QUANTUM COMPUTING

QUANTUM COMPUTING

From Concepts to Code

by Andrew Glassner

no starch press®

San Francisco

Printed in the United States of America

First printing

29 28 27 26 25   1 2 3 4 5

ISBN-13: 978-1-7185-0400-4 (print)
ISBN-13: 978-1-7185-0401-1 (ebook)

 Published by No Starch Press®, Inc.
245 8th Street, San Francisco, CA 94103
phone: +1.415.863.9900
www.nostarch.com; info@nostarch.com

Publisher: William Pollock
Managing Editor: Jill Franklin
Production Manager: Sabrina Plomitallo-González
Production Editor: Miles Bond
Developmental Editors: Annie Choi and Suzanne Olivier
Cover Illustrator: Josh Kemble
Interior Design: Octopod Studios
Technical Reviewer: Ronald T. Kneusel
Copyeditor: Rachel Head
Proofreader: Daniel Wolff

Library of Congress Control Number: 2025010328

For customer service inquiries, please contact info@nostarch.com. For information on distribution, bulk sales, corporate sales, or translations: sales@nostarch.com. For permission to translate this work: rights@nostarch.com. To report counterfeit copies or piracy: counterfeit@nostarch.com. The authorized representative in the EU for product safety and compliance is EU Compliance Partner, Pärnu mnt. 139b-14, 11317 Tallinn, Estonia, hello@eucompliancepartner.com, +3375690241.

[S]

About the Author

Andrew Glassner is a distinguished research scientist at Wētā FX, where he applies computer graphics and machine learning to help artists produce amazing visual effects for film and television.

Glassner served as Papers Chair of the SIGGRAPH '94 Papers Committee, founding editor of the *Journal of Computer Graphics Techniques*, and editor in chief of *ACM Transactions on Graphics*. His many books include the Graphics Gems series, the textbook *Principles of Digital Image Synthesis*, and *Deep Learning: A Visual Approach* (also from No Starch Press).

Glassner has written and directed live-action and animated films, and was the creator-writer-director of an online multiplayer murder-mystery game for The Microsoft Network. He has written three novels and several short and feature screenplays. He has carried out research at labs such as the NYIT Computer Graphics Lab, Xerox PARC, Microsoft Research, and Wētā FX.

Andrew holds a PhD in computer science from the University of North Carolina at Chapel Hill. In his spare time, Andrew paints, composes music, plays jazz piano, writes fiction, and hikes. He can be followed on Bluesky as @glassner.com.

About the Technical Reviewer

Ronald T. Kneusel is a computer scientist, an expert in machine learning, and a lover of fine craft beers. Kneusel has been working with machine learning in industry since 2003 and has a PhD in machine learning from the University of Colorado, Boulder, and a master's in physics from Michigan State University. He's the author of several books from No Starch Press: *Math for Programming* (2025), *How AI Works* (2023), *Strange Code* (2022), *Practical Deep Learning* (2021), and *Math for Deep Learning* (2021).

BRIEF CONTENTS

CONTENTS IN DETAIL

PART I
STATES, OPERATORS, AND SYSTEMS

1
A CURIOUS DECK OF CARDS   5

2
QUANTUM STATES 27

3
OPERATORS 83

4
WORKING WITH QUBITS 99

5
SYSTEMS 109

6
MEASUREMENT 157

PART II
QUANTUM ALGORITHMS

7
TELEPORTATION
207

8
DEUTSCH'S ALGORITHM
227

ACKNOWLEDGMENTS

Nobody writes a book alone. Many people offered me their time, support, expertise, and insight during the development and writing of this book.

For encouragement and help in ways big and small over the years it took to write this book, it gives me great pleasure to thank (in alphabetical order): Mike, Jenn, and Georgia Ambrose; Eric Braun and Wendy Meg Siegel; Steven Drucker and Lourdes Romao; Luca Fascione; Adam Finkelstein; Bruce Glassner and Lisa DeGisi; Eric and Cathy Haines; and Tom Rieke.

Several people generously offered to read an early version of this book. For their valuable time and thoughtful suggestions for improvement, I thank Paul Beardsley, Amir Ebrahimi, John Kalaigian, Jarno Mielikäinen, Del Rajan, Oliver Rosten, and Ramon Montoya Vozmediano. Of course, any errors or problems that remain are my own.

My employer, Wētā FX, supported this project in the midst of its wonderful and creative atmosphere. Big thanks to Navi Brouwer, Peter Hillman, Daniel Hodson, Julia Whyman Jones, Joe Letteri, Millie Maier, Joe Marks, Melissa Roberts, Kimball Thurston, and all my other colleagues and friends in the beautiful cities of Wellington, New Zealand, and Vancouver, Canada.

The fine folks at No Starch Press brought this book to life. Thank you to Miles Bond, Annie Choi, Jill Franklin, Rachel Head, Suzanne Olivier, and Bill Pollock, and everyone else who helped bring this book to life.

Thanks also to many of the fine independent coffee shops in Seattle and Vancouver, where I worked on this book: Diva Espresso, Cafe Ladro, Cafe Vita, Distant Worlds Coffeehouse, Firehouse Coffee, The Fremont Coffee Company, Fuel Coffee, and JJ Bean on Alberni Street. Somehow, wrestling with a tricky bit of math is more enjoyable in a warm and friendly coffee shop.

INTRODUCTION

A hundred thousand welcomes!
—William Shakespeare, *The Tragedy of Coriolanus*, 1605 [192]

 Welcome! This book's purpose is to give you all the information you need to write and run programs on quantum computers.

You'll find that the journey is mind-stretching, exhilarating, challenging, and rewarding. Your brain will thank you for giving it a treat, and by the end of this book you'll have developed skills that put you ahead of the curve for the next big revolution in computer programming.

This stuff is fantastic, which is why I was drawn to it in the first place. Because I love to share cool stuff with cool people, I've written this book to help you join the party and discover and master these exciting new ideas.

What Is Quantum Computing?

The term *quantum computing* sure sounds exotic, but what does it mean? Quantum computing is an entirely new way to think about and execute algorithms. It involves a kind of programming that runs on a *quantum computer*, an electronic device specifically designed to let you manipulate objects at the incredibly tiny quantum scale. Objects at this scale behave in ways that are nothing like what we're used to. We can write programs that use these quantum behaviors to carry out tasks that are wildly beyond the range of those that we can perform on conventional computers.

For example, we can evaluate an algorithm using billions, trillions, or even trillions of trillions of different inputs *simultaneously*. The catch is that each time we run the program, only a single one of these results is available to us as output. But if we're clever, we can make it likely that the output we want is the one we get.

Popular articles on quantum computing sometimes imply that these systems will replace conventional computing, essentially making it all go faster. That's not the case! This new type of hardware runs new types of programs, often targeted at new classes of problems that are beyond the reach of conventional computers. It's a whole new way to think about computing.

Why Should You Read This Book?

You should read this book if quantum computing sounds intriguing to you, or if you want to write your own programs to realize your own ideas. This is an especially exciting time to get into this field. It's still barely a blip on most people's radars, so there's a good chance that whatever you want to do with quantum computing, you'll be the first person to do it.

Right now, there are only a few real quantum computers in the world, and most of them don't have a whole lot of computational power. But these computers are real, and they work. Some are even on the web, where you can use them for free! People are working hard on the formidable engineering efforts that will enable us to scale up these machines so that they're more powerful, more reliable, and easier to use.

Many people (including me) believe that the quantum computing field is on a trajectory much like that charted by classical computers. The computers we're used to today started out as physically hulking devices that were incredibly finicky and expensive to operate. Now they've evolved into the microchips that are in everything from your favorite cloud server to your wristwatch and toaster.

Whether we have to wait a few years, or a few years more, quantum computers will emerge that are powerful, reliable, and affordable. They will have a tremendous impact not just on computing but on society at large, because of the scope and scale of problems they can help us solve. When that happens, many people will suddenly turn their focus to quantum computing, and the field will blossom. By getting familiar with the subject now, you'll be ahead of the curve and ready to lead the way for others.

Quantum computing is radically different from conventional programming: There are no loops, no functions, and no data structures. We can't even make copies of variables! Almost everything that makes up classical programming gets thrown out the window, and we need to start over with entirely new concepts organized in entirely new ways.

It's wonderful and challenging stuff that engages both your intellect and your imagination. If you love new ideas, and seeing the world in new ways, you're going to love this stuff. It's not just forward-looking technology, it's *fun*.

Who This Book Is For

This book is for people who want to understand quantum computing, and even design and write their own quantum algorithms. If you flipped through the book and were put off by what seemed like a whole lot of intimidating math, please, don't worry! As I'll discuss next, I've tried hard to make it comfortable, even if you haven't had the greatest relationship with math and its notation in the past. The whole point of the math is to communicate ideas, so I focus on an informal approach designed to make that communication work for you.

We'll take it slow, and you'll see that once you get familiar with pulling apart mathematical expressions into their components, everything fits together smoothly and without mystery.

Although the symbols used in quantum computation, and therefore in this book, might look unfamiliar, most of them are shorthands for different lists of numbers. If you remember some high school math, you're all set.

By the time we're done, you should be able to glance at almost any of the math in the book, state it in words, and immediately nod and think, "Yup, sure, that's reasonable." It's really just a matter of knowing what the symbols mean and how to parse them in context.

What You'll Learn

This book will show you how quantum algorithms work and how to write your own.

Quantum computers can seemingly calculate billions of solutions to a problem *simultaneously*. They enable us to teleport the state of a quantum object to another quantum object anywhere in the universe, instantly. And they can crack the encryption software used today that protects most of the world's online secrets.

They also have unique limitations. For example, we can't make a copy of an arbitrary piece of quantum information. We sometimes get incorrect outputs. And while a quantum computer can evaluate vast numbers of inputs simultaneously, we can only get a single output as the result of any run of the program, and we usually can't even choose which of the many possible outputs we'll get (though, as mentioned earlier, with some effort we can increase the likelihood of getting a useful answer).

It's a whole new world out there!

How We'll Do It

Let's address the elephant in the room right away: I do not expect that you're a math whiz or a physicist.

If you've flipped through the pages of this book, you've probably seen lots of stuff that looks like unfamiliar and strange math. It may seem off-putting, particularly if you're not comfortable with math and its unusual notation.

Don't worry. Let's first see why the math is necessary for this book, and then I'll explain how I've organized things so that the math will be a friendly companion for you, rather than a challenge.

Using Metaphors

One way to bypass any math is to describe everything using words and pictures, expanding on related ideas we're already familiar with. That is, metaphors.

I love metaphors, because they're a great way to understand something new and unfamiliar. In fact, all of Chapter 1 is one big metaphor.

Metaphors are a wonderful way to share experiences and get comfortable with new ideas. They're the soul of much poetry, song, and other arts. A great metaphor can open new doors of understanding, help people connect, and even help us understand our place in the universe.

Some books on quantum computing embrace these wonderful qualities and use metaphors to describe most or even all of the ideas unique to quantum computing. They describe quantum objects and their behavior using pictures, stories, and references to everyday objects like baseballs or rain clouds. For people who only want a rough understanding of the quantum world, these approaches can be both appealing and helpful, as long as they're not taken too seriously or literally.

But my goal in this book is to help you learn how to read and write quantum programs that actually run and produce useful results. Since I love metaphors and visual explanations, I tried hard to use these techniques in this book rather than math, but I couldn't find a way.

The problem is that metaphors are, unavoidably, approximations. This means they are sometimes incomplete, misleading, or just plain wrong. Figure 1 illustrates these problems.

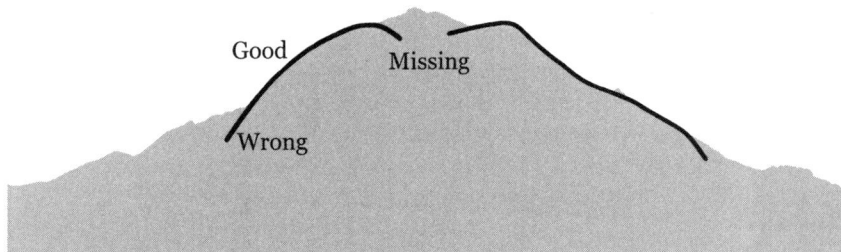

Figure 1: A silhouette of Mount Rainier, along with two curves acting as a metaphor for its shape

This image shows a silhouette of Mount Rainier, whose shape we want to understand. The two curved lines are a metaphor for the mountain's shape. This metaphor is a good match to the silhouette in some places, but in other places it's wrong. Some parts of the silhouette aren't captured by the metaphor at all. In short, this metaphor, like most metaphors, is *incomplete* and sometimes *incorrect*. If we reason about the mountain's shape based just on these curves, we're likely to be misled.

The problems shown in Figure 1 aren't surprising. After all, if a metaphor were an exact match to the subject matter, it would *be* the subject matter, not a metaphor!

The limitations of metaphors for understanding quantum computing are a serious problem. The core issue is that the behavior of quantum-scale objects is deeply unfamiliar to us, so we tend to be bad at dealing with the problems in any metaphor that tries to describe it.

For example, where a metaphor is incomplete, we usually fill in the gaps using our intuition and experience. Unfortunately, this is a quick ticket to misunderstanding in the quantum realm. We simply cannot use our common sense and reason when dealing with quantum objects and their behavior, because they're fundamentally different from what we're used to. When we fill in the holes inevitably left by a metaphor, we draw conclusions that make sense to us but don't match the real world at the quantum scale.

Just as bad are the places where metaphors are incorrect, because they don't accurately describe the phenomena they reference. Here, we're misled by believing the metaphor when it's actually a poor description of what it represents.

Figure 2 illustrates both of these problems. The dotted lines show conclusions we might reasonably draw about the shape of the mountain, based on our sometimes incomplete and incorrect metaphor.

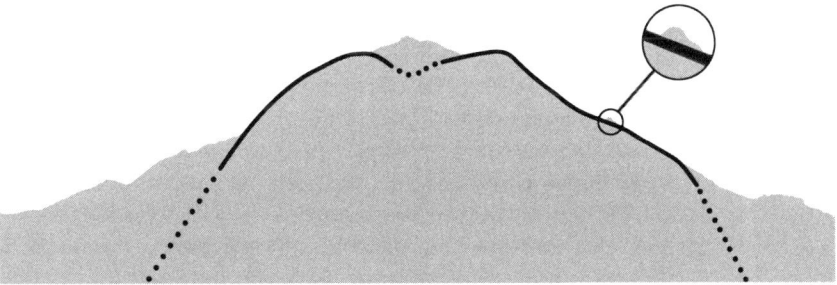

Figure 2: Drawing incorrect conclusions from the metaphor of Figure 1

Where the metaphor is incomplete, it's too easy to guess incorrectly at what's left out and develop the wrong idea of the mountain's shape, as we can see at the peak and on the sides. When the metaphor is just wrong at matching the shape it represents, as shown in the inset, we can miss what could be critical details, without any idea that we're overlooking something essential.

In quantum computing, these problems can have disastrous consequences: Our programs won't work, and we won't know why. Our mental model of the quantum world is simply wrong, leading us astray.

We need some other way to describe the quantum world that is comprehensible, accurate, and complete. The good news is that we have such a description. The bad news, perhaps, is that it's written in the language of mathematics, with its terse notation.

Using Math

For our present work, where complete generality based on exquisitely precise terminology is not the goal, we only need a light dose of some of the fundamental notions.

—Brad Osgood, *The Fourier Transform and Its Applications*, 2007 [153]

At its core, mathematical notation is really no different from a recipe for baking lemon cookies, the instructions for assembling a bookshelf, or a program listing in a computer language like Python. Each of these examples is described by a sequence of steps and relationships, using its own language, conventions, and symbols. A Python programmer needs to know what def means, a baker needs to know that tbsp refers to a tablespoon, and someone assembling a bookshelf has to know that "23x7x1" refers to the dimensions of a shelf. Practically every field of human endeavor has its own special language and symbols.

Math is no different. But math notation can look a lot less approachable than the special notations used by photographers or people keeping score at a bowling alley. Long ago, mathematicians decided that brevity mattered more than almost anything. They wanted the most compact possible way to write down the relationships between different objects.

This choice persists, and it can make written math much harder to read and comprehend than it needs to be.

Other fields that aren't obsessed with this kind of minimalism often produce notations that are easier for people to deal with. For example, consider conventional programming. If we're keeping track of the number of apples in a store, we might use a variable called numApples. That's just a label, so it could have been numberOfApples, or appleCount. What it would almost never be is the single letter a. But if you were reading an equation that referred to the number of apples in a store, that variable might well be written as just a. If a were already being used, say for the number of apricots, then the author writing that equation would likely pick some other letter, such as g. There would be almost no way to guess at the meaning of g, and if you missed the spot where the author told you that it stood for a quantity of apples (assuming that they remembered to tell you), you'd probably never figure it out. The equation using g, and everything that came after it, would be incomprehensible.

Now throw in weird symbols like + and ⟩ and ⊗ and Δ and Σ, and it can turn into a hopeless jumble that becomes so frustrating you just walk away from the whole mess. I get it. I've been there. So I've been careful to lessen the frustration.

In this book, I'll introduce every bit of the math when it appears, with discussion and context. If you flip through the book, you'll see that I annotate almost every line of a series of expressions with a note explaining what's going on. There's a glossary of all the notation in the Appendix. I'll also use many letters consistently. For example, d, i, and N always mean the same thing throughout the book.

So why use this problematic, super-compressed math notation here? Why not replace it with something better?

The reality is that math is how everyone, the world over, discusses quantum computing and writes quantum programs. This book is meant to introduce you to the world of quantum computing. If I used some custom approach instead of math, I would ultimately not be doing you any favors. I want you to be able to read other quantum computing books and resources if you want, and to be equipped to read other people's programs and write your own. It's also important to be able to communicate with other people. Everyone doing any kind of quantum computing uses mathematical language and its dense notation, so if you're not familiar with it, those resources will be unavailable to you.

The other reason we use math is that it *works*. It doesn't work like a metaphor, with gaps and mistakes. The math of quantum computing has been refined to the point where it is both *accurate* and *complete*. Simply put, it has always correctly predicted nature. This is astonishing! The math in this field has *never been wrong*, meaning that it has always perfectly predicted how a quantum computer will behave.

What this completeness and accuracy gives us is something that makes the effort of learning the math and its notation worth it: If the math tells us that manipulating a quantum particle in a particular way will give us a particular result, then that's what will happen. We can count on it. No surprises, no mysteries, no approximations, and no gaps. We don't have to guess, either by reason or by analogy. The math matches the reality, without omissions, without exceptions. That's amazing.

So, we use the math because we must. It's the only description of the quantum world that is so accurate and so complete that we can unquestioningly rely on it when writing quantum algorithms.

Why this math is such a perfect match to the real world is a puzzle that nobody really understands (we'll come back to this a couple of times in the book). For our purposes, it's enough to know that it simply works.

This is a book that *uses* math, but it's not a math book. We're not studying math for its own sake (though much of it is beautiful in its own way, and offers its own rewards). I've used math here because *not* using math would be a disservice to you.

Therefore, I've taken a careful but informal approach to the math in this book. I use the ideas and notation that help us get our work done, but I skip going into lengthy definitions, proofs, and edge cases. For those who want to go more deeply and broadly into all the math I discuss here, I've provided lots of references that fill in all the details.

My hope is that as you work your way through this book, regardless of your previous experience with math and its notation, you'll come to actively appreciate and even enjoy (yes, enjoy!) the values of this language. Equations can be concise to a fault, but they can also be exquisitely precise and eloquent. In addition to the utility of a well-expressed mathematical expression, there is often beauty, and I want to help you appreciate both its utility and its beauty.

Reading Math

When I go to a performance of a play by Shakespeare, I often find it's initially hard to understand the language. It takes a few minutes until I catch the rhythm, and his unfamiliar syntax starts making sense again.

It can also take some time to catch the rhythm of mathematical writing. If you're not familiar with the kind of math writing in this book, I promise you that you can master it. The good news is that there's a pleasant and effective process for reading and understanding math that almost everyone uses. I follow this process myself, and I recommend that you do, too.

Generally speaking, most big batches of equations are meant to change one kind of expression into another. Most of the time, the exact mechanics that get us from start to finish are less important than the new result. So on first exposure, just skim the steps (really, just look them over without too much attention), and focus on the first and last lines.

Those steps are important, though, because they demonstrate the kind of thinking you'll follow yourself when you write new quantum algorithms. So later, return to the math, and this time read it more carefully. Try to describe in words what the math is talking about. Break it down, explain it to a friend (or imagine doing so), draw pictures, or whatever helps you to translate the math notation into meaning inside your own head. When you feel like you've had enough, move on. If you feel like there's more to learn, come back again later.

Each return visit can be fun and rewarding, because each time you'll see more depth and more connections, and the pieces will begin to lock together in ways that feel natural. You'll develop mastery. You'll be learning how to write quantum programs. It's a great feeling!

To summarize, the first time you see an unfamiliar block of math, I recommend that you don't try to understand it by steamrollering your way through. Just skim it and read on. Then return later for a closer look. You'll find that making a few relaxed revisits is the fastest and most pleasant way to catch the rhythm and meaning of the math, and understand what it's saying.

Another thing to keep in mind when reading the math in this book is that you're seeing the final version of every equation and sequence. Not my many false starts, dead ends, and mistakes. These occupy a shelf full of pads in my office, each filled up in succession as I worked on this book. Some expressions took many attempts over many days until I finally found a clean way to represent an idea or make a well-reasoned chain of steps from one expression to another.

I'd show you these failures if I thought they would be helpful, but your process and difficulties will differ from mine. The best I can do is to show you the clearest math I can write, and encourage you to seek the same goals in your own work. It's mostly about persistence and trying different approaches until one works. Also remember that just as we all edit our writing, it's important to also edit our math. Revise and rethink your equations to make them as clear as you can.

Just as it takes practice and time to learn how to throw a baseball, play an instrument, or draw a likeness of a person, it also takes time and practice to learn how to read and write math, particularly when the ideas and notation are new to you. When you write math, don't expect to produce clean and useful results on your first try, even when you're expressing ideas you already know. Thinking in new abstractions can be hard. Be patient and gentle with yourself, practice, and embrace the struggle, and I promise that you'll develop increasing fluency, reading and writing the math of quantum computing with confidence.

Proving Things

Many times throughout this book, we'll look at the mathematical relationships between objects, written in symbolic form as equations.

You might wonder where these equations come from. Are they supposed to be obvious? No, they're rarely supposed to be self-evident; if they were, I wouldn't be presenting them to you! Usually, these relationships were originally developed by someone who had an insight, or maybe just a hunch, and worked on the idea until they found a way to verify it. Then they worked out a way to prove their result, other people read the description (called a *proof*) and agreed it was valid, and the idea became part of the fabric of the subject.

Since these relationships will be important to us, I want you to understand them. However, because this is an informal book, I won't show you rigorous proofs [203]. Instead, my goal is to help you develop an intuitive understanding of why these statements are true. So, I'll usually approach each new idea in one of three ways:

- Deriving the relationship from concepts we've already covered
- Presenting a non-rigorous argument suggesting its validity
- Demonstrating that it holds by showing an example

I'll choose the method that's clearest and most convincing for each case. While I hope this will be satisfying for most readers, those interested in more detail will be able to find it in the citations.

Is Quantum Weird?

In popular writing, it's common to see the behaviors of objects at the quantum scale described as "weird," "strange," or even "bizarre." Those descriptions carry a negative connotation. They are unfair and even misleading. These phenomena are simply unfamiliar to us.

There's no hard rule for what size we mean when we refer to the quantum scale, but it's usually incomprehensibly smaller than we're used to. The smallest physical size that most people can directly perceive visually is about 0.1 millimeters, which is roughly the thickness of a piece of paper [184]. Let's compare this to an electron, which is typically considered a quantum object. Objects that small don't have a well-defined size, but we can informally say that since an electron is part of every atom, it's surely smaller than a hydrogen atom. That puts the size of an electron somewhere less than 5.3×10^{-8} (or 0.000000053) millimeters [253].

It's a different world down there. It might have turned out that these ultra-tiny things behave just like apples, trees, mountains, and other objects we can directly perceive. These are the objects that gave rise to all of our common sense and intuition.

But that's just not how the universe operates. At the quantum scale, objects behave in ways that are unusual to us. I don't think it's fair to call them "bizarre" just because they're outside of our experience. They're different, sure. But it's a big universe, with room for everyone.

In quantum computing, we make use of four particular properties of these quantum objects. Chapter 1 will show you a metaphor that illustrates how these four essential properties work. They will feel reasonable because we'll simulate them in software. And we'll use no math. That chapter will make it easier for you to get comfortable with the quantum world, so that when we later discuss these ideas in the context of quantum computing, they will already be familiar.

What You'll Need

Of course, you'll need to have a quantum computer on your desk. Just kidding!

Once you start writing quantum programs, you'll have two ways to execute them.

The first is to use a quantum *simulator*, a piece of software you can run on your home computer (or on a cloud service, if you prefer) that simulates a real quantum computer. Written by experts, these simulators are faithful representations of what real hardware would actually do. Free, open source simulators are available in a wide variety of programming languages for different operating systems (Chapter 14 offers some starting points).

The second option is to use someone else's quantum computer. Then it's their job (not yours) to spend millions of dollars to craft these delicate machines and keep them tuned up, calibrated, and otherwise maintained. Using a web interface, you submit your quantum program to a queue, and you wait your turn. When your program gets to the front of the line, it's loaded on the computer and run, the outputs are collected, and they're sent back to you. Some big companies offer a limited amount of free time on their quantum computers each month to anyone who wants to use them.

Between the free simulators and the available free time on real hardware, you can develop and run real quantum algorithms. If you want or need more time or a bigger computer, you can get these additional resources online as well, for a fee.

We'll return to these topics when we start looking at complete quantum programs in Part II.

Who Wrote This Book?

As of 2025, AI-based systems that are capable of producing text (in addition to generating other media) are changing the landscape of writing. People are using chatbots to write office memos, term papers, business proposals, newspaper articles, technical reports, and probably anything else that's based on text, from love letters to birthday cards.

It can be hard to know whether you're reading something written by a person or the output of a program trained to produce text that is statistically similar to that written by a person. The result is that it can be challenging to know how much trust we can place in anything we read: Are we reading something produced by a real person, or a clever simulation?

So, I think it's important for you to know where this book came from. It came from me, Andrew, the guy whose name is on the cover. Every word, every letter, every comma and parenthesis came from me (or one of my wonderful human editors). With the exception of plots showing actual data and circuit diagrams produced by a library following my explicit instructions, I drew every figure by hand. I used a wide variety of software tools to help me write text, create images, run quantum programs, check the math, and provide other support, but none of them wrote any of the book.

I conceived of, outlined, wrote, rewrote (and rewrote), illustrated, and otherwise created this book. It was then produced by other skilled humans who turned my manuscript into a final, printed book. We're all people. We made this book for you.

Overview

This book is organized into two parts. The first part builds up all the ideas, terminology, pictures, and math you'll need to understand and write quantum programs. The second part teaches you quantum programming by going through some of the most famous and important quantum algorithms. Once you know those algorithms, you'll be ready to write your own quantum programs.

Here's an overview of what each chapter covers.

Part I: States, Operators, and Systems

Part I starts at ground level and gradually presents the ideas, language, graphics, and math that make up quantum computing.

This foundation is vital, because it will serve as our bedrock when we start doing things with quantum objects that are far beyond our intuition.

By the end of Part I, you'll know all the math and core concepts of quantum computing.

Chapter 1: A Curious Deck of Cards Before we get into the details of quantum computing, I'll present a metaphor to explain what makes this field different from classical computing. We'll cover the four critical ideas, called superposition, entanglement, interference, and measurement. This informal introduction will help you see what's special about these ideas in a comfortable, everyday setting.

Chapter 2: Quantum States Just as the core of classical programming is the bit, the core of quantum programming is the qubit (pronounced CUE-bit). The description of the information in a qubit is called its state. We'll see how states are represented and what rules they obey. Along the way, we'll cover some of the key ideas and math that underlie the description of a qubit.

Chapter 3: Operators Programming is about change: turning inputs into outputs. In quantum programming, we apply a series of transformations to qubits to turn them from their default input states into useful output states. These transformations are performed by objects we call operators. We'll see how to describe operators and how they change the state of the qubits they transform.

Chapter 4: Working with Qubits Now that we know what qubits are and how operators transform them, we'll put those ideas together. We'll meet a few important operators and see why they're useful for manipulating quantum states.

Chapter 5: Systems Useful classical programs require large numbers of bits. Quantum programs are similar, requiring large numbers of qubits. We'll see how to assemble lots of qubits together into a single representation called a system. Happily, the same techniques will also let us combine multiple operators into their own systems. These new representations will allow us to efficiently describe how groups of operations transform groups of qubits, which is a common way to think about quantum programming.

Chapter 6: Measurement The last step of a program on any kind of computer is to get an output. In quantum computing, we say that we get our output by measuring the states of the qubits at the end of an algorithm. Ultimately, we get back a classical bitstring of 0s and 1s. But even with perfect knowledge of the states of the qubits, we usually can't predict with certainty what we'll measure at the output. We'll explore why this is the case, how to represent what we might measure, and how we can still create algorithms that will produce useful results.

Part II: Quantum Algorithms

In Part II, we'll use the tools from Part I to write quantum programs!

We'll look at several famous quantum algorithms, usually named for the people who invented them. Each of these algorithms will introduce one or more new ideas that have become part of the toolbox of quantum programming.

Chapter 7: Teleportation   This quantum algorithm doesn't compute anything in the usual sense. Instead, it transports, or teleports, a quantum state from one quantum object to another, located anywhere in the universe. And it does this *instantly*, faster than the speed of light. Though this appears to violate one of the basic rules of the theory of relativity, this phenomenon has been experimentally confirmed many times. The quantum world is full of surprises!

Chapter 8: Deutsch's Algorithm   Our first real quantum program solves a small problem. On any classical computer, solving this problem requires calling a function two times, once for each of two inputs. There's just no way around it. But on a quantum computer, we only need to call that function once, because quantum computers can evaluate both inputs *simultaneously*. Mind blown.

Chapter 9: Deutsch–Jozsa's Algorithm   The program in Chapter 8 uses only single qubits as inputs and outputs. We'll see how to extend it to handle multiple qubits.

Chapter 10: Bernstein–Vazirani's Algorithm   We start with an oracle, the name we use for a small quantum program that we can run but are not allowed to examine. This oracle holds a secret bitstring. We'll see how, by querying the oracle only once, we can discover the entire secret string.

Chapter 11: Simon's Algorithm   Now we'll use a more complicated oracle, which we're still not allowed to examine. This oracle also holds some secret information we want to discover. We'll use a quantum algorithm along with a classical algorithm that runs on a conventional computer. Together, this team can discover the oracle's secret far more quickly than any classical algorithm by itself ever could.

Chapter 12: Grover's Algorithm   Searching problems come up all the time in computer science. Suppose we're told that in a database of objects, there's one object that will satisfy some set of conditions. Our job is to find it. This algorithm does just that. It doesn't work every time, but we usually need just a few runs to get the right answer, no matter how large the database is. This algorithm (and its variations) has become a component in many quantum programs.

Chapter 13: Shor's Algorithm   Hey, you know all those secrets on the internet? They're kept safe through a method called encryption, which scrambles a piece of information in a way that turns it into nonsense for everyone except someone who knows the secret method for unscrambling it.

The great majority of the information we use today is protected by algorithms that perform an encryption step that nobody has found a practical way to undo, unless you've been given the secret key. This technique is used to encrypt bank balances, direct messages, industrial secrets, military secrets, love letters, business plans, and just about everything on the internet. This protection would fall apart if we could break that encryption. With quantum computers, we can do just that.

The method is another hybrid algorithm, with a quantum algorithm in the middle of two classical algorithms. The quantum part of this algorithm requires quantum computers that are significantly bigger and more reliable than those we have now. So, for the moment, our privacy is safe. But someday, quantum hardware will become powerful enough to reveal all of the secrets encrypted using today's systems.

Chapter 14: Next Steps   Now that you can write your own quantum programs, what's next? I'll suggest a number of ways you can build on the topics covered in the book.

I'll also give you resources for your next steps, such as references to learn from, free software simulators of quantum computers to download and use, and some online services offering real quantum hardware on which to run your programs.

Then I'll survey some applications, such as maze solving, creating pictures, and even a few quantum games.

Appendix: Notation   The appendix summarizes all the mathematical symbols, variables, and graphics that I use in this book for easy reference. If you ever forget what any symbol, letter, or shape refers to, you can find it here.

PART I

STATES, OPERATORS, AND SYSTEMS

*If I had 10 hours to cut down a tree, I'd spend
the first 9 sharpening my axe.*
—Abraham Lincoln (apocryphal), ~1860

*The universe cannot be read until we have learnt the language and
become familiar with the characters in which it is written.*
—Galileo Galilei, *Opere Il Saggiatore*, 1585 [68]

We're at the start of an exciting journey!

If this were a book on river rafting, we'd talk about rocks and currents and reading the river. If it were about photography, we'd learn about f-stops and exposures. Every field has its language and its notation, and quantum computing is no different.

In this first half of the book, we'll explore the language and symbols that let us talk about quantum computing. In the second half, we'll use what we learn here to write quantum programs!

The first chapter of the book presents a metaphor for the quantum world. Here, we'll meet the four key ideas of *superposition, measurement, interference,* and *entanglement* that make quantum computing special. We'll see them in an everyday setting with equipment that we could build in an evening. With a general sense of these ideas under our belts, we'll then leave the metaphor behind for the real stuff.

We'll begin by seeing how to describe the information that can be held in a quantum computer. Quantum programs are all about creating and manipulating information. We call each piece of quantum information a *quantum state*, and those will be our focus in Chapter 2.

To control these states, we'll use *quantum operations* to gradually change them from their initial values into new values that can help us solve problems. Chapter 3 describes several of these operations and shows how they're used.

In Chapter 4, we'll put these pieces together and see our first quantum program!

Most quantum programs need to manipulate lots of information. We'll see in Chapter 5 how to assemble many quantum states and quantum operations into collections, or systems. This will make it convenient to build up big programs without getting lost in a maze of details.

Finally, the goal of every computer program is to ultimately give us an answer to the problem we created it to solve. In quantum computing, we say that we *measure* the output of the computation to get our answer. We'll look at measurement in Chapter 6, and learn about the many surprises that are part of making quantum measurements.

Figure P-1 shows a more detailed breakdown of the chapters that make up Part I.

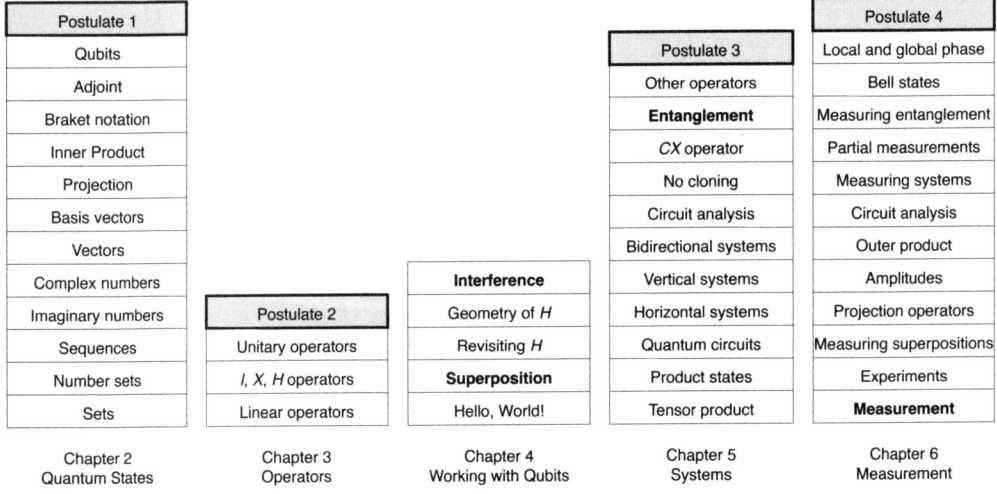

Figure P-1: An overview of Part I

Here, each chapter is represented as a tower, where we start at the bottom and work our way up, building and extending our ideas as we go.

You can see that most chapters build up to a *postulate*. This is a short description of a key quality of quantum computers that details a vital aspect of what makes them different from the classical computers we're used to. The postulates are short, but they each embrace a broad variety of ideas. Each chapter begins where the previous one left off and slowly builds up new pieces to reach another postulate, until we've covered all four of them. Then we'll be ready to start writing programs!

The blocks in each column of Figure P-1 roughly correspond to that chapter's outline. Many of the terms in the blocks may be new to you and mean little or nothing at all, and that's fine. After all, if you knew all this stuff, you wouldn't be reading this book! My job is to help you become fluent with every one of these concepts. Eventually, you'll be as comfortable with every one of these ideas as a plumber talking about different pipes and their uses, or an artist talking about different kinds of paints and color systems, or a playwright discussing the tools of story structure and dialog.

Many of the ideas in Part I belong to the field of mathematics, so I'll be writing them using mathematical notation. If this isn't your cup of tea, don't worry! Whatever experience you've had with math in the past, set it aside. In this book, math is our helper and friend, not a mountain to be overcome. We'll take it slowly and carefully, and you'll see that the notation we'll use is little more than a way to express what we could say in words, only much more compactly.

The math in this book is deliberately informal. I use just enough detail to get across each point. I won't go into all the nooks and crannies, and I'll rarely prove anything formally. Again, the math is to help us, and we only need it to the extent that it's helpful.

The math we'll use in this book all belongs to a field called *linear algebra*. My approach to this math is illustrated in Figure P-2.

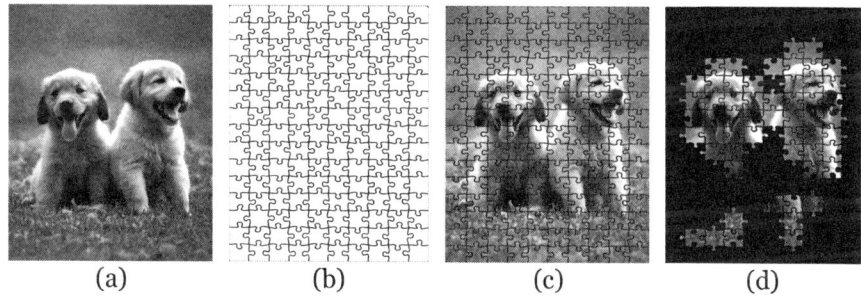

(a)   (b)   (c)   (d)

Figure P-2: (a) The world of linear algebra. (b) The ideas in that world. (c) How the ideas make up the whole. (d) What we'll cover in this book. (Photo by Chevanon Photography from pexels.com.)

In Figure P-2(a), I've represented the world of linear algebra with a picture of two adorable puppies. Figure P-2(b) represents each of the many ideas that make up the field as a jigsaw piece. I'm using a jigsaw puzzle here because over many years, mathematicians have honed each of these ideas so that it interlocks perfectly with the others. In Figure P-2(c), we see how the many aspects of linear algebra combine to make up the whole.

But we won't be covering all of the pieces! We won't even cover most of them. Instead, this book will be like Figure P-2(d). I'll pick and choose just the pieces we need to become quantum programmers, and skip the others. And as I said, I'll take an informal approach and won't even dig into most of those included pieces very deeply.

As Figure P-2(d) suggests, when we're done with our partial jigsaw, you'll clearly understand the parts of linear algebra that will be useful to us as quantum programmers. The essential parts will all be there. If you prefer a more formal approach than the one I use here, or if you crave more details, there are many wonderful resources for linear algebra on the web and in books [211] [117] [220].

My goal in Part I is to give you the tools that will let you dig into the quantum programs in Part II. By the end of this book, you'll be fully prepared to write your own quantum programs!

Now you know what we're up to and what we'll be talking about for the next few chapters.

The big idea is to meet and become comfortable with the ideas, language, and symbols that we use to discuss quantum information and how we manipulate it.

It's going to be pretty wild.

Let's get started!

1

A CURIOUS DECK OF CARDS

Now, let's use this utterly contrived situation that will never occur in order to build intuition about the real world.
—Zach Weinersmith, *Saturday Morning Breakfast Cereal*, 2023 [229]

 This chapter is an optional, math-free introduction to the key concepts of quantum computing. I recommend reading this chapter because the context you'll gain here will help you get comfortable with the formal representations that make up the body of the book. If you're keen to jump right into the mechanics, though, you can go straight to Chapter 2.

Quantum computing is exciting stuff! It stretches our brains with a bunch of new, challenging, and cool ideas. This chapter is all about getting familiar with these ideas.

Our discussion here is deliberately *not technical*. While you're reading this chapter, I encourage you to switch off the part of your brain that focuses on details and give free rein to the wild, free-associating part of your brain that soaks up cool new ideas. We'll come back to the details later.

Today's computers are called *von Neumann machines*, after the mathematician John von Neumann [269]. They're also called *conventional computers* or *classical computers.* The physical implementations of these computers are built around electronic devices that early programmers directly manipulated. Most of us today prefer to use high-level languages that abstract away the details of the underlying electronics. We may have a general sense that there are things happening at the level of electrons, but we usually think about programs in terms of abstract ideas like iteration and subroutines. Those ideas all grew out of the capabilities of the hardware.

Quantum computing is based on a hardware model that has little in common with the hardware of classical computers. This new type of hardware leads to a new philosophy of solving problems that, while still algorithmic, is so radically different from that of conventional languages that it's essentially a new way to think about programming. For example, there are no data structures, no loops, and no subroutines (though there might be, one day).

To keep our wits when working with these devices, we need some kind of mental model that describes what they can do, and how. In this chapter, I'll present a mental model based on a metaphor, where quantum objects are represented by electronic playing cards.

Electronic Playing Cards

We'll use electronic playing cards as stand-ins for the quantum-scale objects that will be our main focus in this book. These imaginary cards are unusual, but they're not radical. You could build them right now with everyday parts if you wanted to.

Our cards will approximate the behavior of the real quantum objects we'll meet later, but because they're only a metaphor, the match won't be perfect.

These cards have four unusual properties, called superposition, measurement, interference, and entanglement. Our goal in this chapter is to become familiar with these four ideas by seeing how the cards work.

Each card is a little computer with a display and some other electronics, as shown in Figure 1-1.

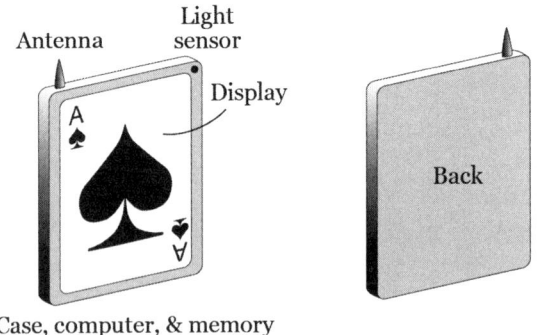

Figure 1-1: An electronic playing card

Every card we'll use in this discussion looks like the one in Figure 1-1. With an exception we'll get to at the end of the chapter, I'll assume that all the cards that are manufactured have identical hardware and software.

Each card is a thin box. Its most obvious feature is a display that takes up almost all of one side of the box. There's also a light sensor next to the display, and an antenna on top for communications. Inside the box, there's a computer, some memory, and a battery.

The cards are shipped from the manufacturer in standby mode, where the computer is turned off and the display is blank.

This display is unusual because it's *write-once*: It starts out blank, but once we draw a picture on the display, it retains that picture forever.

When we first open up the box containing our cards, each card's light sensor notes that its environment has gone from dark to light. When we later place the card face down on a table, the light sensor detects that it's no longer receiving light.

At some point, when the card is turned face up, the sensor detects that it's receiving light again, and it sends a signal to the computer to tell it that the card has been turned over. The computer responds by determining the value of the card (using a process we'll discuss soon), then drawing the corresponding picture on the display.

To prevent cheating, we can never directly access the card's memory. Only the card's computer can read from, or write to, its internal memory. But we get some limited, one-way access through the card's antenna, with which we can send instructions to the computer telling it to manipulate the memory on our behalf. The computer will carry out the instructions we send it, but aside from the picture it draws on the display when the card is turned face up, there's no way for the computer to report back to us anything else, including what's in the memory.

After it has received the light sensor's signal that the card has been turned face up and drawn a picture on the display, the computer erases its internal memory and turns off all the electronics. All the circuitry inside the card is now useless. The image on the display remains that way, fixed forever. From then on, the card is functionally no different from one printed on card stock.

States

To play with these cards, we start by placing them on a table with their displays down. As we discussed, when any card is then turned face up, the light sensor detects the change in light, and it triggers the computer inside to draw a picture on the display.

The computer determines what picture to draw based on the information in its memory. For example, if upon being triggered the computer determines that the card should be a three of diamonds, it will draw the picture for a three of diamonds on the display, and the display will stay that way forever.

Figure 1-2 shows a visual summary of these steps. Note that everything happens immediately as soon as any light hits the sensor, so by the time we can see the display it's already holding its final picture.

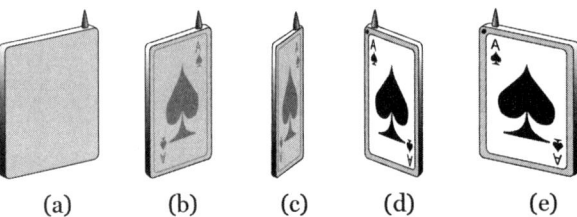

Figure 1-2: Turning a card over so we can see its face

In Figure 1-2(a), the card is face down and the display is blank. In step (b), we imagine that we have lifted one edge and started to turn the card over. The light sensor detects this event and turns on the electronics. Immediately, the computer consults its internal memory and chooses one card. The computer then draws the corresponding picture on the display, clears the memory, and turns off all electronics, shutting itself down. We can think of all the parts of step (b) as happening instantly. In parts (c) through (e), we finish turning over the card, so we can clearly see the picture of the chosen card on the display.

Since the computer decides what to draw on the display based on what's in the memory, let's look more closely at what gets stored there.

The memory holds a list of card values, like the ace of spades, three of clubs, and nine of diamonds. Each such value has an associated *probability*, which is a real number from 0 to 1. This identifies how likely it is that, when the card is turned over, the computer will pick that value for the card.

There's a nice, more general word that we can use to describe the value of any object: It's called a *state*. We say that an object that has the value of a state is *in* that state.

For example, suppose the list holds the three of clubs and the nine of diamonds, each with a probability of 1/2, as in Figure 1-3.

State	Probability
3♣	1/2
9♦	1/2

Figure 1-3: A list inside the memory containing two states

Because the probability of each state tells us the chance that the computer will select it, and these two states have equal probabilities, the card is equally likely to end up as a three of clubs or nine of diamonds. Alternatively, suppose the list looks like Figure 1-4.

State	Probability
3♣	2/3
9♦	1/3

Figure 1-4: A different list inside the memory containing the same two states we saw in Figure 1-3, but with different probabilities

Here, the three of clubs has a probability of 2/3 and the nine of diamonds has a probability of 1/3. That means the computer is twice as likely to choose the three as the nine.

Superposition

A list of one or more states, each with a probability, is called a *superposition*. The concept of a superposition is critical to quantum computation, and it's used by almost every quantum algorithm. A superposition is never empty. In our cards, the memory holds a superposition that contains at least one state, but possibly many more, each with a probability. The memory in each card is used to store a single superposition (that is, a list of states and their probabilities), and nothing else.

Though we usually write a superposition as a list, the order in which the states are listed is irrelevant, as shown in Figure 1-5.

State	Probability		State	Probability
3♦	1/3		K♣	1/6
5♠	1/2		5♠	1/2
K♣	1/6		3♦	1/3
(a)			(b)	

Figure 1-5: The order of the states (and their probabilities) in the superposition doesn't matter.

The superposition in part (a) of Figure 1-5 is the same as the superposition in part (b). Writing the superposition as a list requires that the states are listed in some order. To get closer to the idea that the order doesn't matter, it might be helpful to think of a superposition not as a list, but as a bag full of objects representing states. These objects can have any shape we like, such as cubes, balls, or whales. To choose a state, we close our eyes and reach into the bag. The greater a state's probability, the larger its physical representation is, making it more likely that we'll select it.

Because the computer inside the card chooses a state using the probabilities in the superposition when the card is turned over, we are fundamentally unable to predict which card value will end up on the display. The best prediction we can make is to write out the superposition itself, identifying each possible state the card can end up in, along with the probability of that state being selected by the computer.

We sometimes say that before the card is turned over, its state *is* the superposition in its memory. That is, the superposition is itself the state of the card before the computer's choice is made.

Since only one picture ends up on the display, it's natural to want to unwind this process and recover what the superposition was *before* we turned the card over.

But this is impossible! Remember that I said that after the computer chooses a state, it draws the corresponding picture on the display, erases the memory, and shuts itself down. So there's no way to unravel the process to figure out what the superposition was, based just on what is displayed after we turn the card over.

Let's return to a face-down card. If the card's memory holds, say, a superposition of three states (with their probabilities), it might be tempting to think of this as the card being in "all three states at once," but that would be a misleading description. If the card were in all three states, it would be showing all three pictures on the display. Instead, it's in the superposition state, which is a hybrid idea composed of these three states and their probabilities. It's a collection of possible outcomes.

When all the probabilities in a superposition are the same, we say that we're working with an *equal superposition* or *uniform superposition*.

Initialization

When the manufacturer makes our cards, they have to determine how to *initialize* them, or set the starting contents of each card's memory.

If they wanted to emulate traditional cards, then the manufacturer could make a deck of 52 cards where each card's memory is a superposition of just one state, with a probability of 1. If each card is initialized with a different single state in its superposition, then when we turn over all the cards we'll get one of each of the 52 cards in a normal deck.

What a waste of potential! We might as well have just bought a deck of normal cards. Let's make this more interesting. By changing the initial memory, we'll make our cards unsuitable for many existing games, but we open the door to new games.

We could initialize every card to a uniform superposition that contains each of the card's 52 possible states, each with the same probability of $1/52$, as shown in Figure 1-6.

A♥	2♥	⋯	K♥	A♠	⋯	K♠	A♦	⋯	K♦	A♣	⋯	K♣
$1/52$	$1/52$		$1/52$	$1/52$		$1/52$	$1/52$		$1/52$	$1/52$		$1/52$

Figure 1-6: Each of the 52 states has a probability of $1/52$.

This is unlike shuffling a traditional deck: There's nothing preventing the same state from being chosen by multiple cards, since each card makes its own random selection from its own memory. For example, it's perfectly possible that if we turned over the first five of these cards, we could get the jack of clubs for each one.

We'd definitely need to come up with new rules for games played with a deck initialized this way. Imagine a poker game where you reveal your winning hand is five aces of hearts, as in Figure 1-7, only to find your opponent has the identical hand!

Figure 1-7: Even if you get this terrific hand, your opponent might also hold the same cards.

This illustrates the point that until a card is turned over, it has no definite value (and thus nothing on the display). The value ultimately shown on each card when it's turned face up is independent of the value shown on any other card.

Measurement

We say that when we turn over a card, we're *observing* the card, or *making an observation*. If we use a camera or some other device to observe the state of the card, we refer to this process as *measuring* the card, or *making a measurement*. In practice, we usually treat observation and measurement (and their related terms) as synonyms and use them interchangeably.

When we measure a card (that is, turn it over), we're triggering an automatic and unavoidable process where the computer in the card consults the superposition in its memory, chooses a single state based on its probability, draws that card on the display permanently, erases the memory, and finally shuts itself down. This all happens essentially instantaneously. We can't stop the process, pause it, peek at intermediate results, or modify the steps in any way.

Drawing on the mathematical language for this process, we say that making a measurement or observation causes the superposition to *collapse*. One way to think of this term is to imagine all of the states in the superposition stacked up like a delicate tower of blocks. The act of making a measurement causes the tower to collapse. When the dust settles, all the blocks have crumbled and disappeared except for one.

If we're given a card that hasn't yet been observed, its superposition will be a mystery to us, because the computer is prohibited from ever telling us what's in its memory. In the next section we'll see commands that we can send to the computer to get it to change the memory on our behalf, but we get no feedback on those requests.

Measurement is the only way we ever get to learn *anything* about the actual contents of a card's memory. And even then, we learn only one thing, and it's about the past: We learn that the state shown on the display *was* in the memory before it was wiped clean. We don't learn about the identities of any other states in the superposition, or even if there were other such states. And we don't learn about the probabilities of any states at all, except that the one that appears on the display had a probability that wasn't 0.

The inability to read a card's memory makes it hard to check if the cards were properly initialized, or if the computer inside each card has been correctly executing our instructions. For example, suppose we are given a card and we're told nothing of its current state. Then we turn it over, and it shows the three of hearts. For all we know, that card's superposition could have actually been just that single state, with a probability of 1. Or the three of hearts could have been one of several states in the card's superposition. There's absolutely no way to know.

There is a way to check up on the manufacturer, though. We could gather up lots of cards that they assure us are all initialized to the same superposition, and turn them all over. We'd expect that after enough cards have been turned over (or measured), the relative populations of each card value will reveal the common superposition that all the cards started with. This is illustrated in Figure 1-8.

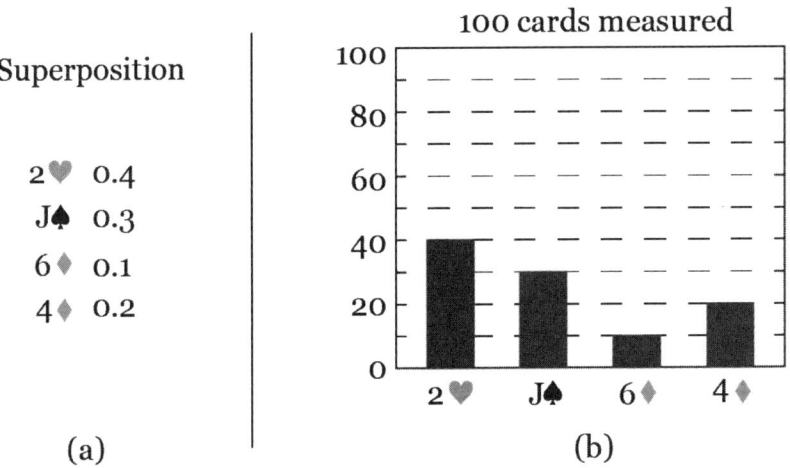

Figure 1-8: (a) The starting superposition of every card. (b) The number of cards of each value observed when we turn over 100 such cards.

This time-consuming process is the only way we can learn more about a card's superposition.

Operating on Cards

So far, I've discussed what happens when we turn over a card that was initialized by the manufacturer.

But we'd like to actually do things with our cards, like play games with them. Since we can't read the memory inside a card, we can imagine betting games based on what state will show up when a card is turned over.

The manufacturer may tell us the superposition that they put into each card's memory at the time it was made. We could use this information to create interesting games. Let's assume that some particular card was initialized with an equal superposition for every value, but we've placed a bet that we'll see a seven (of any suit) when we turn it over. To increase our odds of winning, we need to make the states corresponding to the sevens more probable and the other states less probable.

To change these probabilities from their initial values, we'll use the antenna on the top of the card. The antenna is connected to the card's computer, so we can send messages to the computer instructing it to manipulate the superposition in the card's memory for us.

Imagine you and a friend are playing a game with three cards. The cards are fresh from the manufacturer, who has told you that every card's memory was initialized with an equal superposition of all 52 possible card values. To play the game, you and your friend each secretly change each card's superposition just a little on each turn.

Maybe your bet is that, when the cards are turned over, all the values will be less than five, and your friend is betting that they will all be diamonds. In this case, it's conceivable either or both of you could win.

But if you're betting that all three cards will be less than five, and your friend is betting that they will all be greater than seven, at least one of you is bound to lose.

Each message that either of you sends to a card will consist of the name of a routine that the computer knows how to execute, perhaps followed by a list of numbers and states that tell that routine exactly what to do.

Using the language of mathematics, we call each of these routines an *operator*, and we say that when we tell the computer to execute one of these operators, we're *operating* on the card, or *applying an operator*. Each of the numbers and states that accompanies an operator is called an *argument*.

Let's start with an operator I'll name include. This takes two arguments: a state and a probability.

When the computer receives this message, it first looks in its memory to see if that state is in the current superposition. If the state isn't there, then the computer appends the state and its probability to the superposition. Parts (a) through (c) in Figure 1-9 illustrate this process.

2♥ 0.4	include	2♥ 0.4	2♥ 0.25
J♣ 0.6	3♦	J♣ 0.6	J♣ 0.375
	with probability	3♦ 0.6	3♦ 0.375
	0.6		
(a)	(b)	(c)	(d)

Figure 1-9: Using include to introduce a new state and probability into the superposition

Let's examine what's happening in each part of this figure:

(a) This is the starting superposition, before the message arrives.

(b) The include message introduces the state three of diamonds with a probability of 0.6.

(c) Since the three of diamonds isn't already in the superposition, the computer appends it to the list with the probability 0.6.

(d) The probabilities are then uniformly scaled so that they add up to 1.

Part (d) is an additional step that we perform after executing every operator: We divide all the probabilities by their sum after the operator is applied. The result is a new list of scaled probabilities that add up to 1 (since I'm showing only the first few digits of each floating-point number in this figure and those to follow, their total might be slightly more or less than 1).

In this example, the sum in part (c) is 0.4 + 0.6 + 0.6 = 1.6, so we divide each probability in (c) by 1.6 to get the final probability values in part (d). This means that the final probability of the two of hearts is $0.4 / 1.6 = 2/8 = 1/4 = 0.25$, that of the jack of clubs is $0.6 / 1.6 = 3/8 = 0.375$, and that of the three of diamonds is also $0.6 / 1.6 = 3/8 = 0.375$. Adding up these results, we have $2/8 + 3/8 + 3/8 = 1$.

This process is called *normalization*, and it's important because the definition of probabilities demands that in any complete list of outcomes, the sum of their probabilities must add up to 1. Since we want the computer to treat the numbers in the superposition as probabilities, we must make sure we satisfy that definition, and therefore we scale the numbers so they add up to 1.

Every operation we perform from now on will automatically be followed by a normalization step, which I'll explicitly show in the figures.

Now let's consider what happens if the state is already present. Since we're just making things up right now, we can define include to behave in any way we like. For this discussion, I'll say that when we give it an existing state, the computer should replace the current probability for that state with the sum of that current probability and the probability in the argument. This process is shown in Figure 1-10.

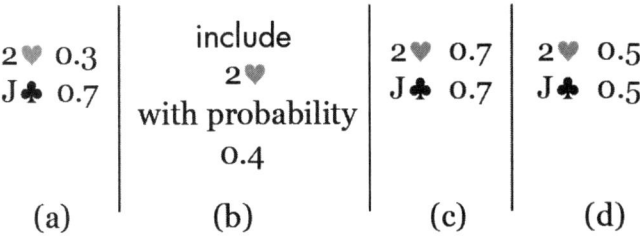

Figure 1-10: An include message referring to an existing state

Let's look at Figure 1-10 step by step:

(a) This is the superposition before the message arrives.

(b) The `include` message introduces the state two of hearts with a probability of 0.4.

(c) Since the two of hearts is already in the superposition with a probability of 0.3, the computer replaces that probability with 0.3 + 0.4 = 0.7.

(d) The probabilities are then uniformly scaled so that they add up to 1.

Let's look at a few more possible operations we might ask our cards to carry out for us. The more operators we have, the more different kinds of games we can invent.

Suppose we have an operator called `flip`. Given a single state as an argument, this operator causes the computer to replace the current probability for that state with 1 minus that current probability (and then the whole superposition is normalized, as always). If the state isn't already in the superposition, then we pretend it was there with a probability of 0, so we append that state to the list with a probability of 1 − 0 = 1.

So, if the argument to `flip` is the two of hearts, and the current probability for the two of hearts is 0.4, then the computer will replace that 0.4 with 1 − 0.4 = 0.6, as shown in Figure 1-11 (in this figure and those that follow, I've rounded numbers to 2 digits of precision).

2♥ 0.4	flip	2♥ 0.6	2♥ 0.5
4♦ 0.1	2♥	4♦ 0.1	4♦ 0.08
Q♦ 0.3		Q♦ 0.3	Q♦ 0.25
7♥ 0.2		7♥ 0.2	7♥ 0.17
(a)	(b)	(c)	(d)

Figure 1-11: Carrying out the `flip` operator

Let's look at this operation step by step:

(a) These are the initial probabilities for a card.

(b) The `flip` message arrives, with the argument two of hearts.

(c) The probability of the two of hearts becomes 1 − 0.4 = 0.6.

(d) As usual, we follow up the command with a normalization step.

The `flip` command specified a single state to modify. Now let's imagine a new operator called `swap`. This command takes two arguments that identify two states and exchanges their probabilities, as shown in Figure 1-12.

2♥ 0.4	swap	2♥ 0.3	2♥ 0.3
4♦ 0.1	2♥, Q♦	4♦ 0.1	4♦ 0.1
Q♦ 0.3		Q♦ 0.4	Q♦ 0.4
7♥ 0.2		7♥ 0.2	7♥ 0.2
(a)	(b)	(c)	(d)

Figure 1-12: Carrying out the swap operator

Here, since we've only moved the probabilities around, normalizing the superposition doesn't have any effect. In this figure:

(a) These are the initial probabilities for a card.

(b) The swap instruction with the arguments "two of hearts" and "queen of diamonds" arrives.

(c) These are the probabilities after executing the instruction.

(d) After normalizing step (c), there is no change.

The key thing to remember about all of the manipulations we've discussed so far (and others we can imagine) is that while the computer can read from its memory and write to it, we can't do either. There is no way for us to read back the values of these probabilities, ever. Our one and only way to learn anything about the memory is to turn the card over. Then, we can infer that the value shown on the display had a nonzero probability before the card was turned over. Otherwise, we just have to trust that our operations are being carried out correctly.

In Figures 1-9 through 1-12, I assumed that we knew the card's superposition before the message arrived, and I assumed that each operation was carried out perfectly. If we were mistaken about the starting superposition, or the computer messed up the operation somehow, the final column in each figure wouldn't represent the superposition in the card's memory, and there would be no way for us to know. Again, all we can learn by turning over the card is that the state shown was in the superposition.

Amplitudes and Probabilities

Over the course of play, we'll often want to increase or decrease the probabilities of different states. We can measure any card after any operation because we know that, thanks to performing normalization after every operation, the probabilities in the superposition are guaranteed to add up to 1. Thus, the computer can legitimately use the rules of probability to randomly select one state.

But what if we decrease the value of a probability so much that it becomes negative? The normalization step we've seen doesn't do anything to address that possibility. Is that going to be a problem?

A negative probability is just fine while we're applying operators to the card, because eventually we might change that probability again and make it greater than 0 when we get around to measuring it. But what if the probability is still negative at the time of measurement?

If the measurement step encounters a number that's supposed to represent a probability, but it's less than 0 or greater than 1, then that number is not actually a probability. As a result, calculations with that number would probably create nonsense, perhaps ultimately causing the card's computer to crash. That's not good!

We can be confident that the numbers that are supposed to represent the probabilities of states won't be greater than 1, thanks to our normalization step. But there's nothing right now to stop those numbers from dropping below 0. We would like to be able to freely change probabilities without constantly worrying about whether they become negative.

We can solve this problem by changing the numbers we store in our superpositions. Rather than storing a probability with each state, we instead store a different number called the *amplitude*. Unlike a probability, an amplitude can be negative.

To find the probability of a state, we multiply the state's amplitude (which we'll see will always be between −1 and 1) with itself. The result, sometimes called the *square* of the amplitude, will always be a real number between 0 and 1. This two-step mechanism means that we don't have to constantly check if the number associated with a state in a superposition is going negative and, if so, do something about it.

For example, if some command causes a state's amplitude to become −0.3, it's no problem, because $(-0.3)^2 = 0.09$, which is a valid probability. So, this mechanism of storing the amplitude and squaring it to get the probability ensures that all the values used by the measurement will be positive, and we're back in business. The downside is that we can't just read the probabilities directly from a superposition anymore. Instead, the superposition gives us the amplitude of each state, which we need to square to find that state's probability.

We'll need to make a corresponding change to our normalization step, so from now on, we'll say that normalization scales the amplitudes in a superposition so that their squared values add up to 1. That is, it's the probabilities that add up to 1, and not necessarily the amplitudes. From this point forward, when I refer to normalization and show it in figures, it will always mean this new version.

Now we can manipulate the amplitudes freely. Normalization will make sure that the squares of the amplitudes in a superposition are all between 0 and 1 and all add up to 1. The measuring process will always work, and the card will never crash.

Whew.

Figure 1-13 shows an example of this new convention.

4♠ 0.4	include	4♠ 0.4	4♠ $0.4/\sqrt{0.65} \approx$ 0.50
2♦ 0.1	2♦	2♦ −0.6	2♦ $-0.6/\sqrt{0.65} \approx -0.74$
J♥ 0.3	with amplitude	J♥ 0.3	J♥ $0.3/\sqrt{0.65} \approx$ 0.37
9♦ 0.2	−0.7	9♦ 0.2	9♦ $0.2/\sqrt{0.65} \approx$ 0.25
(a)	(b)	(c)	(d)

Figure 1-13: Subtracting some probability from a state

Here, we use the include instruction as before, only now with an amplitude of −0.7. The amplitude of the two of diamonds goes from 0.1 to −0.6. To check if the numbers in column (c) are normalized, we square each value and then add them together. That gives us 0.65. Since that's not 1, we need to scale each number in (c) by $\sqrt{0.65}$. This gives us the results to the right of column (d). Note that the new amplitudes don't add up to 1. But if we square these numbers to turn them into probabilities, and add up those probabilities, we get 1 as desired.

Interference

When two amplitudes are added together, we call this process *interference*, using language from physics that describes how waves interact.

When the result of interference is an amplitude for a state that's less than the amplitude it started with, we call this *destructive interference*. You can think of the term *destructive* as telling us that the starting amplitude is reduced, causing it to become less than it was. In Figure 1-13, we saw how the amplitude can become negative.

The opposite effect is where the sum of the old and new amplitudes creates a new value of increased amplitude. That's called *constructive interference*. You might think of the word *constructive* as meaning that the values combine to construct a new, larger value.

By adding positive and negative amounts to the amplitudes for every state in the superposition, we can adjust them so that the state we want to measure has a large probability and all other states have small probabilities. That way, when we turn over the card, there's a good chance we'll see the card we want.

Carried to an extreme, we can completely remove a state from a superposition by using destructive interference to set that state's amplitude to 0. The computer could literally then remove that state from the superposition in memory, or leave it there with an amplitude of 0. Either way, it will have a probability of 0 and thus will never be chosen and will never show up on the display.

The trick to removing a state from a superposition is to send the card an include command with an amplitude that's exactly the opposite of the value it currently has, as in Figure 1-14.

4♠ 0.4	include	4♠ 0.4	4♠ $0.4/\sqrt{0.26} \approx 0.78$
2♦ 0.1	9♦	2♦ 0.1	2♦ $0.1/\sqrt{0.26} \approx 0.20$
J♥ 0.3	with amplitude	J♥ 0.3	J♥ $0.3/\sqrt{0.26} \approx 0.59$
9♦ 0.2	−0.2	9♦ 0	
(a)	(b)	(c)	(d)

Figure 1-14: Removing a state from a superposition

The process is as follows:

(a) This is the superposition before the message arrives.

(b) The include message says to introduce the state nine of diamonds with an amplitude of −0.2.

(c) Since the nine of diamonds is already in the superposition with an amplitude of 0.2, the computer adds the old amplitude and the new to get 0.2 + (−0.2) = 0.

(d) The amplitudes are then uniformly scaled so that their squared values add up to 1.

Pulling off this maneuver requires knowing the value of the state's amplitude, so we can cancel it exactly. If we know how the card was initialized, and every change we've made to it since then, then we can work out the right message to send the card to send a state's amplitude to 0. But when our friend is making secret changes in addition to our changes, all we can do is guess at the appropriate amplitude. That's part of the fun of the game.

When we measure our card at the end of a calculation, if our desired state has a much larger probability than any other state, there's a good chance we'll observe that answer on the display.

Entanglement

Let's give our cards one last ability, called *entanglement*. Entanglement refers to a special link between objects (in our case, pairs of cards).

This relationship involves the antenna on each card. Until now, we've only used each card's antenna as a way to give its computer commands to execute. Now, pairs of cards will use their antennas to communicate with one another.

Entangled Cards

Let's revisit Figure 1-6, which illustrated an initial state for a deck where every card has a superposition of all 52 states, each with a probability of 1/52.

Let's say that before sealing up this deck of cards, the manufacturer picks two cards from the deck (which two cards doesn't matter, since they're all the same), and *links them together*.

It creates this linkage by including two new routines in the computer software for each card: one for sending a message, and the other for receiving it.

The sending routine is triggered when either card is turned over (for this discussion, I'll assume that one card is always turned over before the other card, even if only by a tiny fraction of a second).

As usual, the first card's computer collapses its superposition down to a single state and draws the corresponding picture on the display. But just before it turns off its power, it sends a secret message via its antenna (so, I'll call this card the *sender*). This message says only that the sender has been observed, and names the value that its superposition collapsed to. As soon as that message makes it out of the antenna, the sender's computer shuts down completely and permanently.

This message is completely secret, and the only card that can read it, or even know it was sent, is the other half of the pair.

When this other card (the *receiver*) receives this message, it's still face down. Nevertheless, the instant the message arrives, the receiver's computer collapses its superposition to the *same state* as that in the message, draws the corresponding picture on the display, and turns itself off permanently. As always, this all happens essentially simultaneously and instantly. If and when we turn the receiver over, we are guaranteed to see the same picture that was on the sender.

Figure 1-15 shows the idea visually.

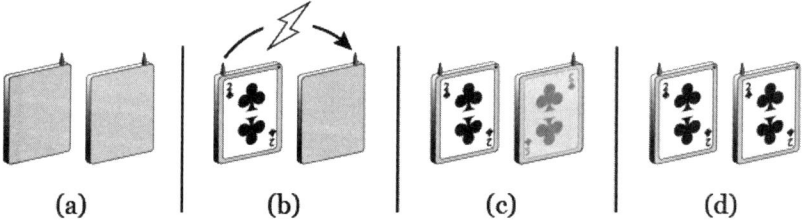

(a)   (b)   (c)   (d)

Figure 1-15: When one card of an entangled pair is observed, the other card's state collapses immediately to the same value.

In part (a), we start with a pair of linked cards, both face down. Suppose the left card is turned over first. In part (b) we see that it collapsed to the two of clubs. This card is now called the sender, and in part (b) it sends a message to the other card in the pair, the receiver, carrying the fact that the sender has been turned over and is showing the two of clubs. Immediately, the receiver collapses its own superposition to the two of clubs, draws this on its display, and shuts down. In part (c) I've shown a kind of X-ray view of the card, so we can see what's now on its display even though the card is still face down. As we see in part (d), if and when the receiver is turned over, it will show the two of clubs as well.

We say that two cards linked together in this way are *entangled*.

The manufacturer might mark entangled pairs of cards by printing a shared mark on the back of each. For example, they could put a letter A on the backs of the first pair that they entangle, a letter B on the next pair, and so on. The order in which these labels are applied doesn't matter. Their only purpose is to let us identify pairs of entangled cards. If we don't want to know about which cards are entangled, there's no need to mark them.

The manufacturer could create entangled groups of any number of cards, but we'll stick with pairs for this discussion.

Every card in the deck could, in principle, be entangled with another card, so a deck of 52 cards could contain up to 26 entangled pairs. Not all cards must be part of an entangled pair.

Recall that when both cards in an entangled pair are face down, our initialization means that every state is in the superposition with a probability of 1/52. So, when the first card is turned over, it can be in any of the 52 possible states. The effect of entanglement is to then change the other card so that, when it's turned over, it will show the same state as the first card.

Entanglement is an intimate connection. Although each card has its own computer and memory, we can think of the pair as conceptually *sharing a single superposition*. When either card is measured, because they're connected, that shared superposition collapses. This is a strange idea: two separate physical devices that are described by a shared state, such that if that state changes, both devices are immediately affected.

If we take this point of view, then we can't really talk about the two cards as having separate identities, or separate superpositions. Because we can't talk about the superposition of one card without including the other, we think of the two cards as a single unit, or a pair that shares a superposition.

Entanglement in Action

Let's use entanglement to get out of a tricky situation. Suppose you run a small casino that specializes in the card game 21, or Blackjack. If you're not familiar with the game, here's a brief recap of a typical version of the core gameplay.

Initially, the dealer has a face-down deck of cards and you (the player) have none. Play proceeds in independent rounds called *hands*. Let's suppose it's just you and the dealer. You start a hand by placing a bet. The dealer gives you two *private cards*, dealt face down, and then gives themselves two cards, one face-down and one face-up. You can look at your private cards, but you don't reveal them to anyone else. Your goal is to keep asking for cards to reach a total as close as you can get to 21 without going over (every number card is worth its numerical value, royalty cards are worth 10 points, and each ace can be individually either 1 or 11 at your choice, which you don't have to declare until the hand is over).

To play, you can announce either that you want one more card or that you're stopping. If you ask for another card, it's dealt face up, so we call it a *public card*. You can keep asking for more public cards, one at a time, until you decide to stop or your total (including your private cards) is at or over 21.

If you hit exactly 21, you win immediately and get back your bet and more. If you go over 21, you lose immediately and the dealer collects your bet. If you stop before reaching 21, the dealer turns over their private card and then gives themselves more cards according to fixed rules, until the rules tell them to stop, they hit exactly 21, or they go over 21. If the dealer goes over 21, or your total is greater than the dealer's total, you win and get back your bet and some more.

There are rules for ties, and lots of variations, but this is all we'll need for our discussion. Usually the game is played with one dealer and multiple players, each of whom is playing independently against the dealer. For this discussion, I'll say that when the hand is over, everyone turns over their two private cards for all to see.

Unlike your competitors, your casino uses our electronic cards. You buy the cards initialized with an equal superposition of all 52 values, so every card has an equal probability of showing any value when it's looked at. These unusual cards give your casino an extra spice nobody else can offer (it also means that your customers can't improve their odds, and reduce your profits, by counting cards). We'll say that when a player peeks at each of their private cards, that lets enough light hit the sensor to trigger the collapse into a single state that's drawn on the display. As usual, only the player is allowed to view their private cards.

One day, you get a call from a famous, rich, but elusive player. They'd like to play at your casino next week and spend a whole lot of money. Great news! But they've recently had some bad press, and they don't want to go out in public. You promise them every kind of privacy, but they aren't interested. Instead, they suggest playing from home with a two-way video system. Both you and they will have cameras pointed straight down at the cards.

You really want this player, and you have no reason not to trust them, but you've seen too many spy movies to not be suspicious of this kind of video setup, particularly when large sums of money are involved. There are just too many opportunities for foul play.

And then it hits you: Use entanglement! You can entangle the cards so that nobody has to trust anyone else. It's not a big deal to include this, since you're already using electronic cards.

You suggest to the player that you order two decks of cards from the manufacturer, just like your usual cards, except that the manufacturer entangles the top two cards of each deck, and the next two, and so on. The result is that each deck of 52 cards is card-by-card entangled with the other deck, as in Figure 1-16. You can order multiple such decks if you expect the game to go on for a while.

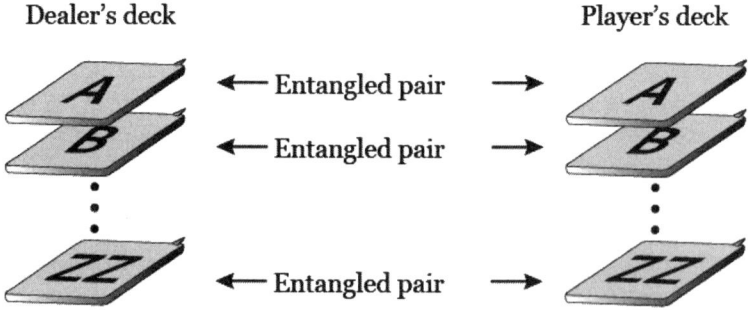

Figure 1-16: Two decks of cards where the pair of cards at the same position in each deck are entangled with each other

You propose that your dealer and the player each play with one of these decks. Every time the dealer deals a card from their deck, the player does the same thing with their own deck. Otherwise, the game plays as usual.

Happily, the player agrees, the cards are ordered, the manufacturer ships one deck to you and the other deck to the player, and the game is on!

Let's see why neither party now needs to trust the other. Each hand begins with the dealer dealing two cards for the player from their deck, which they place face down. The player mirrors this action with their own deck, dealing themselves two face-down cards. Because they're electronic cards where every value is equally likely, at this point nobody can say what the values of the cards are. Now the player privately looks at their cards, causing their superpositions to collapse and the displays to show their final, randomly chosen values. At that moment, the two face-down cards on the dealer's table *also* collapse to those same states!

The game continues this way. Each time the player asks for another card, the dealer takes one from their deck and places it on the table face up. At the same time, the player deals one card from their deck and places it face up. Thanks to entanglement, while nobody can predict what these cards will show, once turned face up they will always show the same value.

The real benefit of this comes at the end of the hand, when everyone turns over their private cards. We're *guaranteed* that they will match! If they don't, either the cards have malfunctioned or someone cheated. Neither party has to trust the other.

We've solved the trust problem!

Well, not quite. While we've prevented one way of cheating, we've created another.

Suppose that the player, upon receiving their deck of cards in the mail, immediately opens it up and looks at every card. Their cards all collapse to specific values, and so do all the cards in the dealer's deck, which might still be in the mail. Now the player stacks their cards back up and carefully puts them back in the box. Later, in the actual game, the player can play perfectly, since they always know what card is coming next. It's a clean cheat because there's no way for the dealer to know that the player peeked at the cards beforehand.

In other words, when the player looked at their deck, they not only caused their cards to collapse to individual values, they also caused the dealer's cards to collapse to those same values. There's no way for the dealer to detect this because they can't ever get any information from a card except by turning it over and looking at it. There's no way to even check that the display is blank without looking at it, which would instantly collapse the card (if it wasn't already collapsed).

The moral of this tale is that entanglement is powerful because it lets the observation of one card immediately affect another card, but its implications can be subtle.

We can make an interesting variation on entanglement: Instead of both cards showing the same state, they could show *opposite* states. Playing cards don't have natural opposites, so for the moment let's say that the opposite of any card is the card with the same number, but with the other suit of the same color. For example, if we have the three of clubs, its opposite is the three of spades, since spades is the other black suit. In the same way, the opposite of the eight of hearts is the eight of diamonds.

The idea is shown in Figure 1-17.

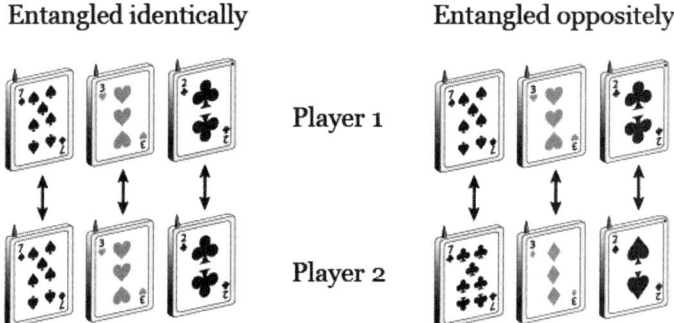

Figure 1-17: Two types of entanglement

Assume that Player 1 turns over their cards first. If the cards are entangled identically, then Player 2 will find their cards are the same. If the cards are entangled oppositely, then Player 2 will see the opposite (the other suit of the same color) of each of Player 1's cards.

Whether both cards of an entangled pair will show the same or opposite states is a decision made by the manufacturer when they create the cards.

We can make up lots of cool new games for these electronic playing cards. If you're feeling inspired, you could build and program these cards tonight. Or you could write software to simulate them.

Summary

If you come across an analogy that improves your understanding, by all means, embrace it. But don't love it too much, or too long.

—Willy McAllister, "Basic Electrical Quantities:
Current, Voltage, Power" [134]

The metaphor of electronic playing cards has done its job, and now we'll bid it farewell.

The purpose of the cards was to help us get a sense of the four features of superposition, measurement, interference, and entanglement. These are the fundamental pillars of quantum computing. I had to adapt these ideas a little from their quantum forms to make them fit into our playing card metaphor. In the following chapters, I'll revisit these principles and replace the approximations in this chapter with the real things.

Our goal from now on will be to construct a careful description of the quantum world as revealed by experiments and explore how we can use that information to create algorithms.

We need this formal approach because quantum objects don't behave the way objects at a human scale behave, so our intuitions about how objects work are initially unreliable at the quantum scale. The precise and logical formalism that we'll develop, expressed in mathematical notation, will help us stick to the experimentally discovered laws that describe the behavior of quantum objects.

From here on out, we'll forget about playing cards and focus on real quantum objects.

In Chapter 2, we'll see how we represent a quantum bit. To fully understand this description, we'll meet a range of ideas, terms, and mathematical tools that form the core of quantum computing, and these will all become good friends as we use them throughout the rest of this book.

Strap in, because it's going to be an exciting ride!

2

QUANTUM STATES

For the great doesn't happen through impulse alone, and is a succession of little things that are brought together.
—Vincent van Gogh, *Letter to Theo van Gogh*, 1882 [223]

Quantum algorithms work by manipulating quantum information. This chapter is about how we represent and describe that information.

Much of our discussion will refer to mathematical ideas. To ease our way into this, I'll introduce the background math that we'll use in this book. To keep things focused, I'll cover only the math we'll need. I'll keep it accurate but informal, and I won't go into details that won't matter to us. My intention is to bring you up to speed as quickly and clearly as I can. If some or all of this material is already familiar to you, I suggest at least skimming the text so that you'll know what notation and conventions I use in this book.

Getting Started

A traditional computer (also called a *conventional* or *classical* computer) is based on manipulating the simplest unit of information, called a *bit*. A bit can represent one of two possible states, which have been given the names 0 and 1.

A quantum computer follows the same general idea and manipulates the simplest unit of quantum information, called a *qubit* (pronounced CUE-bit).

Quantum algorithms run on quantum computers, which are explicitly designed to take advantage of the physics of objects at incredibly small scale, called the *quantum scale*. Here, the word *quantum* refers to both the tiny physical size of the objects we're discussing and their unusual behavior, which is described by the theory of *quantum mechanics*.

The core concepts of quantum mechanics can be captured in a set of basic principles. Each of these ideas is variously called a *law*, *principle*, *axiom*, or, my favorite, an *ansatz*. All of these words are used, but most commonly they're called the *postulates* of quantum mechanics. Different authors express these principles in different ways.

To cover just enough quantum mechanics to let us understand and design quantum algorithms, I'll state these ideas in the form of four postulates, which summarize what we've learned from experiments. These postulates can't be derived from some simpler set of principles. Rather, they *are* the principles, based on physical observations, from which everything else is derived.

We'll begin with the postulate that defines the qubit, and in the following chapters we'll take on each of the remaining postulates one at a time.

Postulate 1

There are many different ways to phrase the postulates of quantum mechanics, even when we narrow them down to quantum computing [71] [146] [293]. They vary based on their generality, their mathematical language, and how much physics they refer to. In this book, I'll write the postulates using ideas and terms that are the most helpful when discussing quantum computing (if you go on to study quantum mechanics, you'll see that our postulates are usually special cases of larger ideas).

Let's begin with Postulate 1 [71]:

> **Postulate 1** An isolated quantum system, or qubit, is described by a two-element complex vector of magnitude 1.

Don't worry if this has little to no meaning for you right now. This chapter is all about unpacking this statement and building it back up, so that it makes perfect sense to you. Along the way, we'll meet terminology and notation that we'll use throughout the book.

We'll start with different types of numbers and then combine them into new abstractions, along with some new notation that was custom-designed for quantum objects. When we revisit this postulate at the end of this chapter, you'll know exactly what it's saying.

Numbers

In this book, we'll use several different types of numbers. Each type of number is organized into a structure called a set. In this section, we'll first look at sets and the related idea of lists. Then we'll discuss the different types of numbers we'll use in this book. We'll look closely at the types of numbers called imaginary and complex, because they are at the heart of quantum computing.

Sets and Lists

A *set* is an unstructured collection of objects [246]. The objects are called the *elements* of the set.

In a set, there's no order or sequencing associated with the elements. It's like a bag full of marbles, where we can pull out the marbles one at a time, but we can't predict which one we'll get. If there are two or more identical elements in a set, we usually ignore any duplicates and pretend there's just a single instance of each.

I'll usually write a set by placing its elements between curly brackets and separated by commas. To name a set, I'll use a capital, italic roman letter (a full list of all the notation I use in this book can be found in the Appendix). For example, Equation 2.1 shows two sets, each of which contains the names of three fruits as elements.

$$S = \{\text{kiwi}, \text{watermelon}, \text{apple}\}$$
$$T = \{\text{watermelon}, \text{kiwi}, \text{apple}\} \tag{2.1}$$

The sets S and T are the same, because the order in which we write the elements of a set doesn't matter.

Speaking generally, if a set has a specific number of elements, we call it a *finite set*. If there's a limitless number of elements, we call it an *infinite set*.

Unlike a set, a *list* is a collection of objects in a specific order [242] [117]. A list may also be called a *sequence* or *ordered list*.

I'll usually write the objects in a sequence between square brackets and separated by commas. Depending on the context, I'll name our lists with different typography. For now, I'll name a list with a lowercase roman letter in boldface. Equation 2.2 shows two different lists, $\mathbf{v}$ and $\mathbf{w}$, with the same elements as the sets in Equation 2.1.

$$\mathbf{v} = [\text{kiwi}, \text{watermelon}, \text{apple}]$$
$$\mathbf{w} = [\text{watermelon}, \text{kiwi}, \text{apple}] \tag{2.2}$$

These lists, or sequences, are distinct, because the order in which we write their elements matters.

Remember that curly brackets describe an unordered set, and square brackets describe an ordered list.

When I refer to a particular element of a sequence, I use the name of the sequence (but without the boldface), subscripted by the entry index, numbering from left to right, starting with 0. For example, in Equation 2.2, v_0 is "kiwi" and w_2 is "apple." Some authors retain the boldface when referring to list elements, but dropping the boldface for elements is a common convention. I like this approach because it reminds us that names such as v_0 refer to elements of a sequence and not sequences themselves.

Unfortunately, the world has yet to agree on whether we should start numbering objects from 0 or 1. As a general rule, if an author is discussing computing, they start numbering with 0, while everyone else starts with 1. But there are so many exceptions that unless an author states their choice explicitly, it's the reader's job to infer which convention is being used. This is primarily a computing book, so I'll start numbering with 0. This is important enough to highlight:

Numbering from 0   In this book, I always number items starting with 0.

Note that I'm referring to *numbering* here, or giving numerical labels to elements. When we're *counting* objects, I start with 1, as we usually do.

If a set or list is finite, we don't refer to its number of elements as a "length." That word already has too many meanings! To prevent confusion, we call the number of elements in a set A its *dimensionality*. We write this as $\dim(A)$ or $\dim A$ (another term for this is the *cardinality* of the set A, written $\#A$). These same terms and notations are also used when describing lists.

In this book, dimensionality is always an integer, and I will always represent dimensionality with the letter d. I won't use d to mean anything other than dimensionality.

Types of Numbers

We'll use five different types of numbers in this book. Some of these will match our everyday idea of a "number," while others may be new to you.

Each type of number belongs to its own set, with its own name. Four of these sets are *nested* inside one another like matryoshka dolls, so that each set also contains all the elements of every set it contains. Figure 2-1 shows these five sets (there are other common sets of numbers, but we won't be using them).

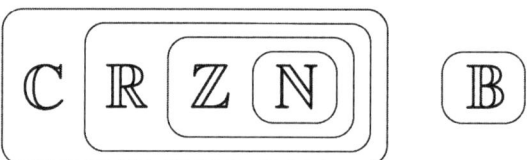

Figure 2-1: The five sets of numbers we'll use in this book

Let's start with the nested sets on the left. The innermost set is called the set of *natural numbers*, also called the *counting numbers*. We write this with the special symbol $\mathbb{N}$ (said out loud as "blackboard N," or just "N"). Curiously, mathematicians don't agree on whether the set $\mathbb{N}$ should include 0 or not [272] [243]. For this book, I'll say that $\mathbb{N}$ is an infinite set that starts at 0, then continues with 1, 2, 3, and so on. We can write this as in Equation 2.3, where the three dots mean that this pattern continues forever.

$$\mathbb{N} = \{0, 1, 2, 3, \ldots\} \tag{2.3}$$

Sometimes we say that $\mathbb{N}$ contains the *non-negative integers*, meaning all the positive whole numbers and 0.

Suppose we want to say that some variable x is a natural number, or equivalently, that x refers to an element of $\mathbb{N}$. To express this symbolically, we use the symbol $\in$, which means "is a member of." So, $x \in \mathbb{N}$ is a compact way to say "the variable x is a member of the set $\mathbb{N}$," "x belongs to $\mathbb{N}$," or "x is in $\mathbb{N}$." All of these mean "x is an integer 0 or greater."

Working outward, the next set contains all the *integers*, including the negative ones. This set is named $\mathbb{Z}$ (from the German word *zahlen*, meaning integer numbers [129]). This set is also infinite.

Continuing outward, we reach the set $\mathbb{R}$, an infinite set containing all the *real numbers*. This set contains every integer as well as every number with a fractional part. For example, the number of ladybugs on a patch of grass might be 17, while their total weight in grams might be 0.357 [8]. These are both real numbers. Classical programmers often refer to real numbers as *floating-point* numbers.

Finally, we reach the most general type of number we'll be using. The set $\mathbb{C}$ is an infinite set that contains the *complex numbers*. The word *complex* here doesn't mean "complicated," but rather is a technical term that describes a generalized version of the real numbers. We'll explore complex numbers in the next section.

On the right side of Figure 2-1, we have the set $\mathbb{B}$ all by itself. This is the set of *bits*, which contains only 0 and 1. That is, it's the finite set $\{0, 1\}$. You might expect this set to be nested inside the natural numbers, but there are a few reasons why most mathematicians keep it separate. One reason is that all the sets on the left are said to be *closed under addition*, which means that if you take any two elements from the set (you can choose the same one twice) and add them together, the result is also in the set. For example, consider the natural numbers. The numbers 3, 4, and 8 are all in $\mathbb{N}$, and the sums $3 + 3 = 6$ and $4 + 8 = 12$ are also in $\mathbb{N}$. But $\mathbb{B}$ does not have this property, because while $1 \in \mathbb{B}$, the integer sum $1 + 1 = 2$ is *not* in $\mathbb{B}$. For this and other reasons, we usually don't place $\mathbb{B}$ inside the hierarchy on the left [92].

In this book, I'll write names for bits, natural numbers, integers, and real numbers with lowercase, italic roman letters, like a and b, and I'll write complex numbers with lowercase Greek letters, like α and β.

We can use the $\in$ symbol for any of these sets, so $x \in \mathbb{B}$, $x \in \mathbb{N}$, $x \in \mathbb{Z}$, and $x \in \mathbb{R}$ mean "x is a bit," "x is a natural number," "x is an integer," and "x is a real number," respectively. I'll write $\alpha \in \mathbb{C}$ when I want to emphasize that α is a complex number.

Postulate 1 tells us that the basic unit of quantum information, a qubit, is described by a complex vector. This is a reference to complex numbers, so let's see what these complex numbers are all about.

Complex Numbers

> *All such expressions, as $\sqrt{-1}$, $\sqrt{-2}$, $\sqrt{-3}$, $\sqrt{-4}$, etc. are consequently impossible, or imaginary numbers, since they represent roots of negative quantities; and of such numbers we may truly assert that they are neither nothing, nor greater than nothing, nor less than nothing; which necessarily constitutes them imaginary, or impossible.*
>
> —Leonhard Euler, *Elements of Algebra*, 1765 [57]

> *That this subject [of imaginary magnitudes] has hitherto been considered from the wrong point of view and surrounded by a mysterious obscurity, is to be attributed largely to an ill-adapted notation. If for instance, +1, 1, and $\sqrt{-1}$ had been called direct, inverse, and lateral units, instead of positive, negative, and imaginary (or even impossible), such an obscurity would have been out of the question.*
>
> —Carl Friedrich Gauss, *Theoria Residiorum Biquadraticorum, Commentario Secunda*, 1832 [70]

A *complex number* has two defining characteristics. First, it involves a special number that doesn't show up explicitly in the sets of numbers in Figure 2-1. Second, we'll see that it involves *two* real numbers.

The special number I just referred to is a unique object that is *defined* to have the value $\sqrt{-1}$. Don't worry if it isn't clear to you intuitively what the square root of -1 might mean. When we're getting used to new types of numbers, they can seem awkward or strange. In fact, many people resisted the idea of negative numbers, thinking them meaningless and confusing [178], and now we use them every day. So for now, just take this as a definition. Later, we'll see a nice way to interpret $\sqrt{-1}$ geometrically.

In this book, I'll refer to $\sqrt{-1}$ with the label i. While almost everyone uses this letter for $\sqrt{-1}$, some engineers use the letter j instead. Adding to the confusion, many math and computer books use the letter i as the name for a variable holding a generic or temporary integer. I'll never use the label i in this book to mean anything other than $\sqrt{-1}$.

I'll mark definitions using a triangle over an equal sign. Equation 2.4 uses this notation to define i.

$$i \stackrel{\Delta}{=} \sqrt{-1} \tag{2.4}$$

We call i an *imaginary number*. Don't read too much into the word "imaginary," though. Mathematicians have a history of giving different types of numbers interesting and colorful names. We've already seen real numbers (though they're no more real than any others). Some other names for types of numbers include perfect, amicable, social, and surreal [174]. So, while the name "imaginary" originated because these numbers seemed strange to the people doing the naming, they're now considered a perfectly fine kind of number, no more or less "imaginary" than any other numbers. The name, however, has stuck.

Mathematicians also have a long history of asking hypothetical questions and seeing where they lead. In this case, they started by assuming that $\sqrt{-1}$ had meaning and saw what might come of that. Later, engineers discovered that many difficult problems they faced could be solved using $\sqrt{-1}$. In short, whatever $\sqrt{-1}$ might mean, using it in calculations was of enormous practical value.

As you may have guessed, when people tried to describe the results of experiments that probed the nature of quantum objects, they found that using i in the equations let them write those results compactly and elegantly. This idea is now deeply embedded in quantum computing, so let's get more familiar with numbers that involve i and how we work with them.

Working with i

Because i will be such an important tool for us, let's look at it a little more closely.

We can treat i like a more familiar real number in many ways. For example, we can multiply it by a real number, so $3i$ and $-17.45i$ are both fine imaginary numbers. We can add and subtract imaginary numbers, too; for example, $3i + 2.2i = 5.2i$.

We'll often use powers of i, or values created by multiplying i with itself one or more times. Exponentiation is defined so that any number (except 0) raised to the power of 0 is 1, so by that definition, $i^0 = 1$ [285]. Since any number raised to the power of 1 is defined to be itself, we also have $i^1 = i$. From $i = \sqrt{-1}$, it follows that $i^2 = -1$. Multiplying i by itself one more time, we find $i^3 = (i^2)i = (-1)i = -i$. Going forward, we find that the powers of i form a repeating cycle of four values. One full cycle and the start of a second are shown in Equation 2.5.

$$
\begin{aligned}
i^0 &= 1 \\
i^1 &= i \\
i^2 &= (i^1)i = i\,i = -1 \\
i^3 &= (i^2)i = (-1)i = -i \\
\hline
i^4 &= (i^3)i = (-i)i = 1 \\
i^5 &= (i^4)i = 1i = i
\end{aligned}
\tag{2.5}
$$

It's fine to use i inside of equations, but we need to eventually get rid of it for final answers that have physical meaning. For example, suppose a calculation uses imaginary numbers as part of the process of predicting how many liters of root beer we need to buy for an upcoming party, and the next to last result is $-5i$. We can't buy $-5i$ liters of anything. But if the final step of the calculation is to multiply that value by i, we'll get $(-5i)i = -5(i^2) = -5(-1) = 5$, and we *can* buy 5 liters of root beer.

Negative numbers are similar to imaginary numbers in this way. We can use negative numbers while performing calculations, but we'd be in trouble if our final result was that we needed to buy -5 liters of root beer. Negative and imaginary numbers are useful, but not for describing physical amounts.

Let's now use imaginary numbers to form a *complex number* (the word "complex" is just another label and doesn't mean complicated).

We can think of a complex number as a list of two values. The first is a real number, which for the moment I'll call a. The second is formed by taking another real number, which I'll call b, and multiplying it by i, creating the imaginary number $b \times i$. Following convention, two variables that are next to one another with no explicit operation should be multiplied, so bi is short for $b \times i$.

Rather than write this pair of values a and bi in typical list form (that is, $[a, bi]$), the convention is to write them as a summation. Thus, the complex number γ built from a and bi is written as in Equation 2.6.

$$\gamma = a + bi \tag{2.6}$$

Equation 2.6 describes a complex number for any values of the real numbers a and b. In symbols, $a \in \mathbb{R}$ and $b \in \mathbb{R}$, while $\gamma \in \mathbb{C}$.

The first element in this list, a, is called the *real part* or *real component* of γ, written as $\mathrm{Re}(\gamma)$. The second element, bi, is called the *imaginary part* or *imaginary component* and is written $\mathrm{Im}(\gamma)$. Some authors use a special typeface and write the real part as $\Re(\gamma)$ and the imaginary part as $\Im(\gamma)$.

When adding real numbers, we can reduce $3 + 5$ to its equivalent, 8. But a single complex number can't be reduced this way. The sum $a + bi$ represents a combination of two fundamentally incompatible objects.

A big reason for writing a complex number using the plus sign rather than as an explicit list is pragmatic: When we use complex numbers in calculations, treating the two parts as elements that have been added works out just as it would if we were working with the sum of two real numbers. For example, multiplying $3 + 2i$ by 5 gives us $5(3 + 2i) = 15 + 10i$. So, we can treat a complex number just like any other sum of two numbers.

Visualization

There are two popular ways to visualize complex numbers, and they lead to different but equally useful ways to think about and compute with them.

Both visualizations rely on using a two-dimensional plane (like the surface of this page). We call this plane the *complex number plane*, the *complex plane*, or, most commonly, just the *number plane*.

One way to start is to draw two perpendicular axes on the page and label the horizontal axis Real and the vertical axis Imaginary (or R and I). Then we can plot a complex number as though it were a point, using the real and imaginary values as the distances along these axes. Figure 2-2 shows an example.

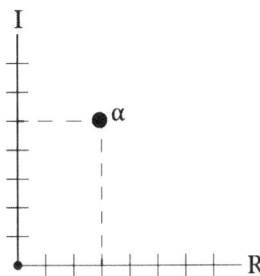

Figure 2-2: Plotting a complex
point $\alpha = 3 + 5i$ along the real (R)
and imaginary (I) axes

This version of the number plane is also called a *rectangular coordinate system*, a *rectangular diagram*, an *Argand diagram* [231] (after Jean Robert Argand [151]), or a *Cartesian diagram* (after René Descartes [90]). The point where the real and imaginary axes cross is called the *origin*.

Alternatively, we can locate points on the number plane by using a *polar coordinate system* in which lines radiate out from a common origin, with circles of increasing radii around that origin, as shown in Figure 2-3.

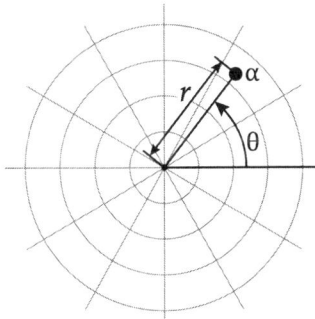

Figure 2-3: The complex point
$\alpha = (r, \theta)$ plotted on a polar
coordinate system

We describe a point in a polar system using the notation (r, θ), where r is a real number that gives us the distance of the point from the origin and θ is a real number representing an angle (in this section, I'll use θ to refer to an angle, which is a real number). The historical convention for these points, like the Cartesian points (x, y), uses parentheses, but they're lists where the order matters.

We call r the *magnitude* of the complex number and θ its *phase*. The (r, θ) notation is called the *polar form* of a complex number. Because r is a distance, it's always 0 or positive. In this book, we'll always measure angles counterclockwise from the three o'clock axis.

When we're doing geometry, we often think of the r in a number written in polar form as "radius," as in the radius of a circle that includes that point.

But since we're using these points to visualize complex numbers, I'll always refer to r as the magnitude.

When you're drawing the number plane, you can choose to draw the rectangular form in Figure 2-2, the polar form in Figure 2-3, or both at the same time. In this book, I'll usually draw the rectangular diagram, since it is graphically more sparse than the polar diagram and so makes it easier to highlight whatever objects I'm drawing on that diagram. If it's helpful, you can always imagine the circles and radii of the polar diagram drawn on top of the rectangular axes.

We can convert between polar and rectangular forms using the trigonometry of a right triangle, as shown in Figure 2-4 (if you're not feeling on solid ground with trigonometry and sines and cosines, there are great introductions on the web [113] [148]).

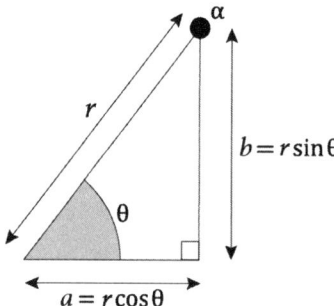

Figure 2-4: How the polar form (r, θ) and rectangular form $a + bi$ are related

We can also use the formulas in Equation 2.7 to convert between polar and rectangular forms.

Given $a + bi$	Given (r, θ)	
$r = \lvert \sqrt{a^2 + b^2} \rvert$	$a = r\cos\theta$	(2.7)
$\theta = \tan^{-1}(b / a)$	$b = r\sin\theta$	

In Equation 2.7, $\tan^{-1}$ is the *inverse tangent* (or *arctangent*) function from trigonometry that takes as input the tangent of an angle and returns the angle itself.

In practice, computing θ with $\tan^{-1}(b / a)$ can suffer numerical problems when a is 0 or nearly so. Modern computer libraries avoid that problem by providing a routine (usually named atan2) that takes the values of b and a as separate arguments.

Around 1740, Leonhard Euler (pronounced OY-ler) discovered something amazing. He wrote out the formulas for sine and cosine in a special way called a *power series*, and he recognized that these expressions looked similar to the power series for a different expression that he already knew [212]. He found that he could make things match by writing the relationship as in Equation 2.8, which is now known as *Euler's formula* [260].

$$r e^{i\theta} = r\cos\theta + i\, r\sin\theta \tag{2.8}$$

Here, e is a real number called *Euler's number*, and it has a value of about 2.718 [259]. In this book, I will never use an e by itself as a name for anything other than Euler's number (sometimes we call this *Napier's number*, in honor of John Napier, who wrote about this number decades before Euler did [144]).

Writing a complex number as $re^{i\theta}$ is called writing it in *exponential form* (sometimes we casually also call this the polar form). The term $e^{i\theta}$ may look strange. Here we are just getting used to imaginary numbers, and now we're using one as an exponent. What could we possibly mean by raising the real number e to an imaginary number $i\theta$? Getting into this would take us too far away from our focus on math for quantum computing, but if you dive into this question, you'll find that you can rewrite such things as more normal looking complex numbers without exponents [217].

What's important for us is that Equation 2.8 gives us a way to turn the polar coordinates r and θ (that is, two real numbers) into a single mathematical object that combines both values. We'll see that the exponential form is often the most convenient way to work with complex numbers when we're performing calculations.

We can write any complex number as either $re^{i\theta}$ or $a + bi$. Both are correct and fine, and we can go back and forth with Equation 2.7. We'll use both forms in this book.

For the rest of this book, I'll use the word *number* to refer to a complex number. Note that if we have a complex number $\alpha = a + 0i$ with no imaginary part, then that's a real number. And if this a has no fractional part, then α is an integer. On the other hand, if $\alpha = 0 + bi$, and b is not 0, then there is no real part and α is an imaginary number. So, the word *number* includes the reals, integers, and counting numbers, as we saw graphically in Figure 2-1. When I want to refer specifically to a real number, an integer, or a counting number, I'll say so explicitly unless I think it's clear from context.

Conjugation

People found that when they worked with complex numbers, such as the number $a + bi$, the related complex number $a - bi$ frequently popped up as well. This happened often enough that they gave this expression its own name and symbol.

Given any complex number $\alpha = a + bi$, the related complex number $a - bi$ goes by the polysyllabic name of *complex conjugate*, or *conjugate* for short. I'll indicate the action of turning a number into its complex conjugate by drawing a little line, or bar, over the original complex number. Thus, if $\alpha = a + bi$, then $\overline{\alpha}$ stands for $a - bi$, and vice versa. If the imaginary part is negative this still holds, so if $\beta = c - di$, then $\overline{\beta} = c + di$ (some authors use an asterisk as a superscript for conjugation, writing α^* instead of $\overline{\alpha}$). We call $\overline{\alpha}$ the conjugate of α, and vice versa.

When we write a complex number in polar form, we can see from Equation 2.7 that the conjugate is formed by replacing θ with $-\theta$. Thus, if $\alpha = re^{i\theta}$, then $\overline{\alpha} = re^{-i\theta}$.

Equation 2.9 summarizes conjugation in notation for both the rectangular and polar forms.

$$\overline{a + bi} = a - bi$$

$$\overline{r\,e^{i\theta}} = r\,e^{-i\theta}$$

(2.9)

Figure 2-5 shows conjugation graphically on the number plane. We can interpret the operation as reflecting, or mirroring, the point around the horizontal axis. Conjugating a complex number twice in a row returns us to where we started.

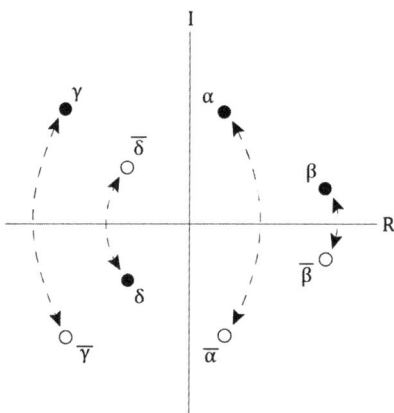

Figure 2-5: Plotting complex numbers (solid dots) and their complex conjugates (empty dots)

A handy property of complex numbers is that the product of any complex number α with its conjugate is *commutative* (that is, the order of the arguments doesn't matter). In symbols, $\alpha\overline{\alpha} = \overline{\alpha}\alpha$. This is because the order in which we multiply any two numbers (including complex numbers) doesn't affect the result. More familiarly, $3 \times 5 = 5 \times 3$.

An important quality of a complex number is its magnitude, which we've seen is the distance from the origin to the complex point when it's plotted on the number plane. The magnitude of a complex number α is written $|\alpha|$ (or sometimes $\|\alpha\|$). The single vertical bar form sure looks like an absolute value operation, and that's no accident, because the magnitude, being a distance, must always be 0 or positive.

If we write a complex number in polar form, then we know the magnitude right away, since it's r. If we have the Cartesian form, we saw in Equation 2.7 that we can find the value r from $r = |\sqrt{a^2 + b^2}|$. We need the absolute value signs here because we know that any positive number has two square roots [247]. For example, the square roots of 25 are 5 and −5, since $5^2 = (-5)^2 = 25$. Because r is a distance and therefore by definition must be positive, we use the absolute value $|\sqrt{a^2 + b^2}|$ for r. From now on, I'll assume we're always using the positive square root.

In this book, we'll often work with complex numbers of magnitude 1. That means $r = 1$, so we can represent these complex numbers as $e^{i\theta}$ for some value of θ and leave off the r term.

Every complex number with a magnitude of 1, when plotted on a polar coordinate system, will lie on a circle of radius 1 around the origin. This is shown in Figure 2-6.

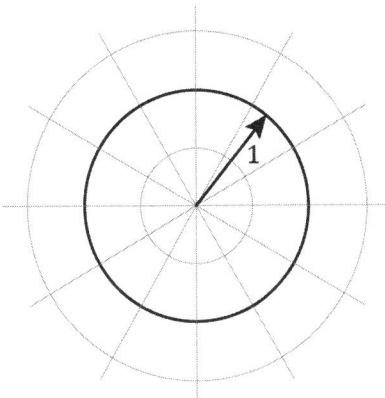

Figure 2-6: All points $e^{i\theta}$ have a radius $r = 1$, so they all lie on a circle of radius 1 around the origin.

The complex conjugate gives us a compact way to compute the squared magnitude of a complex number. All we have to do is multiply that complex number with its conjugate! Let's check that this is true. Equation 2.10 shows the steps to find the squared magnitude of an arbitrary complex number, α, in Cartesian form.

$$
\begin{aligned}
\alpha\overline{\alpha} &= (a + bi)(a - bi) && \text{Expand } \overline{\alpha} \text{ using Eq. 2.9} \\
&= a^2 - abi + bai - (bi)^2 && \text{Multiply the terms} \\
&= a^2 - b^2 i^2 && \text{Replace } (ba - ab)i \text{ with } 0 \\
&= a^2 - -(b^2) && \text{Replace } i^2 \text{ with } -1 \\
&= a^2 + b^2 && \text{Since } --(b^2) = b^2 \\
&= r^2 && \text{Eq. 2.7 tells us } a^2 + b^2 = r^2 \\
&= |\alpha|^2 && \text{Because } r^2 = |\alpha|^2
\end{aligned}
\tag{2.10}
$$

Keep in mind that $\alpha\overline{\alpha}$ gives us the *squared* magnitude of α, not its magnitude. If we want its magnitude, we need to take the square root of this result.

We can summarize what we've just seen as shown in Equation 2.11. We'll often use these properties to calculate $|\alpha|$ and $|\alpha|^2$.

$$
\begin{aligned}
|\alpha|^2 &= \alpha\overline{\alpha} \\
|\alpha| &= \sqrt{\alpha\overline{\alpha}}
\end{aligned}
\tag{2.11}
$$

Since $\alpha\overline{\alpha} = \overline{\alpha}\alpha$, the relations in Equation 2.11 also hold for $\overline{\alpha}\alpha$.

I placed Equation 2.11 in a box because it's particularly important. I'll use boxes like this throughout the book as the mathematical equivalent of setting text in boldface, to signal math that is particularly worth slowing down and paying attention to.

We'll often multiply two complex numbers together, as we just did. Let's look more closely at this operation for the general case of any two complex numbers. I'll start with the Cartesian forms, as in Equation 2.12.

$$\alpha\beta = (a + bi)(c + di) \qquad \text{Expand the complex numbers}$$
$$= ac + adi + bci + bd(i^2) \quad \text{Multiply the terms} \qquad (2.12)$$
$$= (ac - bd) + (ad + bc)i \quad \text{Group real and imaginary}$$

That's nice to know, but to me it's just a big mess of letters that doesn't offer intuitive insight into how the result is related to α and β. Maybe the result will be easier to interpret if we try it again in exponential form, so let's do that, as in Equation 2.13.

$$\alpha\beta = \left(r_\alpha e^{i\theta_\alpha}\right)\left(r_\beta e^{i\theta_\beta}\right) \quad \text{Expand in polar form}$$
$$= \left(r_\alpha r_\beta\right)\left(e^{i\theta_\alpha} e^{i\theta_\beta}\right) \quad \text{Gather similar terms} \qquad (2.13)$$
$$= \left(r_\alpha r_\beta\right) e^{i(\theta_\alpha + \theta_\beta)} \quad \text{Because } e^a e^b = e^{a+b}$$

Now that is a lot more informative! It says that when we multiply complex numbers together, their magnitudes multiply and their angles add. Figure 2-7 shows the process visually.

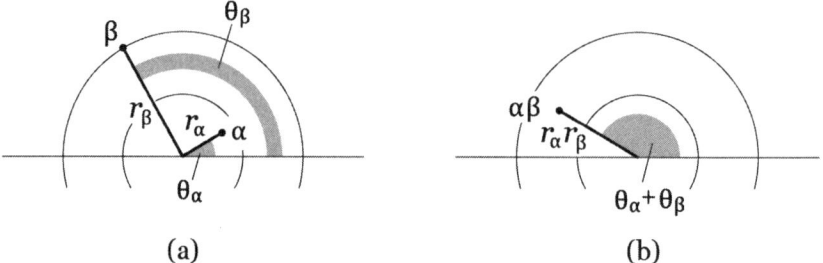

Figure 2-7: Multiplying complex numbers α and β. (a) The starting α and β. (b) Their product, $\alpha\beta$.

In Figure 2-7(a), the magnitude of α is 0.75 and the magnitude of β is 2. When we form $\alpha\beta$ in Figure 2-7(b), the magnitudes multiply to produce a new magnitude of 1.5. We can also see that the angles are added together.

Conjugation has a property we'll find useful: The conjugation of a product is the product of the conjugates. That's a mouthful, but it's simpler in the notation shown in Equation 2.14.

$$\overline{\alpha\beta} = \overline{\alpha}\,\overline{\beta} \qquad (2.14)$$

It's almost always worthwhile to check these kinds of claims. Doing so gives us practice with manipulating the notation, building our confidence and fluency. This skill will help you become a talented and creative quantum programmer, so I encourage you to try to prove these sorts of things for yourself (even informally) before seeing how I do it.

I'll show that Equation 2.14 is true by writing everything out, using the complex numbers $\alpha = a + bi$ and $\beta = c + di$. Equation 2.15 shows the steps. In the last line, I used some temporary variables, s and t, to simplify the results.

$$\overline{\alpha}\,\overline{\beta} = \overline{\alpha\beta} \qquad \text{The goal}$$

$$(a - bi)(c - di) = \overline{(a + bi)(c + di)} \qquad \text{Expand numbers}$$

$$\underbrace{(ac - bd)}_{s} - \underbrace{(ad + bc)}_{t}\,i = \overline{\underbrace{(ac - bd)}_{s} + \underbrace{(ad + bc)}_{t}\,i} \qquad \text{Group and name groups} \quad (2.15)$$

$$s - ti = \overline{s + ti} \qquad \text{Definition of conjugation}$$

Now that we have a handle on what complex numbers are, let's look at how to work with them.

EULER'S IDENTITY

Many people feel that an expression known as Euler's identity is one of the most beautiful and elegant equations in all of mathematics. I agree with them.

Euler's identity is a special case of Euler's formula from Equation 2.8, where we set $r = 1$ and $\theta = \pi$. The result is shown in Equation 2.16.

$$e^{i\pi} + 1 = 0 \qquad (2.16)$$

This relationship is Euler's identity.

Just to check that it's correct, we can expand out the $e^{i\pi}$ term using Equation 2.8, giving us $\cos \pi + i \sin \pi = (-1) + 0i = -1$, and that gives us $(-1) + 1 = 0$. That's the whole thing. But oh, what a thing.

Euler's identity ties together five fundamental branches of mathematics in one short equation, with nothing missing and nothing left over. Reading left to right, we have Euler's number, e, which comes to us from calculus; the number i, which is the essence of imaginary numbers; the number π, which is fundamental to geometry; the integer 1, which is the identity element from algebra (that is, $1x = x$ for every x); and the number 0, which is not just the identity element of arithmetic (that is, $0 + x = x$ for every x) but is unique from all other real numbers because it is neither positive nor negative.

This just scratches the surface. You could write an entire book about just this one compact equation. Or two books [143] [286].

For me, Euler's identity is like a waterfall. It's both simple and sophisticated. This is what people mean when they say that mathematics can be beautiful.

Working with Complex Numbers

Most vagabonds I knowed don't ever want to find the culprit
That remains the object of their long relentless quest
The obsession's in the chasing and not the apprehending
The pursuit, you see, and never the arrest

—Tom Waits, "Foreign Affair," 1977 [227]

Postulate 1 tells us that a quantum bit is described by a "complex vector," which we'll see refers to a list of complex numbers. Manipulating these numbers is therefore central to quantum programming.

To become quantum programmers, we need to be as comfortable with complex numbers as we are with integers and real numbers. This section is intended to help you develop that comfort by working with complex numbers in a few basic ways. My goal is to help you feel at ease with these numbers, rather than to derive specific results for later use.

I'll illustrate much of this section with pictures drawn in the two-dimensional number plane. This lets us think about algebraic operations on complex numbers as geometric operations. A nice reward for this geometric approach is that we'll find a way to think about i in strictly geometric terms, removing any mystery around the expression $\sqrt{-1}$.

To get into the right frame of mind, let's visit the classical real number line, shown in Figure 2-8.

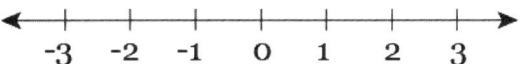

Figure 2-8: The number line

This picture may seem trivial to you, but the number line is a conceptually deep and elegant construction that unifies a long series of gradual expansions of the concept of a number [172] [135]. Roughly speaking, people in the Neolithic era got the number ball rolling with their use of counting numbers. Around the year −1800, people in Egypt developed reciprocals (like $1/2$ and $1/3$). The Greeks introduced irrational and real numbers around −500. The negative numbers were developed in China around −200. The idea of 0 was introduced in India around 600, but it didn't get firmly established as a number until around 1600. Uniting these separate developments into the simple, geometric number line was a brilliant step of intuition and synthesis. Today we take the number line so much for granted that we teach it in elementary school!

The number line has a few wonderful qualities. Perhaps the most important is that it's not just an approximate visualization of real numbers, but it is exquisitely faithful to them. For example, every real number that we could ever imagine has a single, unique point on the number line, and that point in turn corresponds to that exact number. Structurally, numbers that we consider close to other numbers are also close geometrically to the points associated with those numbers. Thanks to these properties, and many others, points on the number line and their corresponding real numbers are intimate partners.

The number plane has the same qualities, only the plane offers a two-dimensional space for representing complex numbers. These complex numbers are just as much "numbers" as the integers and the reals.

To accommodate complex numbers graphically, we complement the number line with a second, vertical axis. Every complex number we can imagine has a unique corresponding point in the number plane, and that point in turn corresponds to that single number. This relationship is so tight that I will sometimes refer to points in the number plane *as* numbers, and vice versa.

The value of drawing complex numbers in the number plane is that it gives us another way to think about them. We've seen that we can write complex numbers as expressions, like $a + bi$ or $re^{i\theta}$, and then manipulate them algebraically. But we can also draw them as points in the complex plane and then think about algebraic operations in geometric terms. The beauty is that these geometric operations are perfect analogs to the algebraic operations.

We've seen this once already, when in Figure 2-5 we observed that complex conjugation corresponds to reflection across the real axis.

The trick to seeing the geometry of this operation, and those we'll look at next, is to think about multiplication in a particular way. Consider multiplying two real numbers, such as $a \times b$. Think of that expression as starting with a and then changing it by multiplying it with b, producing a new value of a, which we can call a'. Then, we could write this as a process that produces a transformed value of a, or $a' = a \times b$. We could equally well think of $a \times b$ as starting with b and transforming it by a if that better fits what we're doing, so we could write $b' = a \times b$. For this discussion, I'll always think of the first term as an input that gets modified to produce a new version of it. I'll also drop the explicit multiplication sign, so the product of two complex numbers α and β will be $\alpha\beta$.

Suppose we start with a complex number α, and we'd like to double its magnitude (that is, its distance from the origin) without changing its angle. The idea is shown in Figure 2-9(a).

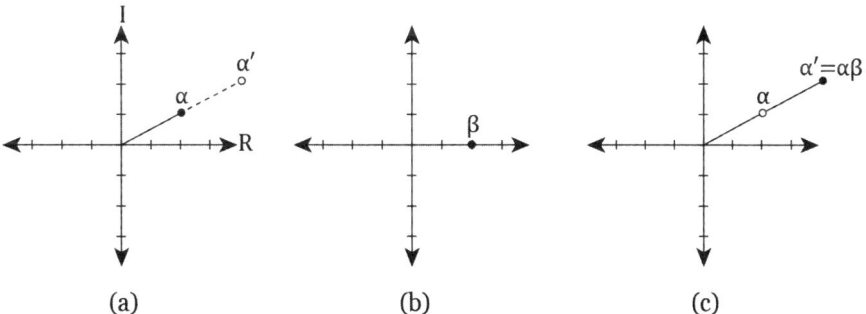

(a) (b) (c)

Figure 2-9: Extending α. (a) Our starting point α and its intended new version, α'. (b) The point $\beta = 2 + 0i$. (c) The product $\alpha\beta$ is the α' we want.

We can create this transformed version of α by multiplying it with the complex number $\beta = 2 + 0i$, shown in Figure 2-9(b). Note that β forms an angle of 0 with the real axis (because it's on the real axis), and it has a magnitude of 2. Thus, when we form $\alpha' = \alpha\beta$, the magnitude of α doubles, but the angle remains the same, as shown in Figure 2-9(c). Bingo!

There was nothing special about the choice of 2 here. Multiplying any α with the complex point $\beta = c + 0i$ scales the magnitude of α by c. This holds for any real number c, including values that are between 0 and 1 (which pull α closer to the origin) and even negative values (which move α to the other side of the origin), as shown in Figure 2-10. An important special case is when $\beta = -1 + 0i$. Then $\alpha\beta = (-1)\alpha = -\alpha$, as shown in Figure 2-10(c).

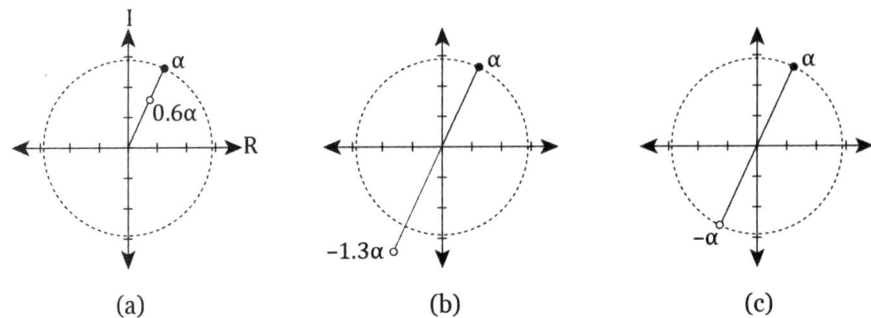

(a)   (b)   (c)

Figure 2-10: Scaling α. The dashed circle shows the starting magnitude of α for reference. (a) A value α scaled by 0.6. (b) Scaling α by −1.3. (c) Scaling α by −1.

Next, let's look at the part of Equation 2.13 that tells us that when complex numbers are multiplied, their angles add. Another way to say this is that complex multiplication creates rotation!

To see this, suppose we want to rotate some arbitrary complex number α counterclockwise around the origin by 90 degrees, or $\pi/2$ radians, without changing its magnitude. To prevent the magnitude of α from changing, any number β we'll multiply with α will need a magnitude of 1. To rotate α by $\pi/2$ radians (that is, to add $\pi/2$ radians to the angle of α), that β needs to form an angle of $\pi/2$ with the real axis. Those requirements completely determine that $\beta = 0 + 1i$, as shown in Figure 2-11(a).

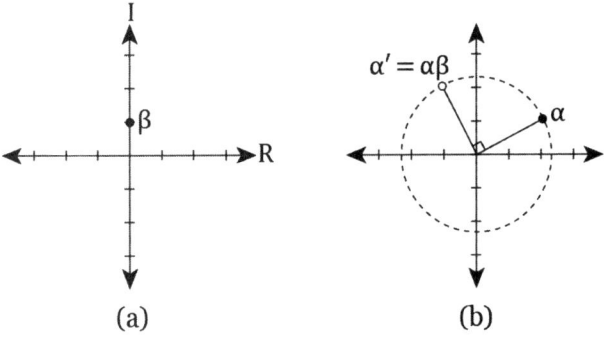

(a)   (b)

Figure 2-11: Rotation of α by complex multiplication. (a) The point $\beta = 0 + 1i$, or i. (b) The point α and the new point $\alpha' = \alpha\beta$, rotated 90 degrees from α.

We can more casually refer to $0 + 1i$ simply as i, so we can view the multiplication of any α with i as a rotation of α by 90 degrees. This works for any α anywhere on the number plane, as shown in Figure 2-11(b).

This geometric viewpoint has a simply marvelous property: It puts to rest the problem of trying to interpret $i = \sqrt{-1}$. We've just seen that multiplying a complex number α by i (or $0 + 1i$), creating $\alpha' = i\alpha$, has the effect in the number plane of rotating α by 90 degrees. If we multiply α' by i again, we'll form $\alpha'' = i\alpha'$, or the original α rotated by a total of 180 degrees. That gives us $-\alpha$, or exactly the same result as $-1 \times \alpha$, as shown in Figure 2-12.

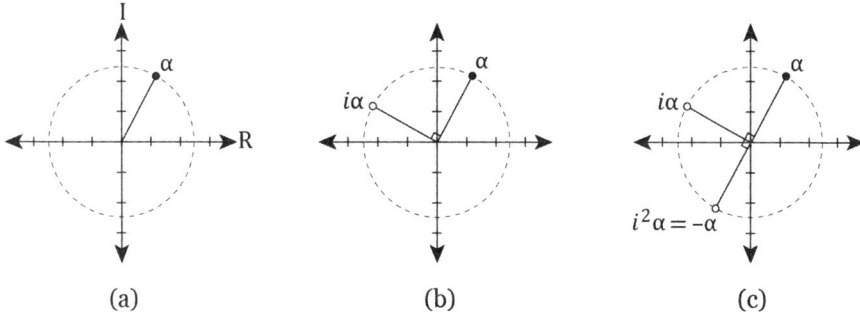

(a)   (b)   (c)

Figure 2-12: Rotating α. (a) The starting α. (b) The product $i\alpha$ is α rotated 90 degrees. (c) Rotating α by another 90 degrees by forming $i^2\alpha$ gives us $-\alpha$.

The result of this operation, α'', can be written as $i\alpha' = i(i\alpha) = i^2\alpha$. That is, applying the rotation twice means multiplying by i twice, or once by i^2.

We've found that $i^2\alpha = -\alpha$, so $i^2 = -1$. When we take the positive square root of both sides of $i^2 = -1$, we get $i = \sqrt{-1}$. This is just a way to write down the observation that multiplying any complex number by i twice gives us a 180-degree rotation of that number in the number plane, which is the same result we get from multiplying that number by -1.

In short, i is that number that we can multiply with any other number α to cause α to rotate by 90 degrees in the number plane.

For this reason, you'll sometimes hear people say something like "multiplying by i causes a 90-degree rotation." That's true!

More generally, it's also true that we can rotate a complex number by *any* angle θ. We need only to multiply that number with another complex number that has a magnitude of 1 and forms an angle θ with the real axis.

If you've worked in computer graphics, or you've done a lot of traditional 2D geometry, you're probably used to thinking of rotating a point around the origin by an angle θ in terms of explicit sines and cosines, which you find by drawing a bunch of right triangles. That approach is perfectly valid, but it can get a little complicated. Multiplying any complex number by a complex number with a magnitude of 1 and at an angle of θ with the real axis, or $e^{i\theta}$, does the same job more simply.

Because complex numbers are vital to quantum computing and will appear throughout this book, let's spend a few more moments with them. I won't introduce any important formulas here; the point is just to illuminate the landscape, not to snag a reward.

Our goal will be to find the square root of i. We have no particular need for this result, but it challenges us to combine what we know about i with what we know about other things, like the square root. It's a good little mental workout, so let's jump in.

We can reason our way to an answer algebraically, but let's first see how nice it is to do it geometrically.

We've seen that multiplying a number α with i produces a version of α rotated by 90 degrees. So, the square root of i would be a complex number β that rotates α by 45 degrees. Then, applying β twice, or applying β^2 once, would be the same as multiplying by i. That is, $\beta^2 = i$, so $\beta = \sqrt{i}$.

If we want the coordinates of β, we can find them by drawing a classical 45–45–90 triangle, as in Figure 2-13. The number β has a magnitude of 1 and makes an angle of 45 degrees (or $\pi/4$ radians) with the real axis.

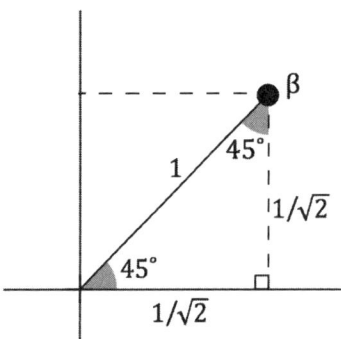

Figure 2-13: The point β is $(1/\sqrt{2}) + (1/\sqrt{2})i$.

Multiplying any α by this β will rotate α counterclockwise by 45 degrees around the origin, as shown in Figure 2-14(a). Multiplying it by β twice, as in Figure 2-14(b), rotates α by 90 degrees, just like multiplying it by i, confirming that $\beta^2 = i$, or $\beta = \sqrt{i}$.

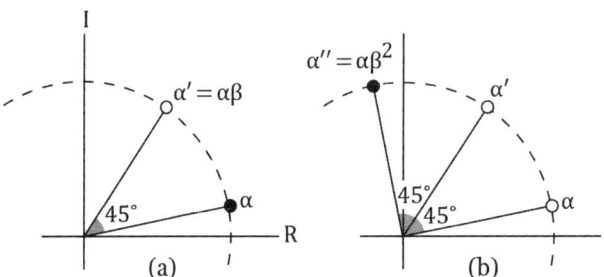

Figure 2-14: Rotating α. (a) The point α and the new $\alpha' = \alpha\beta$, rotated 45 degrees from the original α. (b) Rotating α by 45 degrees again, or forming $\alpha\beta^2$, gives us the original α rotated by 90 degrees.

Note that applying β twice means multiplying by β twice, which is β^2, not 2β.

Because the number plane is such an exquisite match to the complex numbers themselves, this geometry is all we need to be sure that this $\beta = \sqrt{i}$. But for completeness, let's check this result algebraically. The classical trig of the 45–45–90 triangle in Figure 2-13 tells us that $\beta = 1/\sqrt{2} + (1/\sqrt{2})i$. We can find β^2 by multiplying this with itself, and hope that we can manipulate that result to show that it's the same as i. One series of steps that gets us there is shown in Equation 2.17.

$$\beta\beta = \left(\frac{1}{\sqrt{2}} + \frac{1}{\sqrt{2}}i\right)\left(\frac{1}{\sqrt{2}} + \frac{1}{\sqrt{2}}i\right) \qquad \text{Expand } \beta$$

$$= \left(\frac{1}{\sqrt{2}}\right)^2 + 2\left(\frac{1}{\sqrt{2}}\right)\left(\frac{1}{\sqrt{2}}\right)i + \left(\frac{1}{\sqrt{2}}\right)^2 i^2 \qquad \text{Multiply the terms} \qquad (2.17)$$

$$= \frac{1}{2} + 2\left(\frac{1}{2}\right)i - \frac{1}{2} \qquad \text{Collect and simplify}$$

$$= i$$

Great! We've confirmed using algebra what we found by geometry: Squaring $\beta = 1/\sqrt{2} + 1/\sqrt{2}i$ gives us i. In other words, $\beta = \sqrt{i}$.

But wait a second, we're missing something! We know that square roots come in pairs. For example, both 3 and -3 are the square roots of 9, because $3^2 = (-3)^2 = 9$. So, there must be a second choice of β that will also give us $\beta^2 = i$, that comes from following the other square root of i. To find that other β, let's return to thinking geometrically. What other rotation, applied twice, would give us i? The geometry of Figure 2-15 shows us that this is a rotation of -135 degrees, or $-3\pi/4$ radians clockwise from the real axis.

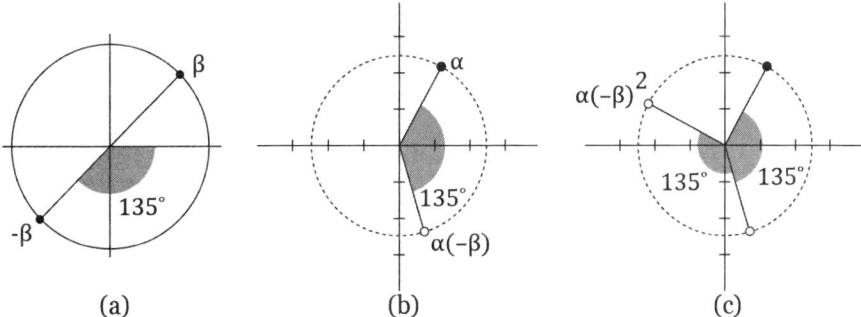

Figure 2-15: The square roots of i. (a) The points β and $-\beta$ are both a distance of 1 from the origin. (b) Rotating an arbitrary point α by multiplying it with $-\beta$. (c) Multiplying α by $-\beta$ twice, or $\alpha(-\beta)^2$, gives us the original α rotated by 90 degrees.

We can see from Figure 2-15 that this rotation, which I'll call β', is the negative of β (that is, $\beta' = -\beta$) at coordinates $-\beta = (-1/\sqrt{2}) + (-1/\sqrt{2})i$. If you multiply out the terms of $(-\beta)(-\beta)$, you'll get the same result as in Equation 2.17, confirming that multiplying any complex number with $(-\beta)^2$ is the same as multiplying it by β^2, with both causing the same 90-degree rotation produced by i. So, β and $-\beta$ are the two square roots of i.

The algebraic approach to this problem gives us the same answers. Let's look at the complex exponential form of the complex numbers involved. I'll start by writing i as the complex point $0 + 1i$, or $e^{i\pi/2}$. To find its square root, we'd like to find a new complex point e^{ib} for some real number b, so that $(e^{ib})^2 = e^{i\pi/2}$. Rewriting $(e^{ib})^2$ as the equivalent $e^{i(2b)}$, we get $2b = \pi/2$, telling us that $b = \pi/4$. The complex number $e^{i\pi/4}$ is the same point β that we saw in Figure 2-13. Phew! The same reasoning confirms that $e^{-i3\pi/4}$ is the other square root of i.

The complex number plane gives us a beautiful way to think about complex numbers and operations with them. It tells us that instead of thinking of i as a mysterious quantity, $\sqrt{-1}$, we can instead think of it geometrically, where multiplying any complex number α with i causes α to rotate around the origin of the number plane counterclockwise by 90 degrees.

List Structure

Throughout this book, we'll create many lists of numbers. For example, the description of a quantum object will be a list of two numbers.

In this section, we'll look at list structure and a standardized way to add up the elements of a list (which we'll do often). Let's start by coming up with a concise way to describe what types of numbers are in a list and how many there are.

We'll start with a 2D point, which we can view as a list of two real numbers. We can describe that list by writing its *structure*, or *format*, as $\mathbb{R} \times \mathbb{R}$. When combining sets, the $\times$ symbol (called the *Cartesian product* in this context) doesn't mean multiplication as we're used to it, but instead stands for a different mathematical operation. For our needs in this book, it can be considered a shorthand for "is attached to" or "is followed by." So, $\mathbb{R} \times \mathbb{R}$ means that we're describing a list containing a real number followed by another real number. If we wanted to describe the structure of a 3D point, we could write it as $\mathbb{R} \times \mathbb{R} \times \mathbb{R}$, or three real numbers in a sequence.

This notation won't scale well for a list of, say, 512 dimensions. So even though $\times$ here doesn't mean multiplication, we'll pretend it does so that we can use exponentiation for repeated multiplication. Just as 5×5 can be written 5^2, so too can our structure of a 2D list of real numbers be written $\mathbb{R}^2$ and our 3D list as $\mathbb{R}^3$. We can use this notation for any kind of list and numbers, so a list of 512 complex numbers could be described as having the structure $\mathbb{C}^{512}$.

We'll often want to process all the items in a set or list one by one and then add up the results. Let's say we're working with a list named $\mathbf{v}$ that has a dimensionality of d (that is, it contains d elements). We'd like to apply some arbitrary function f to each element in the list, then add up those results. We're going to do this a lot! For this discussion, I'll use the letter n for the number of elements, because that's another widespread convention that's worth being familiar with.

Authors write this operation in a few different ways. Let's start with the most traditional form in Equation 2.18, where we use an integer, arbitrarily named k, to step through the values from 0 up to and including $n - 1$ for the n elements in the list. We'll use these values to extract the elements of $\mathbf{v}$ one by one and give them to our function f.

$$\sum_{k=0}^{n-1} f(v_k) = f(v_0) + f(v_1) + \cdots + f(v_{n-1}) \tag{2.18}$$

Since we'll do this so often, I'll introduce a shortcut to reduce some of the clutter. I'll write the sequence containing all the integers from 0 up to and including $n - 1$ as $[n]$. For example, $[4]$ refers to the list of integers $[0, 1, 2, 3]$. The definition is given in Equation 2.19.

$$[n] \overset{\Delta}{=} [0, 1, 2, \ldots, n - 1] \tag{2.19}$$

Note that $[n]$ is a sequence, not a set, because the elements are in a specific order. Using this new notation, we can rewrite Equation 2.18 a little more simply as Equation 2.20.

$$\sum_{k \in [n]} f(v_k) \tag{2.20}$$

In this context, the expression $k \in [n]$ under the big sigma tells us to use each integer from $[n]$ in sequence, one at a time.

I'll use this form frequently throughout the book.

Another way to write these indices is in binary form. A *bitstring* is a sequence of 0s and 1s. Together, they form a binary number. For example, the bitstring 1101 is the binary equivalent of the decimal number 13. We will often work with sets and lists that contain 2^n elements, for some integer n. Such lists are indexed by values from 0 to $2^n - 1$, and every decimal number from 0 to $2^n - 1$ has a unique corresponding bitstring of n elements. We can write the set of all of these bitstrings as $\mathbb{B}^n$, which contains every combination of n elements that are 0 or 1. That means we can write Equation 2.20 as Equation 2.21.

$$\sum_{k \in \mathbb{B}^n} f(v_k) \tag{2.21}$$

I'll often use the capital N for 2^n, so we can also write the range of values from 0 to $2^n - 1$ as $[N]$. Thus, the four expressions in Equation 2.22 are all equivalent.

$$\sum_{k=0}^{2^n-1} f(v_k) \quad = \quad \sum_{k=0}^{N-1} f(v_k) \quad = \quad \sum_{k \in \mathbb{B}^n} f(v_k) \quad = \quad \sum_{k \in [N]} f(v_k) \tag{2.22}$$

I think the rightmost two versions are the easiest to read and comprehend, so I'll use them almost exclusively in this book.

The version in Equation 2.20 is the form I'll use for lists with n elements, for any integer n.

Vectors

I need to clarify the term vector. As you know, we use this term to indicate an object in ordinary space that has a magnitude and a direction. . . . I want you to completely forget about that concept of a vector.

—Leonard Susskind and Art Friedman, *Quantum Mechanics: The Theoretical Minimum*, 2014 [213]

You may be used to using the term *vector* to refer to an arrow in 2D or 3D space. Those arrows are indeed vectors, but the term embraces a much broader and more abstract class of objects [282]. That is, arrows are just one special case of what a vector can be.

Postulate 1 refers to "complex vectors." Because complex vectors are at the core of quantum computing, my goal in this section is to help you feel comfortable working with these objects.

For our purposes, the term *vector* will refer to a list of numbers.

Since Postulate 1 refers to a complex vector, we know that vectors made of complex numbers will be important to us. We call these *complex-valued vectors*, or *complex vectors*, to distinguish them from *real-valued* and *integer-valued* vectors, composed of reals or integers, respectively. For now, I'll continue to write these vectors with bold, lowercase roman letters like $\mathbf{v}$ and $\mathbf{w}$. I'll write the elements of complex vectors with Greek letters like ν and ω, respectively, to emphasize that they're complex.

I'll write each of these objects as a *matrix*, or grid of elements, made of a single column, with the elements written from top to bottom. This is called a *column matrix*. For example, I'll write a three-element complex vector $\mathbf{v}$ as a column matrix of three elements, read from top to bottom, as in Equation 2.23.

$$\mathbf{v} = \begin{bmatrix} \nu_0 \\ \nu_1 \\ \nu_2 \end{bmatrix} \tag{2.23}$$

Like with any other list, the number of elements in a vector is called its *dimensionality*. We write this as dim $\mathbf{v}$, or with parentheses, as in dim($\mathbf{v}$). Be careful not to refer to the dimensionality of a vector as its "length"!

Even though we use vectors as the fundamental data structure for quantum information, we will modify them in only a few ways. Everything else will be built from these operations.

The first operation is called *scalar multiplication*. For our purposes, a *scalar* is a number of any type. For generality, let's use a complex number, σ. We'll *define* the operation of multiplying together a vector $\mathbf{v}$ and a scalar σ as multiplying each element of $\mathbf{v}$ by σ. This definition is illustrated in Equation 2.24, for a vector with three elements. For later reference, I've given this definition the name V1.

$$\sigma \mathbf{v} \overset{\Delta}{=} \sigma \begin{bmatrix} \nu_0 \\ \nu_1 \\ \nu_2 \end{bmatrix} = \begin{bmatrix} \sigma \nu_0 \\ \sigma \nu_1 \\ \sigma \nu_2 \end{bmatrix} \qquad \text{V1} \tag{2.24}$$

The second operation is called *vector addition*. We write this operation using a + sign between two vectors and *define* $\mathbf{v} + \mathbf{w}$ to be the operation of adding their corresponding elements. This definition is illustrated in Equation 2.25 for two vectors with three elements each. I've named this definition V2.

$$\mathbf{v} + \mathbf{w} \overset{\Delta}{=} \begin{bmatrix} \nu_0 \\ \nu_1 \\ \nu_2 \end{bmatrix} + \begin{bmatrix} \omega_0 \\ \omega_1 \\ \omega_2 \end{bmatrix} = \begin{bmatrix} \nu_0 + \omega_0 \\ \nu_1 + \omega_1 \\ \nu_2 + \omega_2 \end{bmatrix} \qquad \text{V2} \qquad\qquad (2.25)$$

This definition requires that $\mathbf{v}$ and $\mathbf{w}$ have the same dimensionality, or number of elements. If they don't, the definition doesn't apply, and $\mathbf{v} + \mathbf{w}$ is a meaningless string of symbols.

Linearity

Chaos should be taught, he argued. It was time to recognize that the standard education of a scientist gave the wrong impression. No matter how elaborate linear mathematics could get. . . . May argued that it inevitably misled scientists about their overwhelmingly nonlinear world.

—James Gleick, *Chaos: Making a New Science, 1987* [77]

The majority of the math in this book belongs to a field called *linear algebra* [211]. Roughly speaking, this is the study of vectors and transformations of vectors.

A key property of linear algebra is in its first word, *linear*, which refers to the idea that all the objects we work with share a characteristic called *linearity*.

To be called linear, an object needs to have (or satisfy) two properties. One property is based on scalar multiplication and the other on vector addition. These properties closely mirror the V1 and V2 properties we just saw.

Let's illustrate these two properties with a function. Assume that we have a function f that takes in a vector $\mathbf{x}$ of any dimensionality as input and produces a vector $\mathbf{y}$ of the same dimensionality as output [241]. We can write this as $\mathbf{y} = f(\mathbf{x})$ for input $\mathbf{x}$ and output $\mathbf{y}$.

The first property of linearity we'll consider tells us that if we multiply an input by some number, that's the same as leaving the input alone and multiplying the output by that same number. I've written this in symbols in Equation 2.26 for some complex number σ and some complex vector $\mathbf{v}$. For later reference, I'll call this property L1.

$$f(\sigma \mathbf{v}) = \sigma f(\mathbf{v}) \qquad \text{L1} \qquad\qquad (2.26)$$

The second property, which I'll call L2, says that if we add two vectors and apply f, we get the same result as we would if we instead applied f to each vector and then added the outputs, as shown in Equation 2.27.

$$f(\mathbf{v} + \mathbf{w}) = f(\mathbf{v}) + f(\mathbf{w}) \qquad \text{L2} \qquad\qquad (2.27)$$

We'll use these two properties so often in this book that they'll become second nature to you. They're that useful.

In Equation 2.28, I've written out these properties by showing all the elements (often called the *tableau form*) for vectors of dimensionality 3.

$$f\left(\begin{bmatrix} \sigma\,\nu_0 \\ \sigma\,\nu_1 \\ \sigma\,\nu_2 \end{bmatrix}\right) = f\left(\sigma\begin{bmatrix} \nu_0 \\ \nu_1 \\ \nu_2 \end{bmatrix}\right) = \sigma f\left(\begin{bmatrix} \nu_0 \\ \nu_1 \\ \nu_2 \end{bmatrix}\right) \qquad \text{L1}$$

(2.28)

$$f\left(\begin{bmatrix} \nu_0 + \omega_0 \\ \nu_1 + \omega_1 \\ \nu_2 + \omega_2 \end{bmatrix}\right) = f\left(\begin{bmatrix} \nu_0 \\ \nu_1 \\ \nu_2 \end{bmatrix}\right) + f\left(\begin{bmatrix} \omega_0 \\ \omega_1 \\ \omega_2 \end{bmatrix}\right) \qquad \text{L2}$$

Let's try out these definitions. I'll pick a tiny function that takes in a single real number (or a real vector of dimensionality 1) and returns a real number. The function is $f(x) = x + 1$. It gets to be called linear only if it satisfies *both* properties L1 and L2. Let's test f using $a = 3$ and $b = 4$, as shown in Equation 2.29. The left column checks property L1, and the right column checks L2.

$$f(x) = x + 1$$
$$a = 3, b = 4$$

$$
\begin{array}{ccc}
\underline{\text{L1}} & & \underline{\text{L2}} \\
a f(b) \overset{?}{=} f(ab) & \qquad & f(a) + f(b) \overset{?}{=} f(a + b) \\
3 f(4) \overset{?}{=} f(3 \times 4) & & f(3) + f(4) \overset{?}{=} f(3 + 4) \\
3 \times 5 \overset{?}{=} f(12) & & 4 + 5 \overset{?}{=} f(7) \\
15 \neq 13 & & 9 \neq 8
\end{array}
$$

(2.29)

Surprise! Despite its simplicity, this function fails both tests. We've found that f is not a linear function.

Let's try another function, g. This function's inputs and outputs are again real numbers (that is, real vectors of dimensionality 1). Let's say $g(x) = 2x$. I'll check for properties L1 and L2 as before, in the left and right columns of Equation 2.30, again using $a = 3$ and $b = 4$.

$$g(x) = 2x$$
$$a = 3, b = 4$$

$$
\begin{array}{ccc}
\underline{\text{L1}} & & \underline{\text{L2}} \\
a g(b) \overset{?}{=} g(ab) & \qquad & g(a) + g(b) \overset{?}{=} g(a + b) \\
3 g(4) \overset{?}{=} g(3 \times 4) & & g(3) + g(4) \overset{?}{=} g(3 + 4) \\
3 \times 8 \overset{?}{=} g(12) & & 6 + 8 \overset{?}{=} g(7) \\
24 = 24 & & 14 = 14
\end{array}
$$

(2.30)

Because both tests are satisfied, g is a linear function.

Well, maybe. While the counterexample for f is a definite proof that f is not linear, I only demonstrated that g was linear for $a = 3$ and $b = 4$. Maybe it's not linear if I use other values. As I mentioned in the Introduction, sometimes an example is all we need to get the gist of an idea, and I think that's the case here. You can develop a formal proof for the linearity of g (or any other function), but it takes a little more work [150].

The moral of these examples is that even if a function appears simple, we can't assume it's linear. If we're not sure that a function is linear, we need to check.

Linearity is so important that the study of almost any physical phenomenon begins by simplifying it to create an approximation where all the equations needed to describe that approximation are linear. Once we've gotten as much out of this simplified version as we can, we gradually include operations that are not linear, or *nonlinear*. Nonlinear operations include taking a square root, or taking the sine or cosine of some value. Anything that doesn't satisfy both conditions L1 and L2 is nonlinear.

We need nonlinear operations to accurately describe the weather, the sounds of a piano, how water waves move, how trees grow, and how human languages spread. Almost any useful description of a real-world phenomenon requires nonlinear operations. This is unfortunate, because nonlinear operations cause our math to become more complicated and harder to understand. But we need these nonlinearities, because most linear approximations are bad at making accurate predictions.

To see how a strictly linear model can fail, consider the time it takes a person to run some distance. An adult male can typically run a 1-kilometer (1K, or 1,000 meter) race in about 4 minutes [181].

Now suppose we want to estimate how long it would take that person to run a marathon. A linear model would say that since there are about 42,000 meters in a marathon, there are 42,000 / 1,000 = 42 segments of 1,000 meters each. In symbols, where $f(x)$ is the time it takes to run x meters, the linear model would say $f(42,000) = 42 \times f(1,000) = 42 \times 4 = 168$, so it would take 168 minutes, or about 2 hours and 50 minutes, to complete the marathon. But that can't be right, because in a 1K race you can burn up all your energy quickly and run fast, while you have to conserve your energy and run at a slower pace to complete a marathon. In fact, the average marathon time for an adult male is about 270 minutes, or about 3 hours and 30 minutes [130].

The linear model underestimated the marathon time by about 40 minutes, which is significant for a race of a few hours.

Cooking is also non-linear. Suppose that a particular recipe tells you to cook a dish for 100 minutes at 450 degrees Fahrenheit. If cooking were linear, you'd get the same results by cooking the dish for 1/10 the time at 10 times the temperature, or 10 minutes at 4,500 degrees. You'd also get the same results by cooking it for 1 minute at 45,000 degrees. That's over four times the heat at the surface of the sun [51]! While nobody would complain that the resulting black lump was undercooked, I don't think it would be nearly as tasty as if you'd followed the recipe's original advice. Cooking times and temperatures are definitely not linear.

There's nothing special about running or cooking. Almost everything in the real world is nonlinear. This makes it astonishing that the core of quantum mechanics is linear—not an approximation, but actually linear. Nobody really knows why this is the case, but it is. The result is that our math is going to be accessible and comprehensible to a degree we'd never have guessed from our experience with describing running, cooking, or almost any other real-world phenomenon. This is a gift we're going to make use of frequently, by using the linearity properties L1 and L2 to simplify our equations.

Bases

Let's use linearity right away to make vectors easier to work with, by representing them as sums of simpler vectors.

Imagine a vector where every element is 0 except for one element, which has the value 1. In this section, I'll call this vector $\mathbf{b}_k$, where the integer index k tells us where the 1 is located, numbering from top to bottom and, as always, starting at 0. I'll assume that the dimensionality of these vectors will be clear from context, matching the dimensionality of any other vectors being discussed. For any dimensionality d, there will be d of these vectors, each one with a single 1. I've shown them in Equation 2.31 for $d = 3$.

$$\mathbf{b}_0 \triangleq \begin{bmatrix} 1 \\ 0 \\ 0 \end{bmatrix}, \qquad \mathbf{b}_1 \triangleq \begin{bmatrix} 0 \\ 1 \\ 0 \end{bmatrix}, \qquad \mathbf{b}_2 \triangleq \begin{bmatrix} 0 \\ 0 \\ 1 \end{bmatrix} \qquad (2.31)$$

We call these vectors *basis vectors* or *bases* (notice that the i changed to an e). Together, they form a *basis set*, or *basis* for short. We call a basis set built from this recipe a *computational basis* or *conventional basis* of d dimensions [49].

Even though our vectors will generally have complex elements, these basis vectors are composed only of the numbers 0 and 1.

Basis vectors are important because they let us break up other vectors into their individual elements, or assemble such elements into a single vector. For example, in the traditional 2D Cartesian coordinate system, the X and Y axes are more formally represented as basis vectors, and every point in that 2D plane is created by scaling those basis vectors and then adding them.

Just as in the Cartesian coordinate system, we can write any vector **v** as a sum of scaled basis vectors. The steps are shown in Equation 2.32, using our vector properties in their most basic form, by assuming that the function f returns its input as its output (that is, $f(\mathbf{x}) = \mathbf{x}$). That means we can ignore the function (and those giant parentheses). Remember that each ν_k is a complex number.

$$\mathbf{v} = \begin{bmatrix} \nu_0 \\ \nu_1 \\ \nu_2 \end{bmatrix}$$

$$= \begin{bmatrix} \nu_0 \\ 0 \\ 0 \end{bmatrix} + \begin{bmatrix} 0 \\ \nu_1 \\ 0 \end{bmatrix} + \begin{bmatrix} 0 \\ 0 \\ \nu_2 \end{bmatrix} \qquad \text{Use property V2} \qquad (2.32)$$

$$= \nu_0 \begin{bmatrix} 1 \\ 0 \\ 0 \end{bmatrix} + \nu_1 \begin{bmatrix} 0 \\ 1 \\ 0 \end{bmatrix} + \nu_2 \begin{bmatrix} 0 \\ 0 \\ 1 \end{bmatrix} \qquad \text{Use property V1}$$

$$= \nu_0 \mathbf{b}_0 + \nu_1 \mathbf{b}_1 + \nu_2 \mathbf{b}_2$$

We can generalize this pattern for vectors of any dimensionality d, giving us Equation 2.33. We call each value ν_k in Equation 2.33 a *coefficient* or *coordinate* of $\mathbf{v}$ in the basis $\mathbf{b}$.

$$\mathbf{v} = \sum_{k \in [d]} \nu_k \mathbf{b}_k \qquad (2.33)$$

I boxed this equation because we'll often write vectors in this form. It lets us manipulate a vector by changing its individual elements ν_k, which is frequently more convenient than modifying the whole vector all at once.

The basis vectors $\mathbf{b}_k$ that I defined in Equation 2.31, generalized for any number of dimensions, will be our default for the rest of the book. From now on, every time I refer to "a basis" or "the basis vectors," I will always mean vectors in the form of those in Equation 2.31 (and their extensions to any number of dimensions). Any time I use some other set as a basis, I'll say so explicitly.

Before we move on, I want to point out that we define the computational basis vectors $\mathbf{b}_k$ in this way for two reasons. First, this is the basis used in almost every discussion of quantum computing, so it's a good thing to get used to. Second, it's a great choice for a basis, because these basis vectors make the math of quantum algorithms particularly clear. If you want, you can create an infinite number of other basis sets, and as long as they satisfy a few rules [283] [248], each one will be just as useful as the computational basis (though perhaps harder for people to interpret).

There are a few more things to learn about vectors and bases that will be of great value throughout the rest of the book. These will be our focus in the next few sections.

The Dot Product

You don't have to be a mechanic (quantum or otherwise) to appreciate a good multitool. A Swiss army knife, Leatherman, or other multipurpose tool seems to solve an unlimited number of everyday problems.

In quantum computing, a function called the *inner product* is our version of a multitool. Much of what we do with this tool is abstract and algebraic, largely because our vectors will be of high dimensionality and thus contain many complex numbers. This makes the actions we perform difficult to visualize, since the pictures we can draw on the page are limited to two dimensions (or three, if we're careful and we agree on a lot of conventions).

Happily, there's a stripped-down version of the inner product that applies nicely to two-dimensional vectors of real numbers. This mini-tool is called the *dot product*.

With 2D vectors and the dot product, we can draw pictures of many of the operations we'll use in this book. Building our intuition with these 2D pictures can help us reason about what these operations will do when we apply their full versions in abstract spaces of higher dimensions.

Thus, in this section we'll see a bunch of useful operations that we can perform in 2D using the dot product.

There are three main reasons for introducing these ideas to you here, even though you might not see the value of learning all of this now.

First, we'll be able to carry out much of our discussion in two dimensions, where we can draw pictures and grasp the geometry directly. Second, by covering these ideas now, together, we can build each idea on the previous ones while they're still fresh in our minds. Finally, having these ideas even roughly in place now will be of great value later. We'll be able to consider more complicated ideas in a smooth and coherent flow, rather than having to constantly stop to introduce a piece of technique, resume the discussion, and then stop again to talk about the next underlying idea. What we do here will pay off later.

The vectors we'll work with in this section will each be made up of only two real numbers. To stress this limitation, I'll call them *arrows*. We can draw arrows by putting their tails at the origin and their heads at a location given by two real numbers representing distances along the X and Y axes, in that order. Because arrows are vectors (though a very special case), I'll name them like vectors, with lowercase, bold roman letters like **v** and **w**.

For just this section, I'll label angles with lowercase Greek letters, like α and θ. Remember that these are real numbers, not just because they're 2D angles, but also because complex numbers don't appear in this section (I'll let you know when we switch back to complex vectors, starting at the section on inner products).

Also for this section only, I'll do everything in the 2D Cartesian coordinate system you learned about in high school. As usual, it has two axes named X and Y, set at right angles to one another. The point where they meet is called the *origin* and has coordinates $(0, 0)$. Figure 2-16 shows an arrow **a** in the plane.

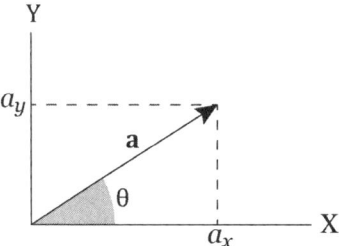

Figure 2-16: An arrow **a** in a typical 2D plane with perpendicular axes X and Y

The components of this arrow **a** are real numbers a_x and a_y.

Our first goal will be to find the angle between two known arrows. Figure 2-17 shows an arrow **a** that makes an angle α with the X axis and an arrow **b** that makes an angle β. We want to find the angle γ between **a** and **b**.

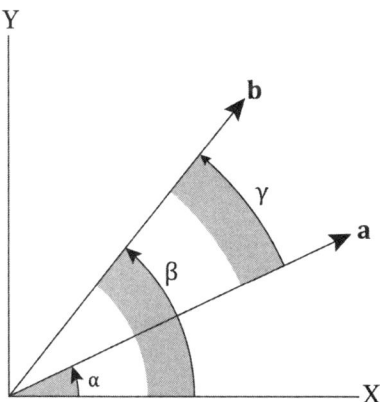

Figure 2-17: Finding the angle γ between **a** and **b**

For now, all we know are the coordinates of **a** and **b**. Let's use these to label the sides of the right triangles formed by these vectors, as in Figure 2-18.

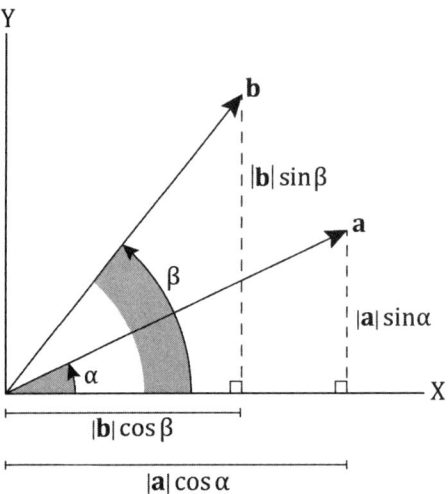

Figure 2-18: Drawing a right triangle for each arrow

The components of **a** and **b** can be read off the right triangles they form with the X axis. I've written the coordinates in Equation 2.34.

$$a_x = |\mathbf{a}| \cos \alpha, \qquad a_y = |\mathbf{a}| \sin \alpha$$
$$b_x = |\mathbf{b}| \cos \beta, \qquad b_y = |\mathbf{b}| \sin \beta \tag{2.34}$$

I've written $|\mathbf{a}|$ for the magnitude of **a**, or the distance from the origin to the tip of **a**. This is a real number. Similarly, $|\mathbf{b}|$ is a real number representing the magnitude of **b**. We can find these magnitudes from the Pythagorean rule, using the coordinates we just identified in Equation 2.34, giving us Equation 2.35.

$$|\mathbf{a}| = \sqrt{a_x^2 + a_y^2}, \qquad |\mathbf{b}| = \sqrt{b_x^2 + b_y^2} \tag{2.35}$$

The next step will come out of nowhere. I'll assert that by forming the sum $a_x b_x + a_y b_y$, we'll get back an expression that tells us how to compute γ.

Having made this claim, let's see that it's true. In Equation 2.36, the first line shows this expression involving the starting combination of coordinates, and the last line gives us an expression involving γ.

$$(a_x b_x) + (a_y b_y)$$

$= (|\mathbf{a}| \cos \alpha \, |\mathbf{b}| \cos \beta) + (|\mathbf{a}| \sin \alpha \, |\mathbf{b}| \sin \beta)$	From Eq. 2.34	
$= |\mathbf{a}| \, |\mathbf{b}| (\cos \alpha \cos \beta + \sin \alpha \sin \beta)$	Rearrange	(2.36)
$= |\mathbf{a}| \, |\mathbf{b}| \cos(\alpha - \beta)$	$(\star)$	
$= |\mathbf{a}| \, |\mathbf{b}| \cos \gamma$	Write $\gamma = \alpha - \beta$	

The line marked with a star uses a rule from trigonometry that is easily forgotten, even if you've seen it before. It's called the *cosine double-angle rule*, and it tells us that the mess on the right end of the second line of Equation 2.36 is just a complicated way of writing $\cos(\alpha - \beta)$. If this isn't familiar, or seems unlikely, you can rederive the rule yourself [60].

The last line of Equation 2.36 gives us an expression involving the angle γ between the two arrows. The expression requires their magnitudes, but since we know the components of the arrows, we can get their magnitudes from Equation 2.35.

The number $|\mathbf{a}|\,|\mathbf{b}|\cos\gamma$ that we get from Equation 2.36 is the dot product I mentioned at the start of this section. Since each line of Equation 2.36 is a different expression for the same number, any of them could be used as the definition of the dot product. Traditionally, we use the expression either on the last line or at the start of the first line. We write this relationship by placing a vertically centered dot between the arrow names, as in $\mathbf{a} \cdot \mathbf{b}$, which we read aloud as "a dot b" (hence the name "dot product"). With these agreements, the definition of the 2D dot product is thus given in Equation 2.37.

$$\mathbf{a} \cdot \mathbf{b} \overset{\Delta}{=} a_x b_x + a_y b_y$$
$$= |\mathbf{a}|\,|\mathbf{b}|\,\cos\gamma \tag{2.37}$$

We can solve for $\cos\gamma$, or γ itself, from Equation 2.37 to give us Equation 2.38.

$$\cos\gamma = \frac{\mathbf{a} \cdot \mathbf{b}}{|\mathbf{a}|\,|\mathbf{b}|}$$
$$\gamma = \cos^{-1}\left(\frac{\mathbf{a} \cdot \mathbf{b}}{|\mathbf{a}|\,|\mathbf{b}|}\right) \tag{2.38}$$

The angle γ between two arrows tells us to what extent they point in the same direction. If $\gamma = 0$, then the arrows point in exactly the same direction (though they might have different magnitudes). If $\gamma = \pi$ (or 180 degrees), then they point in exactly opposite directions. And if the angle is $\pi/2$ or $-\pi/2$ (that is, either 90 or -90 degrees), then the two arrows are perpendicular to one another. A synonym for perpendicular is *orthogonal*.

Figure 2-19 shows several different pairs of arrows with different angles γ and their dot products beneath them. For this figure, I set both magnitudes to 1 (that is, $|\mathbf{a}| = |\mathbf{b}| = 1$), so $\cos\gamma = \mathbf{a} \cdot \mathbf{b}$.

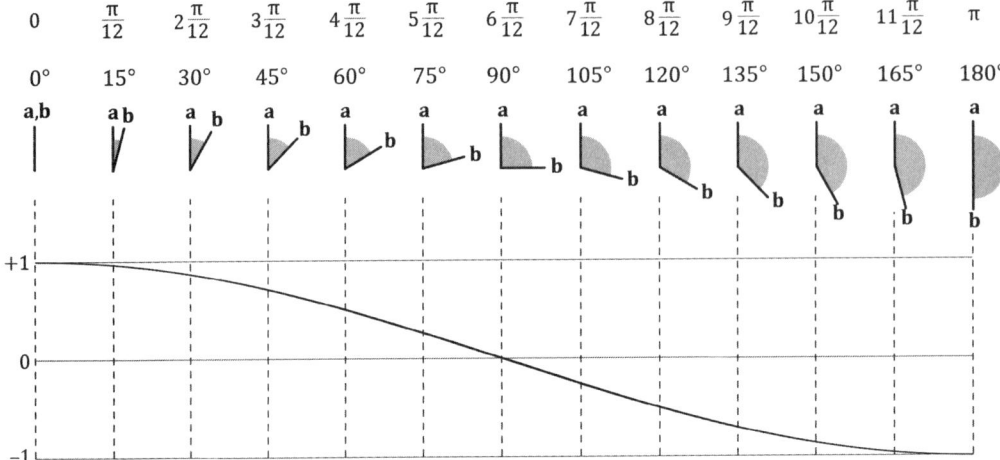

Figure 2-19: Rows 1 and 2: The angle γ for each pair in radians and degrees. Row 3: A pair of arrows, each of magnitude 1, with that γ. Bottom: The cosine curve from 0 to π, showing the dot product for each pair of arrows, along with all the other possible pairs of arrows.

Equation 2.35 shows us how to compute the magnitude of an arrow from its components. We can also find magnitudes with the dot product.

To see this, I'll again just give you the formula that I know will do the job, and then prove that it works. In this case, the method is to find the dot product of an arrow with itself, which will give us the squared magnitude of the arrow.

Let's first run through it using components. Using the definition of Equation 2.37, we can find $\mathbf{a} \cdot \mathbf{a}$ as shown in Equation 2.39.

$$\begin{aligned} \mathbf{a} \cdot \mathbf{a} &= a_x\,a_x + a_y\,a_y \\ &= a_x{}^2 + a_y{}^2 \end{aligned} \tag{2.39}$$

Comparing this to Equation 2.35, we see that $\mathbf{a} \cdot \mathbf{a}$ gives us the square of the magnitude of $\mathbf{a}$, or in symbols, $|\mathbf{a}|^2 = \mathbf{a} \cdot \mathbf{a}$.

To build up our skills at manipulating these kinds of relationships, let's look at a different way to show that $\mathbf{a} \cdot \mathbf{a}$ is the squared magnitude of $\mathbf{a}$.

I'll start with the version of the dot product given by the last line of Equation 2.36. The steps are in Equation 2.40.

$$\begin{aligned} \mathbf{a} \cdot \mathbf{a} &= |\mathbf{a}|\,|\mathbf{a}|\cos\gamma && \text{From Eq. 2.36} \\ &= |\mathbf{a}|\,|\mathbf{a}|\cos 0 && \text{The angle } \gamma \text{ between } \mathbf{a} \text{ and} \\ &&& \text{itself is } 0 \\ &= |\mathbf{a}|\,|\mathbf{a}| && \text{Since } \cos 0 = 1 \\ &= |\mathbf{a}|^2 && \text{Matching Eq. 2.35} \end{aligned} \tag{2.40}$$

We can use the dot product to answer two important questions about arrows. First, given an arrow, does it have a magnitude of 1? This will be important to us because this is a condition required of the vectors that describe quantum states.

Second, given two arrows, are they perpendicular to one another? This is useful to know because perpendicular arrows make the most convenient basis (the basis vectors $\mathbf{b}_k$ we saw in Equation 2.31 are all perpendicular to one another). It also reduces our conceptual overhead when operations get complicated, since we can focus our attention on just one component at a time.

Let's see how the dot product helps us with checking both of these conditions.

Using the Dot Product

Let's first look at how the dot product lets us check if an arrow has a magnitude of 1. Equation 2.40 tells us that the squared magnitude of $\mathbf{a}$ is given by $\mathbf{a} \cdot \mathbf{a}$. So if this dot product has a value of 1, we know that $|\mathbf{a}| = 1$. Boom, done!

Now let's check if the two arrows are perpendicular. I'll start with the form of the dot product on the second line of Equation 2.37. If the arrows each have a nonzero magnitude, but their dot product is 0, then we know that $\cos \gamma$ must be 0. That can only happen if $\gamma = \pi/2$ or $\gamma = -\pi/2$, both of which correspond to a right angle, which we can see visually in the middle of Figure 2-19. So, to check if two arrows are perpendicular, we can compute their dot product. If that result is 0, then we know that the arrows are perpendicular. If the result is anything other than 0, they are not perpendicular (though the closer the result is to 0, the closer they are to being perpendicular).

The dot product of any arrow with itself, given in Equations 2.39 and 2.40, generalizes to arrows of any number of dimensions d. We create the product of every pair of corresponding entries, as we did for the first two entries in Equation 2.37, and then add up all of these products. The resulting formula is given in Equation 2.41. Remember that $[d]$ stands for the sequence of integers from 0 to $d - 1$.

$$\mathbf{a} \cdot \mathbf{b} = \sum_{k \in [d]} a_k \, b_k \qquad (2.41)$$

We'll often want to find the magnitude of a vector. Equation 2.40 showed us how to compute that using the dot product. Let's write the formulas for the squared magnitude, and the magnitude itself, in one place for easy reference. They're in Equation 2.42.

$$|\mathbf{a}|^2 = \mathbf{a} \cdot \mathbf{a}$$
$$|\mathbf{a}| = \sqrt{\mathbf{a} \cdot \mathbf{a}} \qquad (2.42)$$

We've seen how the dot product helps us determine if an arrow has a magnitude of 1 and how to find the angle between two arrows. Recall that we're learning these things in 2D, using the dot product, because all of these operations will be useful when we later generalize the dot product to do similar operations using vectors that describe quantum states, which are made up of many complex numbers. That's a world that we can't draw direct pictures for. Let's continue with another 2D operation that we can draw, and which we'll also later generalize for these larger, more abstract vectors.

Projection

Let's pull another gadget from our dot product multitool. It's a technique called *projection*, which involves a pair of arrows. We want to determine how much of an arrow **a** is, in some sense, a part of, or contained in, another arrow **b**. That's pretty vague, so let's tighten it up and draw a picture of the process.

Think of someone with a flashlight, standing far away from the two arrows. They're shining the light onto **a** from a direction perpendicular to **b**, as shown in Figure 2-20(a). This causes the arrow **a** to cast a shadow, or projection, onto **b**.

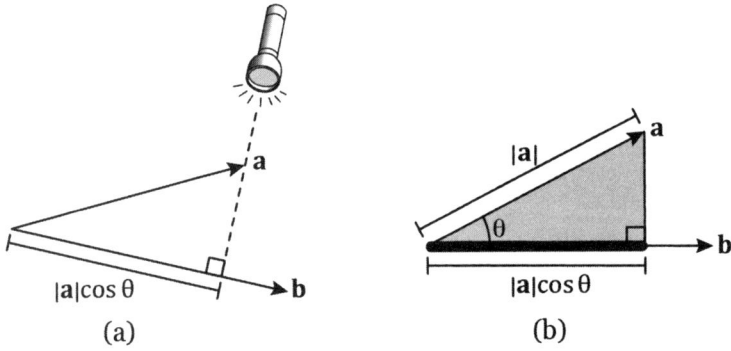

Figure 2-20: Shining a light on **a** from a direction perpendicular to **b** casts a shadow, or projection, of **a** onto **b**.

Figure 2-20(b) shows the shadow, and the thick line shows the projection of **a** onto **b**. The geometry in part (b) tells us that the magnitude of the projection is given by $|\mathbf{a}| \cos \theta$. As we've seen, we can find both $|\mathbf{a}|$ and $\cos \theta$ using the dot product.

What if **b** is pointing the other way, so **a** doesn't fall onto it at all, as in Figure 2-21? We can still compute the magnitude of the projection of **a** onto **b** using the dot product, just as before. In this case, we'll get back a negative real number, since the cosine curve is negative for angles greater than $\pi/2$ (as shown in Figure 2-19).

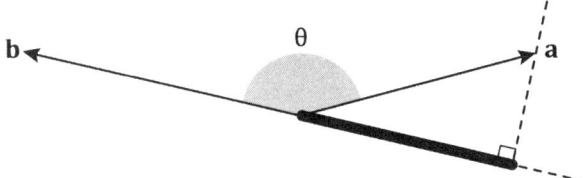

Figure 2-21: Projecting arrow **a** onto **b** when the angle is greater than $\pi/2$

If all we care about is the magnitude of the projection, we're done. But frequently, we'll want to produce a new arrow, $\mathbf{a}'$, in the same direction as **b**, but with a magnitude given by the dot product with **a**. This new arrow is, in some loose sense, the amount of **a** that is also part of **b**. Figure 2-22 shows the idea.

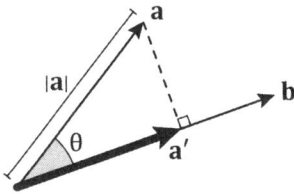

Figure 2-22: Creating $\mathbf{a}'$ from the projection of **a** onto **b**

The ability to find this arrow (well, the generalized version of this arrow) will prove to be useful when we work with quantum states, so let's see how to find it in the 2D world.

We know from Figure 2-22 that the magnitude of $\mathbf{a}'$, or $|\mathbf{a}'|$, is $|\mathbf{a}| \cos\theta$. Using $\cos\theta$ from Equation 2.38, we can write Equation 2.43.

$$
\begin{aligned}
|\mathbf{a}'| &= |\mathbf{a}| \cos\theta && \text{From Fig. 2-22} \\
&= |\mathbf{a}| \frac{\mathbf{a} \cdot \mathbf{b}}{|\mathbf{a}| \, |\mathbf{b}|} && \text{Use } \cos\theta \text{ from Eq 2.38} \\
&= \frac{\mathbf{a} \cdot \mathbf{b}}{|\mathbf{b}|} && \text{Cancel } |\mathbf{a}| \, / \, |\mathbf{a}|
\end{aligned}
\tag{2.43}
$$

How do we scale **b** to this magnitude? The easiest, and usual, way is to make a version of **b** that has a magnitude of 1 and then scale it by the quantity in Equation 2.43. We call an arrow with a length of 1 a *unit arrow*, and we say it has *unit magnitude*. Similarly, we call any vector with a length of 1 a *unit vector*. So in this situation, we want to create a unit arrow in the direction of **b**.

To make a version of **b** with a magnitude of 1, we can scale **b** by dividing it by its magnitude, producing $\mathbf{b} / |\mathbf{b}|$. This will be an arrow pointing in the same direction as **b**, but with a magnitude of 1. We sometimes call this a *unit arrow* (some authors use a hat to mark that an object has a magnitude of 1, so $\hat{\mathbf{b}} = \mathbf{b} / |\mathbf{b}|$). The process is shown in Figure 2-23(a).

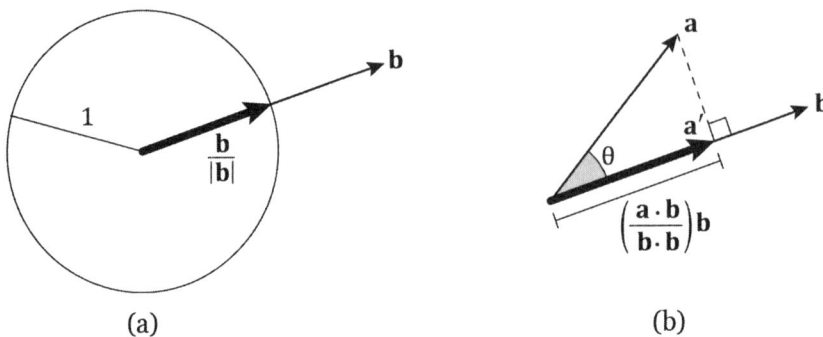

Figure 2-23: (a) The arrow $\mathbf{b}/|\mathbf{b}|$ is an arrow pointing in the direction of $\mathbf{b}$ with a magnitude of 1. (b) Scaling $\mathbf{b}/|\mathbf{b}|$ with the value in Equation 2.43 gives us $\mathbf{a}'$, the projection of $\mathbf{a}$ onto $\mathbf{b}$.

To find $\mathbf{a}'$, we multiply the unit arrow in the direction of $\mathbf{b}$ by the scalar given in Equation 2.43. The result is shown in Equation 2.44.

$$\mathbf{a}' = \frac{\mathbf{a}\cdot\mathbf{b}}{|\mathbf{b}|}\frac{\mathbf{b}}{|\mathbf{b}|} \quad \text{Scale } \mathbf{b}/|\mathbf{b}| \text{ by Eq. 2.43}$$

$$= \frac{\mathbf{a}\cdot\mathbf{b}}{\mathbf{b}\cdot\mathbf{b}}\mathbf{b} \quad \text{Use } |\mathbf{b}|^2 = \mathbf{b}\cdot\mathbf{b} \text{ from Eq. 2.42}$$

(2.44)

In the second line, I replaced $|\mathbf{b}|\,|\mathbf{b}|$ with $\mathbf{b}\cdot\mathbf{b}$ because it saves us the trouble of explicitly computing the magnitude of $|\mathbf{b}|$. The arrow $\mathbf{a}'$ we just made is shown in Figure 2-23(b).

The beautiful thing about this expression for $\mathbf{a}'$ is that it doesn't explicitly involve the coordinates of the arrows. We call this a *coordinate-free* expression. These types of expressions are indispensable when we write quantum algorithms with vectors that have large numbers of dimensions. If we had to write out all the coordinates, the expressions could become gigantic; however, the coordinate-free versions don't change at all, whether our arrows have 2 dimensions or 200.

Suppose that we project $\mathbf{a}$ onto $\mathbf{b}$, and then we project that result onto $\mathbf{b}$ again. I suggest you pause for a moment and think about this. If you were in charge of setting up all of these relationships, what would you *want* the result to be?

Here's how it works out. Start by noting from Equation 2.44 that projecting $\mathbf{a}$ onto $\mathbf{b}$ gives us a scaled version of $\mathbf{b}$. I'll write the scaling term as s, so the projection can be written $s\mathbf{b}$. Now I'll apply Equation 2.44 to project $s\mathbf{b}$ onto $\mathbf{b}$, giving us Equation 2.45.

$$\frac{s\mathbf{b}\cdot\mathbf{b}}{\mathbf{b}\cdot\mathbf{b}}\mathbf{b} = s\frac{\mathbf{b}\cdot\mathbf{b}}{\mathbf{b}\cdot\mathbf{b}}\mathbf{b} = s\mathbf{b} \qquad (2.45)$$

The dot products cancel one another.

So, we've found that if we have the projection of $\mathbf{a}$ onto $\mathbf{b}$, we can project that result back onto $\mathbf{b}$ again as many times as we like, and it doesn't change. In other words, the projection of $s\mathbf{b}$ onto $\mathbf{b}$ is just $s\mathbf{b}$, unchanged. Is this what you'd have wanted it to be?

I find this result reassuring, because to me it makes geometric sense: The shadow of $s\mathbf{b}$ onto $\mathbf{b}$ is just $s\mathbf{b}$ itself.

Let's unfold one final gadget from our dot product multitool.

Change of Basis

Using projection, we can find the coordinates of any arrow as a combination of other arrows. For us, this other set of arrows will usually be the computational basis. The idea is similar to representing a conventional 2D point (x, y) as a combination of the X axis scaled by x and the Y axis scaled by y.

Figure 2-24(a) shows an arrow $\mathbf{p}$ and a pair of orthogonal arrows $\mathbf{a}$ and $\mathbf{b}$. For simplicity, I'll assume that all arrows in this section have a magnitude of 1 (that is, they are of unit magnitude). In symbols, $|\mathbf{a}| = |\mathbf{b}| = |\mathbf{p}| = 1$. This convention will make the math simpler. If you prefer the more general case, you can replace every instance of each arrow, say $\mathbf{a}$, with $\mathbf{a}/|\mathbf{a}|$.

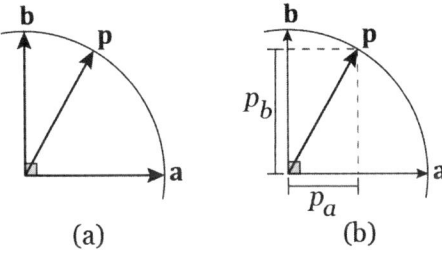

(a) (b)

Figure 2-24: (a) All three arrows have unit magnitude. (b) The coordinates p_a and p_b of $\mathbf{p}$ with respect to $\mathbf{a}$ and $\mathbf{b}$.

In Figure 2-24(b), we see that the coordinates of $\mathbf{p}$ with respect to $\mathbf{a}$ and $\mathbf{b}$ are the projections of $\mathbf{p}$ onto each of these arrows. We can write this in symbols in Equation 2.46.

$$\mathbf{p} = p_a\mathbf{a} + p_b\mathbf{b} = (\mathbf{p} \cdot \mathbf{a})\mathbf{a} + (\mathbf{p} \cdot \mathbf{b})\mathbf{b} \qquad (2.46)$$

Let's rotate the basis arrows $\mathbf{a}$ and $\mathbf{b}$ and call the rotated pair $\mathbf{c}$ and $\mathbf{d}$, as in Figure 2-25(a).

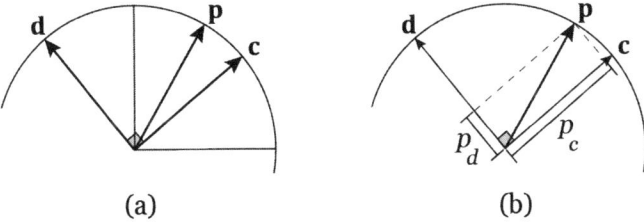

(a) (b)

Figure 2-25: (a) Rotating the basis arrows $\mathbf{a}$ and $\mathbf{b}$ from Figure 2-24, which I've renamed $\mathbf{c}$ and $\mathbf{d}$. (b) The coordinates p_c and p_d of $\mathbf{p}$.

We can find the coordinates of **p**, shown in Figure 2-24(b), from Equation 2.47.

$$\mathbf{p} = p_c\mathbf{c} + p_d\mathbf{d} = (\mathbf{p} \cdot \mathbf{c})\mathbf{c} + (\mathbf{p} \cdot \mathbf{d})\mathbf{d} \tag{2.47}$$

The generalization of this approach is called a *change of basis*. Given the coefficients of an arrow with respect to one set of basis arrows, and a second set of basis arrows we can find the coefficients of that arrow with respect to the second set. This is sometimes a useful step in quantum algorithms, enabling us to represent an operation in a more compact form [211] [117] [220].

This wraps up our discussion of 2D arrows in the plane. Quantum computing generalizes our arrows into complex vectors, sometimes with many dimensions. The wonderful thing about this process is that all the math we've developed will carry over into the new realm with only small changes. The sad thing is that we won't be able to draw pictures of objects and spaces that have huge numbers of dimensions. That's why developing our experience and intuition for these processes in 2D was so important.

Our next step will be to explore and embrace this generalization so that high-dimensional complex vectors will feel nearly as natural and convenient as 2D arrows.

The Inner Product

Now that we've built up some intuition with 2D, real-valued vectors, we'll return to discussing the complex vectors that we use in quantum computing.

Greek letters will now once again refer to complex numbers, not angles. Lowercase bold roman letters like **v** and **w** will refer to vectors, now made up of complex numbers. (You might expect that for consistency we'd use bold Greek letters for vectors of complex numbers, and I think that would make a lot of sense. Alas, that's not how the standard notation developed.)

Postulate 1 tells us that we represent quantum information with vectors of complex numbers, also called *complex-valued* vectors, or *complex vectors* for short.

Since they have the word *vector* in their name, we can be sure that complex vectors obey the scalar multiplication and vector addition properties (V1 and V2) that we met in Equations 2.24 and 2.25. This is good, as those properties will help us keep our math short and comprehensible.

The dot product was a great multitool for real-valued vectors, or arrows. We'd like to generalize it to a version that produces similar results for complex-valued vectors.

That's a pretty vague desire. We can imagine lots of different ways to produce similar results. Let's start with the condition that when we use this new version with real-valued vectors, we get the same results that we get from the dot product.

In computer science terms, we'd like our new operation to *degrade gracefully*. Thus, it will give us useful results for complex vectors and the same results we got from the dot product when we give it real vectors.

Of the many possible ways to define a generalized dot product for complex vectors, one in particular has become typical. It has two great qualities. First, it does indeed match the dot product for real vectors. Second, it makes only a small change to the dot product formula. This means that it will produce results that are much like the dot product's. This is important, because it means we can leverage some of our geometric intuition from the 2D dot product when understanding this bigger and more abstract function.

This is the moment where pictures and geometric reasoning largely end and the symbolic manipulation of algebra pretty much takes over. Now we have to reason not by looking at pictures, but by thinking about *behavior*. Our objects are no longer friendly arrows, but single letters that represent vectors made of many complex numbers. And our operations aren't like shining a flashlight and looking at the shadow, but rather manipulating these letters algebraically. Having 2D experience and intuition gives us helpful analogies to refer to in this world of abstract objects and relationships.

For now, let's advance slowly and carefully.

Defining the Inner Product

Our generalization of the dot product is called the *inner product*. For the moment, I'll write it as a function named "inner product." It takes two vectors as arguments and returns a number.

Let's build things up starting with real numbers. We know that for any real number a, its squared magnitude is a^2, giving us its squared distance from the origin of the number line (for example, $3^2 = (-3)^2 = 9$). When we find the dot product of a real vector with itself, we find the squared magnitude of each element and then add all of these products together.

Let's follow this pattern for complex vectors. For any complex number α, Equation 2.10 tells us that its squared magnitude can be written $\bar{\alpha}\alpha$. So, let's say that to form the inner product (whatever that will be) of a vector $\mathbf{v}$ of many complex numbers, we'll use Equation 2.10 to get the squared magnitude of each element and then add them together.

My provisional version of the inner product will conjugate the first argument, as shown in Equation 2.48.

$$\text{inner product}(\mathbf{a}, \mathbf{a}) = \sum_{k \in [d]} \overline{\alpha_k} \alpha_k \qquad (2.48)$$

If we want to find the dot product of two real vectors $\mathbf{a}$ and $\mathbf{b}$, we multiply all the corresponding terms together and add them up. So, I'll say that we'll do the same thing with our inner product, multiplying together each pair of terms. But as we've seen, one of them will be conjugated. The result is the two-vector version of the inner product shown in Equation 2.49.

$$\text{inner product}(\mathbf{a}, \mathbf{b}) = \sum_{k \in [d]} \overline{\alpha_k} \beta_k \qquad (2.49)$$

Structurally, this looks a lot like the dot product in Equation 2.41, so that's good. And if we give it two real vectors, then the conjugation does nothing, so we multiply together corresponding terms and add them up. Hey, that means we get back the same value as we did from the dot product! We've achieved both of our goals.

In fact, we've just come up with the formula for the inner product that we normally use in quantum computing.

It would be nice to have a more compact way to write the left side of Equation 2.49. What notation should we choose?

We could write the two complex vectors with a symbol between them, like we did with the dot product. Heck, we might even use a dot again! But mathematicians developed a different notation.

The inner product of two complex vectors **a** and **b** is written by naming the vectors in order with a comma between them and placing that sequence between narrow angle brackets, $\langle$ and $\rangle$. With these symbols, the inner product of **a** and **b** is written $\langle \mathbf{a}, \mathbf{b} \rangle$. The expression $\langle \mathbf{a}, \mathbf{b} \rangle$ is read out loud as "the inner product of a and b."

The brackets are not the less-than and greater-than symbols < and > because they are much taller and narrower. Sometimes these symbols are called *chevrons*. The LaTeX typesetting system refers to the symbols $\langle$ and $\rangle$ respectively as *langle* and *rangle*, which seems reasonable to me.

We can now write the inner product as a sum of products, each created from the conjugate of a term from the first vector multiplied with its corresponding term in the second vector. I've summarized this in Equation 2.50.

$$\langle \mathbf{a}, \mathbf{b} \rangle \overset{\Delta}{=} \sum_{k \in [d]} \overline{\alpha_k}\, \beta_k \qquad (2.50)$$

Equation 2.51 expands Equation 2.50 to show its explicit components.

$$\langle \mathbf{a}, \mathbf{b} \rangle = \overline{\alpha_0}\beta_0 + \overline{\alpha_1}\beta_1 + \cdots + \overline{\alpha_{d-1}}\,\beta_{d-1} \qquad (2.51)$$

Since taking the conjugate of a real number causes no change, if we use real-valued vectors, the terms $\overline{\alpha_k}\beta_k$ can be written $a_k\, b_k$. That matches the dot product in Equation 2.41, confirming that the inner product degrades gracefully, just as we wanted!

Finding a Complex Vector's Magnitude

Just as the dot product of a real vector with itself gives us the squared magnitude of that vector, the inner product of a complex vector with itself gives us the squared magnitude of that vector. I've written this in symbols in Equation 2.52.

$$\langle \mathbf{a}, \mathbf{a} \rangle = \sum_{k \in [d]} \overline{\alpha_k}\, \alpha_k = |\mathbf{a}|^2 \qquad (2.52)$$

Equally important is the magnitude of **a**, which is the positive square root of Equation 2.52, shown in Equation 2.53.

$$\sqrt{\langle \mathbf{a}, \mathbf{a} \rangle} = |\mathbf{a}| \qquad (2.53)$$

Choosing Which Term to Conjugate

We've seen that the inner product of two complex numbers is given by Equation 2.50, where the first term in each product is conjugated. We could have conjugated the second argument instead. Why did we choose the first argument?

No good reason, really. The choice is essentially arbitrary. It's like asking why, in a 2D coordinate system, we label the axes X and Y, and point X to the right and (usually) Y upward. It's just tradition.

To everyone's misfortune, mathematicians and physicists have different traditions for which argument gets conjugated when forming the inner product. Historically, mathematicians conjugate the second argument [120] [153], while physicists (including people in the quantum computing field) conjugate the first argument [213] [146]. This is important because the inner product is not commutative; we'll see that the order of the arguments matters.

Most authors conjugate the term that's conventional in their field without explicitly identifying it, so this inconsistency can cause all kinds of confusion and trouble. Keep this in mind when you read any reference material, and make sure that you know whether the author is using the physics-based convention of conjugating the first term or the mathematician's convention of conjugating the second term instead.

Both mathematicians and physicists are mostly consistent in their own fields, so in quantum computing discussions (including this book) it's almost always the first term that gets conjugated when forming the inner product.

This choice has a big impact when we use the inner product to project one complex vector onto another, as we'll see next.

Projection with the Inner Product

The asymmetry of the inner product is important when we use it to project one complex vector onto another. It's essentially the same operation we performed for real vectors, except that if we project one complex vector of d elements onto another we have $2d$ complex numbers, and helpful pictures are hard to create. That's why we looked at the dot product version first.

To set the stage for this projection, recall that Equation 2.44 showed us how to project a 2D arrow **a** onto an arrow **b**, getting back a scaled version of **b**, or s**b** for some real number s. As we saw in Equation 2.45, projecting that result onto **b** made no change, giving us back s**b** again.

We'd like to get the same behavior from the inner product. But because the inner product isn't commutative, we have to choose in which order to write our vectors. Now that we have complex vectors, I'll write the projection of vector **v** onto some other vector **w** as α**w**, or the vector **w** scaled by some complex number α.

To project α**w** back onto **w**, we form the inner product of α**w** and **w**. Should we write this inner product with our scaled vector first, giving us the vector $\langle \alpha\mathbf{w}, \mathbf{w} \rangle \mathbf{w}$, or with the scaled vector second, $\langle \mathbf{w}, \alpha\mathbf{w} \rangle \mathbf{w}$? We know that these two expressions return different vectors, because they conjugate different numbers. For consistency with the dot product, we want the version that, when we project $\alpha\mathbf{w}$ onto $\mathbf{w}$, gives us back $\alpha\mathbf{w}$ again.

Let's try it out both ways, starting with $\langle\alpha\mathbf{w}, \mathbf{w}\rangle$ in Equation 2.54. Since we want to stay close to the dot product, I'll use the form of Equation 2.44 to compute the projection, but I'll use the inner product rather than the dot product. Our hope is to get back our starting vector, $\alpha\mathbf{w}$. As usual, when expanding the inner product I'll assume our vectors are of dimension d (that is, they have d elements, numbered 0 to $d - 1$).

$$\alpha\mathbf{w} \stackrel{?}{=} \frac{\langle\alpha\mathbf{w}, \mathbf{w}\rangle}{\langle\mathbf{w}, \mathbf{w}\rangle} \mathbf{w} \qquad \text{Try } \alpha\mathbf{w} \text{ as the first argument}$$

$$\stackrel{?}{=} \frac{\sum_{k\in[d]} \overline{\alpha\,w_k}\,w_k}{\sum_{k\in[d]} \overline{w_k}\,w_k} \mathbf{w} \qquad \text{Expand the inner products}$$

$$\stackrel{?}{=} \overline{\alpha}\, \frac{\sum_{k\in[d]} \overline{w_k}\,w_k}{\sum_{k\in[d]} \overline{w_k}\,w_k} \mathbf{w} \qquad \text{Pull the constant } \overline{\alpha} \text{ out front}$$

$$\neq \overline{\alpha}\mathbf{w} \qquad \text{The fraction is } 1$$

(2.54)

That's not good. We're getting back $\overline{\alpha}\mathbf{w}$, which is not our starting vector $\alpha\mathbf{w}$ (unless α has no imaginary component).

Let's try the other order, placing $\alpha\mathbf{w}$ in the second position of the inner product, as shown in Equation 2.55.

$$\alpha\mathbf{w} \stackrel{?}{=} \frac{\langle\mathbf{w}, \alpha\mathbf{w}\rangle}{\langle\mathbf{w}, \mathbf{w}\rangle} \mathbf{w} \qquad \text{Try } \alpha\mathbf{w} \text{ as the second argument}$$

$$\stackrel{?}{=} \frac{\sum_{k\in[d]} \overline{w_k}\,\alpha\,w_k}{\sum_{k\in[d]} \overline{w_k}\,w_k} \mathbf{w} \qquad \text{Expand the inner products}$$

$$\stackrel{?}{=} \alpha\, \frac{\sum_{k\in[d]} \overline{w_k}\,w_k}{\sum_{k\in[d]} \overline{w_k}\,w_k} \mathbf{w} \qquad \text{Pull the constant } \alpha \text{ out front}$$

$$= \alpha\mathbf{w} \qquad \text{Just what we started with!}$$

(2.55)

Success! It makes sense that $\alpha\mathbf{w}$ belongs in the second argument, because then α never gets conjugated. Our problem in Equation 2.55 came from placing α in the first argument, so that we ended up with $\overline{\alpha}\mathbf{w}$ rather than our starting vector, $\alpha\mathbf{w}$.

We can summarize this by saying that when we use the inner product to project one complex vector onto another, the vector we're projecting goes in the second position and the vector we're projecting it onto goes in the first position.

So that we can easily refer to it later, I've summarized this rule as Equation 2.56.

$$\text{Write projections as } \langle \textit{onto}, \textit{from} \rangle \qquad (2.56)$$

Forgetting this rule and putting the arguments in the wrong order is an easy error to make when writing and analyzing quantum programs, so I recommend that you commit it to memory and save yourself some debugging sessions!

We've just seen a special case of a more general property of the inner product: If we reverse the order of the arguments, we conjugate the result. In symbols, we can write $\overline{\langle \alpha, \beta \rangle} = \langle \beta, \alpha \rangle$. This property is called *conjugate symmetry*.

Don't forget that we're using the physicist's convention of conjugating the first argument, not the mathematician's convention of conjugating the second [120]. This inconsistency is a sad situation, and it can easily cause all kinds of needless misunderstandings. If you're not certain which convention is being used by an author, check! In this book, I follow the tradition in quantum computing and conjugate the first argument in a complex inner product, so our projections will always follow Equation 2.56.

Braket Notation

Our last big topic in this chapter is a fun one. We'll make all of our math simpler and easier to comprehend with a new piece of notation for complex vectors.

Everyone uses this notation, so it's definitely worth getting familiar with. It may look strange at first blush, but it will soon become second nature.

This style is called *braket notation* (or *bra–ket notation*), pronounced like the word *bracket*. The notation and its name were devised by the physicist Paul Dirac in 1939 [50]. It's also referred to as *Dirac notation*, in his honor [140] [255].

In braket notation, a complex vector is written not as a bold letter like **v** or **w**, but as a lowercase Greek letter with a vertical bar on the left and a rangle on the right, such as $|\alpha\rangle$ or $|\beta\rangle$. Together, this bundle of symbols is called a *ket*. We pronounce $|\alpha\rangle$ out loud as "ket alpha."

Note that this new symbol is not a single complex number, which in this book is represented by a lowercase Greek letter by itself, like α or β. When a lowercase Greek letter is wrapped in ket notation, like $|\alpha\rangle$ or $|\beta\rangle$, it refers to a *vector* of complex numbers, usually with many elements. It's the ket symbols that tell us that we have a vector and not a single number.

I'll use the words *vector* and *ket* interchangeably from now on. When you see a ket, such as $|\alpha\rangle$, you know you're looking at a vector of one or more complex elements. Sometimes authors will place a subscript after the ket to indicate the number of elements, or dimensionality. So, $|\gamma\rangle_8$ is a complex vector with 8 elements. This kind of explicit subscript is rare, however, since the dimensionality is usually clear from the discussion.

Using the language of linear algebra, we represent a ket as a *column matrix*, which we've seen is a grid of numbers that has one column and d rows, one for each of the ket's dimensions. For example, Equation 2.57 shows a ket $|\alpha\rangle$ made up of four complex numbers, named α_0 through α_3. We start numbering at the top and proceed downward. Note that the elements, like α_0, aren't wrapped in ket notation, so they're just complex numbers, not vectors.

$$|\alpha\rangle = \begin{bmatrix} \alpha_0 \\ \alpha_1 \\ \alpha_2 \\ \alpha_3 \end{bmatrix} \tag{2.57}$$

We'd like to be able to multiply kets together, for example to form inner products. But there's a problem. The usual rules of matrix multiplication require that when two matrices are multiplied, the number of columns in the left matrix and the number of rows in the right matrix must be the same. This prohibits us from multiplying a ket with another ket (if matrix multiplication is unfamiliar to you, or you've forgotten the details, check out the many online websites and videos on the topic, such as [271] and [116]).

On the other hand, we *can* multiply a column matrix with a row matrix of the same number of elements. If we could turn a ket into a row matrix, then we could multiply a ket in its usual column matrix form with another ket written as a row matrix and get back a single complex number. In other words, if we conjugated the first ket and also turned it from a column into a row, we could compute an inner product by multiplying the two matrices.

Let's find a way to do that. The effort will return massive rewards. We'll see in later chapters that quantum algorithms are built from treating all of our objects as matrices and combining them with matrix multiplication.

Happily, if we want to write the elements of a ket in a row, rather than a column, there's nothing stopping us from simply writing it that way. This row won't be a ket anymore, but it's something that we can multiply with a ket. Problem solved! To prevent confusion, we need a new piece of notation that helps us keep track of when a list of complex numbers is being written as a column (that is, a ket) or as a row. This is important because, as we'll see, rows and columns will describe different kinds of mathematical objects.

There is a standard name for turning a column matrix into a row matrix: It's called taking the *transpose* of the matrix. The transpose of a column is a row, and the transpose of a row is a column (we'll see in Chapter 3 that we can also transpose matrices of many rows and/or columns).

To indicate the process "take this matrix and form its transpose," we augment the matrix name with a superscript of a capital roman T. So if we have a ket $|\omega\rangle$, its transpose $|\omega\rangle^{\mathrm{T}}$ is a row of the same elements, now written left to right, as shown in Equation 2.58.

$$|\omega\rangle = \begin{bmatrix} \omega_0 \\ \omega_1 \\ \omega_2 \\ \omega_3 \end{bmatrix}, \qquad |\omega\rangle^{\mathrm{T}} = \begin{bmatrix} \omega_0 & \omega_1 & \omega_2 & \omega_3 \end{bmatrix} \qquad (2.58)$$

We can now multiply any ket $|\alpha\rangle$ with any other ket $|\beta\rangle$ by first turning one of these column matrices into a row matrix. For the rest of this chapter, I'll always turn the first column into a row (in Chapter 6, we'll see the result of multiplying a column on the left with a row on the right).

Equation 2.59 shows an example of multiplying two kets $|\alpha\rangle$ and $|\beta\rangle$ in this way, where I've transposed the first ket, $|\alpha\rangle$.

$$|\alpha\rangle \, |\beta\rangle \rightarrow |\alpha\rangle^{\mathrm{T}} \, |\beta\rangle = \begin{bmatrix} \alpha_0 & \alpha_1 & \alpha_2 \end{bmatrix} \begin{bmatrix} \beta_0 \\ \beta_1 \\ \beta_2 \end{bmatrix} = \alpha_0\beta_0 + \alpha_1\beta_1 + \alpha_2\beta_2 \quad (2.59)$$

The most common reason we'll want to do this is to compute an inner product. That means we have to conjugate the elements of the first ket. For example, to find the inner product of $|\omega\rangle$ and $|\alpha\rangle$, we can turn $|\omega\rangle$ into a row by transposing it, and then we'll have to conjugate every element of $|\omega\rangle^{\mathrm{T}}$.

We can indicate that we want to also conjugate the elements of $|\omega\rangle^{\mathrm{T}}$ by including the bar for conjugation along with the T for transposition. Since these two operations don't affect one another, we can perform them in either order.

Let's see this. Equation 2.60 starts with $|\omega\rangle$, then transposes it and conjugates it to form $\overline{|\omega\rangle^{\mathrm{T}}}$. To keep track of the order, read the symbols from the inside out, so we start with the ket $|\omega\rangle$, then transpose it, and then conjugate.

$$\begin{bmatrix} \omega_0 \\ \omega_1 \\ \omega_2 \\ \omega_3 \end{bmatrix}, \quad |\omega\rangle^{\mathrm{T}} = \begin{bmatrix} \omega_0 & \omega_1 & \omega_2 & \omega_3 \end{bmatrix}, \quad \overline{|\omega\rangle^{\mathrm{T}}} = \begin{bmatrix} \overline{\omega_0} & \overline{\omega_1} & \overline{\omega_2} & \overline{\omega_3} \end{bmatrix} \quad (2.60)$$

In contrast, Equation 2.61 goes in the other order, starting with ket $|\omega\rangle$, which is first conjugated and then transposed to get $\overline{|\omega\rangle}^{\mathrm{T}}$.

$$|\omega\rangle = \begin{bmatrix} \omega_0 \\ \omega_1 \\ \omega_2 \\ \omega_3 \end{bmatrix}, \qquad \overline{|\omega\rangle} = \begin{bmatrix} \overline{\omega_0} \\ \overline{\omega_1} \\ \overline{\omega_2} \\ \overline{\omega_3} \end{bmatrix}, \qquad \overline{|\omega\rangle}^{\mathrm{T}} = \begin{bmatrix} \overline{\omega_0} & \overline{\omega_1} & \overline{\omega_2} & \overline{\omega_3} \end{bmatrix} \quad (2.61)$$

Equations 2.60 and Equation 2.61 produce the same result: $|\omega\rangle$ both conjugated and transposed.

Now we can compute the inner product of $|\omega\rangle$ and $|\alpha\rangle$ using matrix multiplication, by multiplying the row matrix $\overline{|\omega\rangle^{\mathrm{T}}}$ (or $\overline{|\omega\rangle}^{\mathrm{T}}$) with the column matrix $|\alpha\rangle$, as shown in Equation 2.62.

$$\langle |\omega\rangle, |\alpha\rangle\rangle = \begin{bmatrix} \overline{\omega_0} & \overline{\omega_1} & \overline{\omega_2} & \overline{\omega_3} \end{bmatrix} \begin{bmatrix} \alpha_0 \\ \alpha_1 \\ \alpha_2 \\ \alpha_3 \end{bmatrix} = \overline{\omega_0}\alpha_0 + \overline{\omega_1}\alpha_1 + \overline{\omega_2}\alpha_2 + \overline{\omega_3}\alpha_3 \quad (2.62)$$

We'll write lots of inner products in the chapters ahead, and every time we do so, we'll want to express it in this matrix multiplication form. That means we'll both transpose and conjugate the first ket. We could write this explicitly as $\overline{|\omega\rangle^{\mathrm{T}}}$ or $\overline{|\omega\rangle}^{\mathrm{T}}$, but linear algebra offers us a ready-made name and symbol that combines both operations.

The conjugated transpose (or transposed conjugate) of a vector is called the *adjoint* of the vector, and we write it using a dagger symbol (†) as a superscript. This is shown in Equation 2.63, where the first term is called the adjoint of the ket. This is easier to read and write.

$$|\omega\rangle^{\dagger} = \overline{|\omega\rangle^{\mathrm{T}}} = \overline{|\omega\rangle}^{\mathrm{T}} = \begin{bmatrix} \overline{\omega_0} & \overline{\omega_1} & \overline{\omega_2} & \overline{\omega_3} \end{bmatrix} \quad (2.63)$$

We pronounce $|\omega\rangle^{\dagger}$ out loud as "the adjoint of ket omega." If we know from context that "omega" refers to a ket, we can more simply say "omega adjoint," "adjoint omega," or "omega dagger."

Now we can form an inner product of two kets by taking the adjoint of the first and then multiplying the matrices. The adjoint takes care of both turning the first column into a row (so we can multiply it with a column) and conjugating each element (because the inner product requires that).

The inner product of two kets $|\omega\rangle$ and $|\alpha\rangle$ as a matrix multiplication is shown in Equation 2.64.

$$\langle |\omega\rangle, |\alpha\rangle\rangle = |\omega\rangle^{\dagger} |\alpha\rangle \quad (2.64)$$

Forming the adjoint of a ket is so common and important that Dirac gave the result its own symbol and name. He called it a *bra* and wrote it with the ket symbols flipped around, so we place a langle on the left and a vertical bar on the right. Thus, we get the definition in Equation 2.65.

$$\langle\alpha| \overset{\Delta}{=} |\alpha\rangle^{\dagger} \quad (2.65)$$

In the same way, if we take the adjoint of a bra, we form its corresponding ket, so $\langle\alpha|^{\dagger} = |\alpha\rangle$. This tells us that if we take the adjoint of a ket $|\alpha\rangle$ twice, we get $|\alpha\rangle$ back again, and the same holds for a bra.

Together, the bra and ket give us bra–ket, or braket, notation.

Looking at the Braket

Let's write the inner product using our new ket and bra notation. The second argument is a ket and the first argument is a bra, or a ket that has been transposed and conjugated (that is, we use its adjoint). The ket is a column matrix and the bra is a row matrix, so if we place them side by side we're implicitly saying that we want to multiply these two matrices together.

We want to project a vector $|\alpha\rangle$ (our starting or *from* vector) onto another vector $|\omega\rangle$ (our target or *onto* vector). Equation 2.56 tells us that $|\omega\rangle$ comes first in the inner product. Equation 2.66 shows five different but equivalent ways to write down this inner product.

$$\langle\, |\omega\rangle, |\alpha\rangle \,\rangle = \overline{|\omega\rangle^{\mathrm{T}}} |\alpha\rangle = |\omega\rangle^{\dagger} |\alpha\rangle$$

$$= \langle\omega|\, |\alpha\rangle = \langle\omega|\alpha\rangle \tag{2.66}$$

In the final form of this expression, I did away with the repeated vertical bars and just mushed the bra and ket together.

The final expression in Equation 2.66, $\langle\omega|\alpha\rangle$, is how we normally write the inner product of any two complex-valued vectors $|\alpha\rangle$ and $|\omega\rangle$. Note that $|\omega\rangle$ is a ket, but when it's the first argument in an inner product we use its adjoint, written as the bra $\langle\omega|$.

And now we can see where the names *bra* and *ket* come from. In math, any pair of matched grouping symbols (like parentheses or square brackets) is called a *bracket* (note the *c* in that word) [233]. Dirac apparently viewed the notation $\langle\omega|\alpha\rangle$ as a kind of bracket itself, and he gave the symbols those names so that, used together to form an inner product, they form a "bra–ket." Thus, we call an inner product like $\langle\omega|\alpha\rangle$ a "braket." I think it's a pretty weak pun, but that terminology is here to stay.

We pronounce $\langle\omega|\alpha\rangle$ out loud as "the inner product of omega and alpha," or "bracket omega alpha."

In Equation 2.67, I've written out the components of the bra and ket matrices in a braket in tableau form. Here, the vectors are of dimension $d = 3$.

$$\langle\omega|\alpha\rangle = \left\langle \begin{bmatrix} \omega_0 \\ \omega_1 \\ \omega_2 \end{bmatrix}, \begin{bmatrix} \alpha_0 \\ \alpha_1 \\ \alpha_2 \end{bmatrix} \right\rangle$$

$$= \begin{bmatrix} \overline{\omega_0} & \overline{\omega_1} & \overline{\omega_2} \end{bmatrix} \begin{bmatrix} \alpha_0 \\ \alpha_1 \\ \alpha_2 \end{bmatrix} = \overline{\omega_0}\alpha_0 + \overline{\omega_1}\alpha_1 + \overline{\omega_2}\alpha_2 \tag{2.67}$$

The more general expression for any dimensionality is shown in Equation 2.68.

$$\langle\omega|\alpha\rangle = \begin{bmatrix} \overline{\omega_0} & \overline{\omega_1} & \dots & \overline{\omega_{d-1}} \end{bmatrix} \begin{bmatrix} \alpha_0 \\ \alpha_1 \\ \vdots \\ \alpha_{d-1} \end{bmatrix} = \sum_{k\in[d]} \langle\omega|_k\, |\alpha\rangle_k = \sum_{k\in[d]} \overline{\omega_k}\alpha_k \tag{2.68}$$

Let's connect the braket explicitly to the inner product defined in Equation 2.50. They are two ways to write the same thing. That is, the braket *is* the inner product we use in quantum computing (and thus, in the rest of this book). In symbols, we can write Equation 2.69.

$$\langle \omega | \alpha \rangle = \sum_{k \in [d]} \overline{\omega_k} \alpha_k = \big\langle\, |\omega\rangle , |\alpha\rangle \,\big\rangle \qquad (2.69)$$

Now let's take our braket out for a spin. I'll use it to compute the magnitude of a ket $|\alpha\rangle$ of dimension d. That's the inner product of $|\alpha\rangle$ with itself. The steps are in Equation 2.70. In the second line, if we know that we're working with ket $|\alpha\rangle$, then we can write each element of $|\alpha\rangle$ more simply as α_k.

$$
\begin{aligned}
\langle \alpha | \alpha \rangle &= \sum_{k \in [d]} \langle \alpha |_k \, |\alpha\rangle_k && \text{Expand using Eq. 2.68} \\
&= \sum_{k \in [d]} \overline{\alpha_k}\, \alpha_k && \text{Conjugate the first argument} \\
&= \sum_{k \in [d]} |\alpha_k|^2 && \text{Use } |\alpha|^2 = \overline{\alpha}\alpha \text{ from Eq. 2.10}
\end{aligned}
\qquad (2.70)
$$

The final result is of the same form as Equation 2.42, which we can now restate in Dirac notation as shown in Equation 2.71.

$$
\begin{aligned}
\big| |\alpha\rangle \big|^2 &= \langle \alpha | \alpha \rangle \\
\big| |\alpha\rangle \big| &= \sqrt{\langle \alpha | \alpha \rangle}
\end{aligned}
\qquad (2.71)
$$

The expressions on the left are a little clumsy to read and write due to those repeated vertical bars. The term $\big| |\alpha\rangle \big|^2$ refers to the squared magnitude of the vector represented by $|\alpha\rangle$.

Let's lock down this new notation using something we've already seen: representing a vector with respect to a set of bases, as we did for arrows in Equation 2.46. That is, we'll project a vector onto the basis vectors.

To keep our equations tidy and easier to read, I'll use a symbol called the *Kronecker delta*, named for Leonard Kronecker [240]. We write it as the lowercase Greek letter δ with two integer subscripts, like this: $\delta_{a,b}$ (sometimes people leave out the comma). The symbol $\delta_{a,b}$ stands for a number that is 1 when $a = b$, and otherwise is 0. I've summarized this definition in Equation 2.72.

$$\delta_{k,m} \triangleq \begin{cases} 0, & \text{if } k \neq m \\ 1, & \text{if } k = m \end{cases} \qquad (2.72)$$

This symbol gives us a convenient way to pick out, or *sift*, one element from a list. Equation 2.73 shows an example.

$$\sum_{k\in[d]} \delta_{a,k}\alpha_k = \alpha_a \qquad (2.73)$$

In Equation 2.73, $\delta_{a,k}$ is 0 for all terms except when $a = k$, when it's 1. This means that every term in the summation is 0 except α_a, which is therefore the result of the summation. This symbol might seem unnecessary, since if we know a, we could just write α_a right away. But we'll see in a moment that the Kronecker delta is useful for simplifying expressions.

We often use the Kronecker delta when we're working with multiple basis states. In the computational basis, every basis state has a magnitude of 1. That is, it has unit magnitude, or it's *normalized* or, more simply, *normal* (note that this is a specialized use of the word "normal" and doesn't imply the vector is typical in some way). We also say the vector has *unit norm*. Every computational basis state is perpendicular to every other, which we express by saying that they are all *orthogonal* to one another. We capture both mutual orthogonality and unit magnitude together in the portmanteau word *orthonormal*.

One reason we use the computational basis exclusively in this book is because it is orthonormal (there are an infinite number of orthonormal bases, but the computational basis is the most convenient for us humans because each state has a single 1 element, and all the rest are 0).

For generality, in this example I won't use the computational basis vectors. I'll only assume that the basis vectors are orthonormal.

I've written out the projection process in Equation 2.74, using vector notation on the left and braket notation on the right. In vector terms, we want to find the coordinates of vector **a** with respect to one of the d basis vectors, **w**. I said earlier that orthonormal basis vectors can help us simplify our equations, and we'll see that in action here.

I've written the vector we want to represent (**a** or $|\alpha\rangle$) as a sum of each basis vector k ($\mathbf{w}_k$ or $|\omega_k\rangle$) scaled by γ_k. Our goal is to find the coordinate (or value of γ_k) for one specific basis vector, $\mathbf{w}_m$ or $|\omega_m\rangle$. We can repeat the process using each basis vector in turn to find its corresponding coordinate.

In Equation 2.74, I've written each step on its own line so you can see the one thing that changes from one step to the next. Both sides of each line mean the same thing. There's no room for comments, so I've numbered the lines and will discuss them right after the math.

Vector Notation		Braket Notation	
$\langle \mathbf{w}_m, \mathbf{a} \rangle$	(1)	$\langle \omega_m \mid \alpha \rangle$	
$= \left\langle \mathbf{w}_m, \sum_{k \in [d]} \gamma_k \mathbf{w}_k \right\rangle$	(2)	$= \left\langle \omega_m \middle\| \sum_{k \in [d]} \gamma_k \mid \omega_k \rangle \right\rangle$	
$= \sum_{k \in [d]} \langle \mathbf{w}_m, \gamma_k \mathbf{w}_k \rangle$	(3)	$= \sum_{k \in [d]} \langle \omega_m \mid \gamma_k \mid \omega_k \rangle$	(2.74)
$= \sum_{k \in [d]} \gamma_k \langle \mathbf{w}_m, \mathbf{w}_k \rangle$	(4)	$= \sum_{k \in [d]} \gamma_k \langle \omega_m \mid \omega_k \rangle$	
$= \sum_{k \in [d]} \gamma_k \delta_{m,k}$	(5)	$= \sum_{k \in [d]} \gamma_k \delta_{m,k}$	
$= \gamma_m$	(6)	$= \gamma_m$	

Since the braket notation on the right is the new stuff, I'll focus on that in this discussion.

Recall that our goal is to find the coefficient, or scaling factor, that results from projecting $\mathbf{a}$ onto $\mathbf{w}_m$ (or $\mid \alpha \rangle$ onto $\mid \omega_m \rangle$). This factor is given by the inner product. Line 1 writes the inner product (or braket) of ket $\mid \alpha \rangle$ onto the basis ket $\mid \omega_m \rangle$. Note that the arguments in the braket are in the order *onto* then *from*, following Equation 2.56.

Line 2 expands $\mid \alpha \rangle$ as a sum of some coefficient γ_k on each basis $\mid \omega_k \rangle$. On line 3, I moved the bra $\langle \omega_m \mid$ inside the summation, since it doesn't depend on the summation index k and thus is constant throughout the summation loop. Note that the expression $\langle \omega_m \mid \gamma_k \mid \omega_k \rangle$ inside the summation is nothing new, though it might look new. It's the bra $\langle \omega_m \mid$, times the number γ_k, times the ket $\mid \omega_k \rangle$.

On line 4, since multiplication doesn't care about the order of its terms, I moved the number γ_k one position left, leaving us with a braket. On line 5, I used the fact that the basis is orthonormal, so the inner product of any basis on any other is 0 unless they're both the same basis, in which case the result is 1. That relationship is perfectly captured by the Kronecker delta. In this situation, $\delta_{m,k}$ is 0 except when $m = k$, when it's 1.

Finally, on line 6, the Kronecker delta sifts out only the coefficient γ_m (since the terms for all other values of k end up as 0), and we're done.

We've confirmed that γ_m, the coefficient of $\mid \alpha \rangle$ projected onto basis $\mid \omega_m \rangle$, is given by the braket (or inner product) $\langle \omega_m \mid \alpha \rangle$.

For the rest of this book I'll use the braket notation in the right column, since that's the notation used in quantum computing.

Conjugating the Braket

Let's get a little practice with the braket by seeing what happens if we conjugate it. The result will prove useful in Chapter 6.

I'll start with a braket, or inner product, of two arbitrary kets $|\phi\rangle$ and $|\psi\rangle$ in that order, written $\langle\phi|\psi\rangle$. Then I'll conjugate it, giving us $\overline{\langle\phi|\psi\rangle}$. Equation 2.75 shows how to derive another way to write this inner product.

$$
\begin{aligned}
\overline{\langle\phi|\psi\rangle} &= \sum_{k\in[d]} \overline{\overline{\phi_k}\,\psi_k} && \text{Expand with Eq. 2.69} \\
&= \sum_{k\in[d]} \phi_k\,\overline{\psi_k} && \text{Because } \overline{\overline{\phi_k}} = \phi_k \\
&= \sum_{k\in[d]} \overline{\psi_k}\,\phi_k && \text{Multiplication of numbers} \\
& && \text{is commutative} \\
&= \langle\psi|\phi\rangle && \text{Definition of inner product}
\end{aligned}
\tag{2.75}
$$

That's quite nice! To find the conjugate of an inner product, we need only reverse the order of the terms.

Qubits

We've covered a lot of material in this chapter! We'll use these mathematical tools throughout the book as we create and modify qubits to perform quantum computation.

With these techniques now under our belts, we can write down what we mean by a qubit, or a unit of quantum information.

Qubit   A qubit is described by a two-dimensional ket, or a column vector with two complex numbers, that has a magnitude of 1.

Why complex numbers? Why two of them? Why must the ket have a magnitude of 1? In Chapter 6, we'll see that these requirements work out nicely for our math, but nature isn't interested in our convenience. We define a qubit this way because it is the simplest and most economical way that anyone has found to match, or model, the experimental data. Most "rules" of nature have exceptions: Some giraffes are born with short necks [119], and some watermelons weigh more than Bernese Mountain Dogs [84] [7]. But some things are fixed, like the speed of light in a vacuum or the charge of an electron. Just as nobody has ever found a counterexample to those values, nobody has ever found a qubit that doesn't match this description.

A qubit is a theoretical abstraction. It's a unit of information. But in practice, we often use the word *qubit* to refer to this unit, and to the physical object it describes, and to the information represented by that physical object. These three ideas are so closely related that there's rarely any confusion, but be aware that the word has multiple interpretations.

Since a qubit is represented by a two-element complex vector, its two coefficients describe the scaling factors on two basis vectors. Thus, we need to know which basis vectors it's referring to.

In this book, we'll always use the computational basis. The first two computational basis vectors are used so widely that they have been given their own labels. These labels could have been anything, but by long-standing convention they're called $|0\rangle$ and $|1\rangle$. They're defined in Equation 2.76.

$$|0\rangle \triangleq \begin{bmatrix} 1 \\ 0 \end{bmatrix}, \qquad |1\rangle \triangleq \begin{bmatrix} 0 \\ 1 \end{bmatrix} \tag{2.76}$$

Note that these bases are both normal (have unit magnitude), and they are orthogonal to one another, as we can see from their inner products in Equation 2.77. Therefore, we say that this pair of basis states is orthonormal.

$$\langle 0|0\rangle = \begin{bmatrix} 1 & 0 \end{bmatrix} \begin{bmatrix} 1 \\ 0 \end{bmatrix} = 1, \qquad \langle 1|0\rangle = \begin{bmatrix} 0 & 1 \end{bmatrix} \begin{bmatrix} 1 \\ 0 \end{bmatrix} = 0$$

$$\langle 0|1\rangle = \begin{bmatrix} 1 & 0 \end{bmatrix} \begin{bmatrix} 0 \\ 1 \end{bmatrix} = 0, \qquad \langle 1|1\rangle = \begin{bmatrix} 0 & 1 \end{bmatrix} \begin{bmatrix} 0 \\ 1 \end{bmatrix} = 1 \tag{2.77}$$

You might be surprised to see that the state labeled $|0\rangle$ has a 1 at index 0 (as always, numbering from the top and starting at 0), but this naming convention will generalize nicely going forward. The rule is that if a ket is described by an integer, like $|0\rangle$ or $|1\rangle$, every element in the ket has a value of 0 except for the element at the integer index given by the ket's name, which has the value 1. So $|0\rangle$ is 0 everywhere except for a 1 at index 0, and $|1\rangle$ is 0 everywhere except for a 1 at index 1.

Now, finally, we can write a qubit!

A qubit is represented by a unit magnitude combination of the two computational basis vectors, each scaled by a complex number, as in Equation 2.78.

$$|\psi\rangle = \alpha \,|0\rangle + \beta \,|1\rangle, \quad |\alpha|^2 + |\beta|^2 = 1 \tag{2.78}$$

Just as we often use the letter θ for some generic angle when doing 2D geometry, we often use $|\psi\rangle$ and $|\phi\rangle$ for generic qubits.

We can also write this $|\psi\rangle$ in tableau form, as in Equation 2.79.

$$|\psi\rangle = \alpha \,|0\rangle + \beta \,|1\rangle = \alpha \begin{bmatrix} 1 \\ 0 \end{bmatrix} + \beta \begin{bmatrix} 0 \\ 1 \end{bmatrix} = \begin{bmatrix} \alpha \\ \beta \end{bmatrix}, \quad |\alpha|^2 + |\beta|^2 = 1 \tag{2.79}$$

Equation 2.79 gives us three equivalent ways to write out the elements of the ket describing a qubit.

While it's common to use α and β for the coefficients of a single qubit, you can also use subscripts on the name. So, for a state $|\psi\rangle$, the coefficients α and β can also be written with ψ_0 and ψ_1, respectively. This latter form generalizes better to larger vectors, like those we'll see later in the book.

We say that the information held within a qubit and represented by a two-element complex vector with unit norm is a *quantum state*, or *state vector*. Sometimes the last term is mushed together into one big word, *statevector*, which is itself often shortened to *state*. I'll treat all of these as synonyms.

Thus, we say that Equation 2.78 is a statevector that describes a *quantum state*, or a qubit.

When describing a quantum state, we can be formal and say something like, "The qubit of information that is held by this physical object may be described by a state vector $|\psi\rangle$ with amplitudes α and β." That's way too much to repeat over and over, though, so we write this informally as "The state of the qubit is $|\psi\rangle$" or even just "The qubit is $|\psi\rangle$." If I think there's ever a chance of confusion, I'll clarify what's being discussed.

I'd love to draw a picture that represented a qubit, but we can't do that in two dimensions. The problem is that a qubit involves two complex numbers, and thus four real numbers, so a qubit would be a point in a four-dimensional space. Despite many attempts, nobody has found an intuitive way to draw such spaces.

All is not lost, because the requirement that $|\alpha|^2 + |\beta|^2 = 1$ lets us remove one of those dimensions. This leaves us with a three-dimensional picture, which we can even draw on the page.

The most common 3D representation of a qubit is called the *Bloch sphere*, named for Felix Bloch [261]. Unfortunately, despite having some attractive features, the Bloch sphere has some unusual properties that can make it tricky to master. It would also be a substantial detour for us to go into, so I won't describe the Bloch sphere here. If you're curious, there are many references (in books and online) that dive into the Bloch sphere and how to interpret it [252].

Summary

Wow. This has been quite a journey!

We've done a ton of work in this chapter. Let's see how it fits together, starting by revisiting Postulate 1, repeated here for reference:

Postulate 1 An isolated quantum system, or qubit, is described by a two-element complex vector of magnitude 1.

Let's interpret this one piece at a time using our new language and notation. We start with "an isolated quantum system, or qubit," which refers to a physical object of quantum scale that is sufficiently well shielded from its environment that it is unaffected by anything around it. We use this condition so that we can discuss quantum states without always explicitly including their environment and its effects.

No such isolated quantum system exists. Every physical object has *some* kind of connection to the world around it, whether it's heat from the ventilation system, the change in humidity caused by someone nearby sneezing, or gravity waves from a collapsing star in a galaxy far, far away.

But thanks to the work of many scientists and engineers, we can now build quantum computers that approach this ideal. For the purposes of this book, we'll pretend that we're working with truly isolated systems, so our quantum systems are independent of the influences of the often unpredictable natural forces around them.

Next up is the phrase "a two-element complex vector." This refers to a list of two complex numbers that obeys the properties of being a vector. We write such a vector as $|\psi\rangle$, which describes a column matrix that we call a ket. Experiments have shown that all the information we need from a quantum object to perform quantum computing is captured in these two complex numbers.

Finally, we have the requirement that every ket must have "magnitude 1." This is included because experiments have shown that, like the speed of light, this is simply how our universe operates. It will also let us easily talk about the probabilities of what we'll see when we measure a qubit.

Let's conclude with a final bit of terminology about vectors and collections of them.

We say that a set of vectors (including complex vectors, such as kets), along with the rules for scalar multiplication and vector addition, form a *vector space* [248]. Here, the word *space* does not refer to a physical space, but rather to an abstract collection of objects and the properties that describe them. This usage has even moved into popular language, where people who build houses are said to work in the construction space, or people who work in banks are said to work in the financial space.

When we pick a specific inner product (as we did when we picked the braket), the vector space and inner product together are called a *Hilbert space* [239], in honor of the mathematician David Hilbert.

The kets in this book qualify as vectors that form a vector space. Along with our inner product, we've defined a Hilbert space. You'll sometimes hear people say that quantum computing is made up of operations in a Hilbert space, or that quantum information is described by objects in a complex Hilbert space. This is a shorthand way to say that we're working with complex-valued vectors, and we've picked a specific formula for finding inner products.

Now that we know what a qubit is, let's look at how to write quantum algorithms to modify its complex numbers.

> *You have gained 1 quantum essence.*
>
> —*Starfield*, Bethesda Game Studios, 2023 [20]

3

OPERATORS

Operator,
Give me in . . . formation.
—William Spivery, "Operator," 1944 [206]

In Chapter 2, we saw that a qubit is represented by a complex-valued vector with two elements and a magnitude of 1.

One qubit is great, but most quantum programs start with lots of qubits. Each qubit begins in some specific initial state, and then the program modifies the state of each qubit until collectively they hold the answer to a problem. We perform these modifications in steps, changing one or more qubits each time.

Physicists have discovered that nature allows us to perform only a single, narrowly defined class of operations on qubits during a calculation. There are an infinite number of operations in this class, so there's plenty of computational power to go around, but we have to be careful when designing algorithms to choose only actions that belong in this category.

This chapter is all about describing that class of operations and becoming familiar with some of the most common and powerful instances. We'll start by looking at matrices, because they will be our standard way of specifying quantum operations throughout the book. Then we'll look at three important operations, named I, X, and H.

We'll draw a lot of diagrams representing quantum circuits. Figure 3-1 shows the first of these diagrams, presenting a graphical version of a tiny quantum algorithm. In this diagram, the qubit starts at the left, where it has been initialized to an initial value (the quantum state $|0\rangle$), then moves rightward into a box marked H that does . . . something, and that's it. In this chapter we'll see what goes on inside that box, and how the qubit that leaves the right side of the box has been changed.

Figure 3-1: A tiny quantum algorithm

By itself, this drawing doesn't give us the answer to any particular problem, but it does something incredible: It creates a superposition, a kind of information that, as we saw in Chapter 1, is utterly unlike anything we're used to from our everyday lives. Prior to a century ago, this was pure science fiction. And it's only in the last few decades that hardware for executing this algorithm has become practical. Now you can design your own quantum superpositions and make them yourself for free using quantum computers available on the internet.

By assembling multiple operations like H artfully, we can use the science fiction–like powers of quantum objects to create algorithms unlike any we've seen before.

Postulate 2

Our motivation for this chapter comes from another postulate of quantum mechanics. This postulate describes a mathematical model for an observed fact of nature. Like the other postulates, it can be expressed in different ways. Here's the version we'll use in this book [71]:

Postulate 2 Quantum states are changed by unitary operators.

When we talk about "changing" a quantum state, we mean altering the complex numbers that describe it. For a single qubit, we're changing the complex numbers that serve as the weights on the basis states $|0\rangle$ and $|1\rangle$.

The new element in this postulate is the reference to a unitary operator, which we'll discuss later in this chapter. Along the way to that discussion, I'll point out some computational tools that we'll use throughout the book to develop quantum algorithms.

Linear Operators

The mathematical definition of an operator is rather abstract, because it can be applied to a wide variety of objects [244]. In a nutshell, we can think of an operator as a process that modifies another mathematical object. For example, the negation sign − is an operator which turns a positive number like 3 into the negative number −3.

We'll use operators to make changes to vectors. Once we've selected a basis for our vectors (like the computational basis introduced in Chapter 2), we can write an operator as a matrix. Since we'll always use the computational basis in this book, we can treat the name of an operator and its matrix representation as two ways to refer to the same thing [34].

For our purposes, we'll use the term *operator* to refer to a matrix that takes in a matrix as input and produces a new matrix as output. [34]. Each such input matrix could be a one-column matrix (a ket like $|\alpha\rangle$), a one-row matrix (a bra like $\langle\alpha|$), or another matrix of multiple rows and columns. In fact, all of our operator matrices will have an equal number of rows and columns. Since writing down the elements forms a square, each such matrix is called a *square matrix*.

When we use a matrix to modify a state, we choose the matrix elements (which are numbers) to manipulate that state in a specific way. The values of these elements also depend on what basis we're using to describe our quantum states. Because we'll always use the computational basis in this book, we can write each matrix using the numbers appropriate for that basis. If you ever want to use another basis, the numbers I'll be showing for our operators will need to change to match the new basis [85].

Matrices are *linear operators*, meaning that they share the same properties of linearity that we saw in Equations 2.26 and 2.27. Let's rewrite those rules using matrix notation.

I'll write matrices using italic, uppercase roman letters like M or P.

Suppose that we have a matrix M and we want to use it to transform a ket $|\psi\rangle$. We write this operation as $M|\psi\rangle$, which tells us to multiply together the square matrix M and the column matrix $|\psi\rangle$. I've drawn $M|\psi\rangle$ in tableau form in Equation 3.1 for a two-element ket.

$$M|\psi\rangle = \begin{bmatrix} \mu_{00} & \mu_{01} \\ \mu_{10} & \mu_{11} \end{bmatrix} \begin{bmatrix} \psi_0 \\ \psi_1 \end{bmatrix} = \begin{bmatrix} \mu_{00}\psi_0 + \mu_{01}\psi_1 \\ \mu_{10}\psi_0 + \mu_{11}\psi_1 \end{bmatrix} = \begin{bmatrix} \psi_0' \\ \psi_1' \end{bmatrix} = |\psi'\rangle \tag{3.1}$$

Equation 3.1 shows how I'll refer to the elements of a matrix. When our matrices have complex elements, I'll use lowercase Greek letters, as I have here. Each element is indexed by two subscripts. The first index indicates the row, numbering as usual from the top down and starting with 0. The second index indicates the column, numbering from left to right and again starting with 0. Normally, we separate these two indices with a comma, but when there are just single-digit numbers of rows and columns and there's no risk of confusion, we sometimes reduce clutter and omit the comma, as I did here.

As I mentioned earlier, matrices are linear operators. We can write the linearity properties from Equations 2.26 and 2.27 in matrix terms as in Equation 3.2. Here, σ is any complex number, and $|\psi\rangle$ and $|\phi\rangle$ are any two arbitrary kets.

$$
\begin{array}{ll}
M(\sigma|\psi\rangle) = \sigma M|\psi\rangle & \text{L1} \\
M(|\psi\rangle + |\phi\rangle) = M|\psi\rangle + M|\phi\rangle & \text{L2}
\end{array}
\tag{3.2}
$$

We'll use these properties so often in this book that it's worth confirming them. Equation 3.3 shows why property L1 holds, by writing out everything in the components.

$$M(\sigma \,|\psi\rangle) = \begin{bmatrix} \mu_{00} & \mu_{01} \\ \mu_{10} & \mu_{11} \end{bmatrix} \begin{bmatrix} \sigma\psi_0 \\ \sigma\psi_1 \end{bmatrix} \qquad \text{Expand the terms}$$

$$= \begin{bmatrix} \sigma\mu_{00}\psi_0 + \sigma\mu_{01}\psi_1 \\ \sigma\mu_{10}\psi_0 + \sigma\mu_{11}\psi_1 \end{bmatrix} \qquad \text{Multiply the matrices} \qquad (3.3)$$

$$= \sigma \begin{bmatrix} \mu_{00} & \mu_{01} \\ \mu_{10} & \mu_{11} \end{bmatrix} \begin{bmatrix} \psi_0 \\ \psi_1 \end{bmatrix} \qquad \text{Extract } \sigma \text{ and write as matrices}$$

$$= \sigma M \,|\psi\rangle \qquad \text{Use definitions of } M \text{ and } |\psi\rangle$$

To demonstrate property L2, I'll take things slowly. I've written out each step in Equation 3.4. It's a lot like Equation 3.3, but with twice as many pieces to keep track of. The first three lines expand out the matrix multiplication, and then the rest of the lines gather things back together. It's really just a lot of bookkeeping.

$$M(|\psi\rangle + |\phi\rangle) = \begin{bmatrix} \mu_{00} & \mu_{01} \\ \mu_{10} & \mu_{11} \end{bmatrix} \begin{bmatrix} \psi_0 + \phi_0 \\ \psi_1 + \phi_1 \end{bmatrix} \qquad \text{Expand the terms}$$

$$= \begin{bmatrix} \mu_{00}(\psi_0 + \phi_0) + \mu_{01}(\psi_1 + \phi_1) \\ \mu_{10}(\psi_0 + \phi_0) + \mu_{11}(\psi_1 + \phi_1) \end{bmatrix} \qquad \text{Multiply the matrices}$$

$$= \begin{bmatrix} \mu_{00}\psi_0 + \mu_{00}\phi_0 + \mu_{01}\psi_1 + \mu_{01}\phi_1 \\ \mu_{10}\psi_0 + \mu_{10}\phi_0 + \mu_{11}\psi_1 + \mu_{11}\phi_1 \end{bmatrix} \qquad \text{Expand the products} \qquad (3.4)$$

$$= \begin{bmatrix} (\mu_{00}\psi_0 + \mu_{01}\psi_1) + (\mu_{00}\phi_0 + \mu_{01}\phi_1) \\ (\mu_{10}\psi_0 + \mu_{11}\psi_1) + (\mu_{10}\phi_0 + \mu_{11}\phi_1) \end{bmatrix} \qquad \text{Gather } \psi \text{ and } \phi \text{ terms}$$

$$= \begin{bmatrix} \mu_{00} & \mu_{01} \\ \mu_{10} & \mu_{11} \end{bmatrix} \begin{bmatrix} \psi_0 \\ \psi_1 \end{bmatrix} + \begin{bmatrix} \mu_{00} & \mu_{01} \\ \mu_{10} & \mu_{11} \end{bmatrix} \begin{bmatrix} \phi_0 \\ \phi_1 \end{bmatrix} \qquad \text{Write as matrices}$$

$$= M \,|\psi\rangle + M \,|\phi\rangle \qquad \text{Use definitions}$$

It was worth confirming that the two matrix operations in Equation 3.2 really are linear because we're going to use those linearity properties frequently when we work with multiple operators.

Speaking of multiple operators, let's see what happens if we apply two operators in a row.

We'll start with a ket $|\psi\rangle$ and apply a single operator A to it, getting back a new ket $|\psi'\rangle$ (in this example, the letter A stands for a generic linear operator, in the same way that $|\psi\rangle$ refers to a generic ket). In symbols, $|\psi'\rangle = A \,|\psi\rangle$. Figure 3-2 illustrates this.

Figure 3-2: Applying A to $|\psi\rangle$

We start with $|\psi\rangle$ on the left, and it moves to the right, where it's modified by an operator A (that is, it's multiplied by a matrix named A), producing the result $|\psi'\rangle$ on the right.

We usually write this operation as $A|\psi\rangle$, but for the moment let's treat A as though it's a function. We can then write this as $A(|\psi\rangle)$, where the parentheses make it clear that $|\psi\rangle$ is the input to A.

Now let's run $|\psi\rangle$ through two arbitrary operators B and C, in that order, as in Figure 3-3.

Figure 3-3: Applying B to $|\psi\rangle$, then applying C

The first operator to be applied is B, so its output is $B(|\psi\rangle)$. Then we apply C to that output, producing $C(B(|\psi\rangle))$. Let's drop the parentheses now, so the output $|\psi'\rangle = CB|\psi\rangle$.

Notice that in the picture, reading left to right, we see B on the left and C on the right. But in the algebraic notation for $|\psi'\rangle$, written $CB|\psi\rangle$, C is on the left and B is on the right. The left-to-right order has been reversed!

This is because the convention in math is to apply operators from right to left, unless other conventions or parentheses explicitly tell us otherwise (this rule comes from an idea called the *composition of functions* [125]). For example, $4 \times 2 + 1$ is $8 + 1 = 9$, because the convention is to perform all multiplications before additions. But $4 \times (2 + 1) = 4 \times 3 = 12$, thanks to the parentheses.

Both the picture and the algebra produce the same result, but this change in the apparent order of operations can definitely be confusing.

When going between an expression (that is, the algebraic form) and a picture (that is, the visual form), it's super easy to mess up the order of the matrices, but it's vital to use the operations in the proper order. The reason is that matrix multiplication is not *commutative*. This means that, generally, $BC \neq CB$. Writing the matrices in the wrong order will almost always give us the wrong result.

Many people have tried to find a way to prevent this kind of apparent (but illusory) inconsistency between the picture and the algebra [137], but each fix seems to introduce new problems of its own. Ultimately, the standard ways to read English, diagrams, and math just inherently conflict in this situation. Remember that when you see multiple operators in an algebraic expression, they are applied right to left. Thus, $CB|\psi\rangle$ means first compute $B|\psi\rangle$, then apply C to that. This will soon become second nature, but until that habit kicks in, read and write these expressions with care.

Operators I, X, and H

Next, we'll look at three operators that are used in almost every quantum algorithm, usually many times. I'll introduce each one, show you what it does, and present its matrix form in the computational basis.

The Identity Operator I

The *identity operator*, written I, is the operator that passes its input to its output unchanged. In other words, multiplying a state by I is like multiplying a number by 1: Nothing changes.

The matrix form of the identity operator is called an *identity matrix*. This is always a square grid of all 0 elements, but with 1 elements on the *main diagonal*, or the line from the upper-left to the lower-right corner. The identity matrix of d dimensions is a square matrix of d elements on each side. Equation 3.5 shows I_2, the identity matrix for $d = 2$, and I_4, the identity matrix for $d = 4$. Normally we leave off the subscript, inferring the dimensionality of the matrix from context.

$$I_2 = \begin{bmatrix} 1 & 0 \\ 0 & 1 \end{bmatrix}, \qquad I_4 = \begin{bmatrix} 1 & 0 & 0 & 0 \\ 0 & 1 & 0 & 0 \\ 0 & 0 & 1 & 0 \\ 0 & 0 & 0 & 1 \end{bmatrix} \qquad (3.5)$$

The identity matrix has this name because of its similarity to other operations that pass their input to their output unchanged. For example, 0 is the identity element for addition because $0 + x = x$ for any number x, and 1 is the identity element for multiplication because $1x = x$ for any number x. In the same way, applying I to any ket $|\psi\rangle$ gives us back $|\psi\rangle$, or in symbols, $I|\psi\rangle = |\psi\rangle$. I've drawn this in tableau form in Equation 3.6.

$$I|\psi\rangle = \begin{bmatrix} 1 & 0 \\ 0 & 1 \end{bmatrix} \begin{bmatrix} \psi_0 \\ \psi_1 \end{bmatrix} = \begin{bmatrix} \psi_0 \\ \psi_1 \end{bmatrix} \qquad (3.6)$$

If we want to multiply a bra and an identity matrix, we need to put the identity on the right, so that we have a legal matrix multiplication. Thus, $\langle\psi| I = \langle\psi|$.

If we multiply the identity matrix with any other square matrix M of the same dimensionality, we get back M unchanged. This is true whether we multiply with the identity on the left or right of M. In symbols, we express this as $IM = MI = M$.

The identity matrix may not seem to be of much value, but it will turn out to be super important in Chapter 5, when we build up systems of multiple qubits and multiple operators.

The NOT Operator X

Our next operator goes by two equivalent names, *NOT* and *X*. For brevity, I'll usually use *X* in this book. Given an input of two elements, *X* swaps those elements. Equation 3.7 shows the *X* matrix.

$$NOT = X = \begin{bmatrix} 0 & 1 \\ 1 & 0 \end{bmatrix} \tag{3.7}$$

Notice that this looks like the identity matrix, only the 1 elements are on the other diagonal, from the upper right to lower left.

Like with *I*, we can keep this pattern going for a square matrix of any size. The symbol X_n tells us to make a square matrix of n elements on a side that are all 0, except for those on the diagonal from the upper-right to the lower-left corner, which are all 1. Without a subscript, the operator *X* always means X_2, the two-by-two matrix shown in Equation 3.7.

When we compute $X |\psi\rangle$, the two elements of the ket are swapped, as shown in Equation 3.8.

$$X |\psi\rangle = \begin{bmatrix} 0 & 1 \\ 1 & 0 \end{bmatrix} \begin{bmatrix} \psi_0 \\ \psi_1 \end{bmatrix} = \begin{bmatrix} \psi_1 \\ \psi_0 \end{bmatrix} \tag{3.8}$$

Similarly, $\langle\phi| X$ swaps the elements of a bra, as shown in Equation 3.9.

$$\langle\phi| X = \begin{bmatrix} \phi_0 & \phi_1 \end{bmatrix} \begin{bmatrix} 0 & 1 \\ 1 & 0 \end{bmatrix} = \begin{bmatrix} \phi_1 & \phi_0 \end{bmatrix} \tag{3.9}$$

If we multiply *X* and a matrix *M*, we swap the rows of *M* if we form *XM*, and we swap the columns of *M* if we form *MX*, as shown in Figure 3-4.

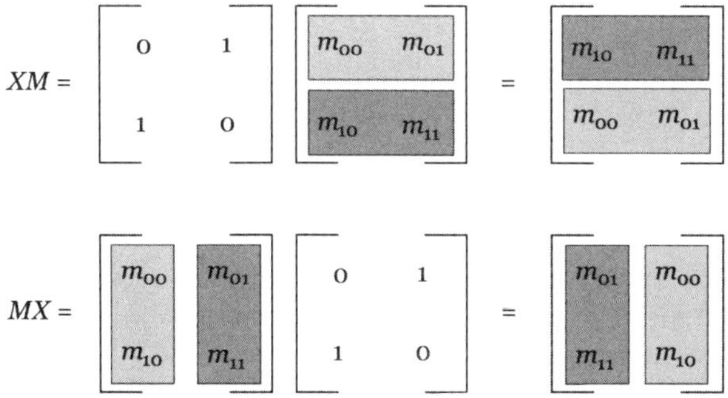

Figure 3-4: The operation XM swaps the rows of M, while MX swaps the columns of M.

These results are different. Generally speaking, $XM \neq MX$, demonstrating the need to take care when writing the products of matrices.

The Hadamard Operator H

Our third operator is the *Hadamard operator*, named for Jacques Hadamard [268]. I'll have much to say about the H operator in this book, but for now let's just scratch the surface.

The matrix for H is shown in Equation 3.10.

$$H = \begin{bmatrix} \frac{1}{\sqrt{2}} & \frac{1}{\sqrt{2}} \\ \frac{1}{\sqrt{2}} & \frac{-1}{\sqrt{2}} \end{bmatrix} \tag{3.10}$$

We'll see a lot of those $1/\sqrt{2}$ factors throughout the rest of the book. To reduce clutter and potential errors, my shorthand is to use the symbol $\vee$ for this value, as defined in Equation 3.11.

$$\vee \overset{\Delta}{=} \frac{1}{\sqrt{2}} \tag{3.11}$$

We pronounce this symbol out loud as "vee," so the expression $\vee^2$ would be pronounced as "vee squared." Using this definition, we can re-write Equation 3.10 as Equation 3.12, which is the form I'll usually use in this book. Here, the $\vee$ multiplies every number in the matrix.

$$H = \vee \begin{bmatrix} 1 & 1 \\ 1 & -1 \end{bmatrix} \tag{3.12}$$

Much better! We'll often see powers of $\vee$, so I've summarized the first few in Equation 3.13.

$$\begin{aligned} \vee^0 &= 1 \\ \vee^1 &= 1/\sqrt{2} \\ \vee^2 &= 1/2 \\ \vee^3 &= 1/(2\sqrt{2}) \\ \vee^4 &= 1/4 \end{aligned} \tag{3.13}$$

From now on, I'll usually write the matrix form of H as in Equation 3.12. It's vital that you don't forget the $\vee$, because without it, states operated upon by H will change in magnitude, violating Postulate 1 of quantum mechanics. Remember to include that $\vee$!

Sometimes it's inconvenient to have $\vee$ dangling out in front of the H matrix, as in Equation 3.12. If that factor is making an expression more complicated rather than simpler, we can explicitly include the $\vee$ factor in the H matrix, as in Equation 3.14.

$$H = \begin{bmatrix} \vee & \vee \\ \vee & -\vee \end{bmatrix} \tag{3.14}$$

Let's take H out for a spin on the quantum state $|\psi\rangle$, as shown in Equation 3.15.

$$H\,|\psi\rangle = \vee \begin{bmatrix} 1 & 1 \\ 1 & -1 \end{bmatrix} \begin{bmatrix} \psi_0 \\ \psi_1 \end{bmatrix} = \vee \begin{bmatrix} \psi_0 + \psi_1 \\ \psi_0 - \psi_1 \end{bmatrix} \tag{3.15}$$

This is a potent result. We've taken a state described by two complex numbers and turned it into a state where both of those numbers are present in each coefficient, once as a sum and once as a difference. We'll see that combining the elements of a state this way opens the door to many of the special properties of quantum computing, making H perhaps the most important matrix in this book.

A Few Matrix Operations

Because we'll use matrices to manipulate qubits, we'll use a lot of matrices in this book. There are a few common ways to manipulate these matrices that will make our expressions simpler and easier to understand, so let's pause for a moment to review these matrix operations. If these are unfamiliar to you, check out any good linear algebra reference [211] [117] [220]. We've already seen these operations applied to our special matrices of a single column (a ket) and a single row (a bra), and now we'll see how they work with a square matrix.

Transposition refers to the process of exchanging every entry $\mu_{j,k}$ of a matrix M with the entry at $\mu_{k,j}$. We can think of this as turning rows into columns, or columns into rows, or reflecting (or mirroring) the matrix along the diagonal line from its upper left to lower right. Figure 3-5 illustrates these three interpretations, which all produce the same result. We call the result of transposition the *transpose* of the original.

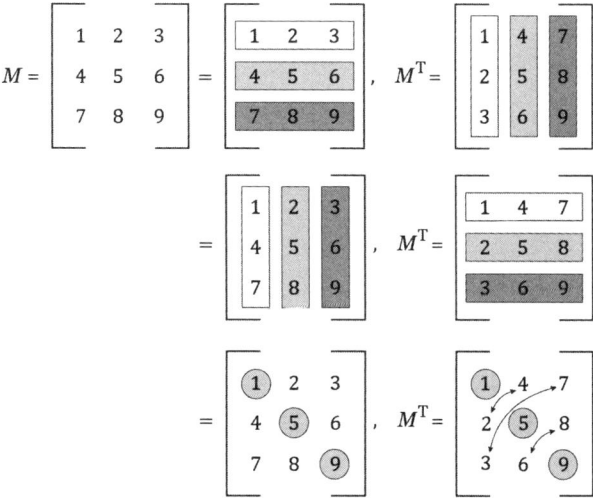

Figure 3-5: Three ways to visualize applying transposition to a three-by-three matrix M

In the top row of Figure 3-5, we think of M in terms of rows; we rotate each row into a column and place them left to right instead of top to bottom. The middle row is like the top, but we start with columns that turn into rows. At the bottom of the figure, we first lock down the elements along the main diagonal, then swap each element with the element on the other side of the main diagonal.

We write the transpose of a matrix M with a superscript of T, or M^T.

There are two important special cases of transposition: Transposing a matrix of a single column produces a matrix of a single row, and vice versa.

A nice property of transposition is that the transpose of the product of two matrices is the product of their transposes in the opposite order. The proof of this takes some effort [162], and I don't think it offers much illumination, so I'll just summarize this in symbols for matrices A and B in Equation 3.16.

$$(AB)^T = B^T A^T \tag{3.16}$$

Conjugation is the operation of replacing every element with its complex conjugate. This is the same operation we discussed in Chapter 2, but now applied to each element in a matrix. If the elements are real numbers, this operation has no effect.

We write the conjugated matrix as $\overline{M}$, with elements $\overline{\mu}_{i,j}$.

The *adjoint* is the result of applying both transposition and conjugation. Since transposition and conjugation have independent effects, applying them in either order produces the same result. So, the conjugate of the transpose, $\overline{M^T}$, is the same as the transpose of the conjugate, $\overline{M}^T$. Either sequence of operations produces the adjoint. We write the adjoint of matrix M with a superscript of a dagger, giving us $M^\dagger$ (some people use an asterisk, writing the adjoint as M^*).

To put everything in one place, Equation 3.17 shows a matrix M and its transpose, conjugate, and adjoint.

$$M = \begin{bmatrix} \mu_{00} & \mu_{01} \\ \mu_{10} & \mu_{11} \end{bmatrix}$$

$$\tag{3.17}$$

$$M^T = \begin{bmatrix} \mu_{00} & \mu_{10} \\ \mu_{01} & \mu_{11} \end{bmatrix}, \qquad \overline{M} = \begin{bmatrix} \overline{\mu_{00}} & \overline{\mu_{01}} \\ \overline{\mu_{10}} & \overline{\mu_{11}} \end{bmatrix}, \qquad M^\dagger = \begin{bmatrix} \overline{\mu_{00}} & \overline{\mu_{10}} \\ \overline{\mu_{01}} & \overline{\mu_{11}} \end{bmatrix}$$

As we saw in Chapter 2, the adjoint is an important tool for us, because it's the mechanism that lets us transform a ket (a column matrix) into its associated bra (the corresponding row matrix with conjugated values), and vice versa. In symbols, $|\alpha\rangle^\dagger = \langle\alpha|$. This is illustrated in tableau form in Equation 3.18.

$$|\alpha\rangle^\dagger = \begin{bmatrix} \alpha_0 \\ \alpha_1 \\ \vdots \\ \alpha_{d-1} \end{bmatrix}^\dagger = \begin{bmatrix} \overline{\alpha_0} & \overline{\alpha_1} & \ldots & \overline{\alpha_{d-1}} \end{bmatrix} = \langle\alpha| \tag{3.18}$$

The adjoint of a bra, $\langle\alpha|^\dagger = |\alpha\rangle$, is shown in Equation 3.19.

$$\langle\alpha|^\dagger = \begin{bmatrix} \overline{\alpha_0} & \overline{\alpha_1} & \ldots & \overline{\alpha_{d-1}} \end{bmatrix}^\dagger = \begin{bmatrix} \alpha_0 \\ \alpha_1 \\ \vdots \\ \alpha_{d-1} \end{bmatrix} = |\alpha\rangle \tag{3.19}$$

In Equation 3.19, I wrote the elements in the bra as conjugated versions of the elements in the ket so that everything matched up with Equation 3.18. The key thing is that the elements in the bra and ket are the complex conjugates of each other, regardless of which one is considered the "starting" value and which is the conjugate.

The transpose, conjugation, and adjoint all undo themselves if applied twice. That is, $(M^\mathrm{T})^\mathrm{T} = \overline{\overline{M}} = (M^\dagger)^\dagger = M$.

The *inverse* of a matrix M is a new matrix, written M^{-1}. It's defined so that when it's multiplied with M we get back the identity matrix I.

In this book, all of our inverses work from both sides, giving us Equation 3.20.

$$MM^{-1} = M^{-1}M = I \tag{3.20}$$

Computing the inverse of a matrix is a mechanical process, but it requires some work (details of the process are given in any linear algebra reference, such as those I mentioned earlier). Not every matrix has an inverse [267]. Such matrices are called *singular, non-invertible*, or *degenerate*.

The process of computing the inverse for a matrix will fail if the matrix has no inverse. We won't need to worry about such conditions, because all of the matrices in this book for which we want to find inverses will be *invertible*, meaning that they all have a well-defined inverse. So, you can be sure that any time I write M^{-1}, that inverse will exist.

A common trick when we see a matrix at the start of either side of an equality is to multiply the start of both sides of the equality by that matrix's inverse. Equation 3.21 shows how this works if we want to isolate the vector $|\psi\rangle$ in the equality $M|\psi\rangle = |\phi\rangle$.

$$
\begin{aligned}
M|\psi\rangle &= |\phi\rangle & &\text{Starting relationship} \\
M^{-1}M|\psi\rangle &= M^{-1}|\phi\rangle & &\text{Multiply both sides by } M^{-1} \\
I|\psi\rangle &= M^{-1}|\phi\rangle & &\text{On left side, replace } M^{-1}M \text{ with } I \\
|\psi\rangle &= M^{-1}|\phi\rangle & &\text{Since } I|\psi\rangle = |\psi\rangle, \text{ remove the } I
\end{aligned}
\tag{3.21}
$$

This is a nice process to keep in mind if we're given the state $|\phi\rangle$ as a result of applying M to $|\psi\rangle$, but we'd really prefer to know what the starting $|\psi\rangle$ was.

We can now show that the adjoint of the product of two matrices is the same as the product of their adjoints in the opposite order. The steps are shown in Equation 3.22.

$$
\begin{aligned}
(AB)^\dagger &= \overline{AB}^T &&\text{The definition of the adjoint} \\
&= (\overline{A}\,\overline{B})^T &&\text{Using Eq. 2.14 for matrices} \\
&= \overline{B}^T\overline{A}^T &&\text{Use Eq. 3.16} \\
&= B^\dagger A^\dagger &&\text{The definition of the adjoint}
\end{aligned}
\tag{3.22}
$$

This property will come in handy in Chapter 6, when we look at how to measure quantum states.

Unitary Operators

In quantum computing, we can only use one particular type of linear operator in our programs. In this section, we'll see why that's the case and what characterizes these operators.

We saw in Chapter 2 that inner products are useful tools for characterizing the relationships between two states. In this section, we're looking at the effects of operators on states. So, let's find the inner product of two arbitrary states $|\phi\rangle$ and $|\psi\rangle$, apply some arbitrary operator M to one of them, and find their new inner product. This will give us some insight about what effect multiplying one state by M has on the relationship between the states.

I'll choose to project $|\psi\rangle$ onto $|\phi\rangle$, so the inner product is $\langle\phi|\psi\rangle$, and I'll choose to apply M to $|\psi\rangle$. This gives us a new inner product, written in Equation 3.23.

$$
\langle\phi|\,(M\,|\psi\rangle)
\tag{3.23}
$$

Like any inner product, this gives us back a scalar. Because matrix multiplication is associative, we can write this expression by bundling together $\langle\phi|\,M$ first. And since the order in which we compute the matrix products doesn't matter, we don't need the parentheses at all, letting us write the same expression in three ways, as in Equation 3.24.

$$
\langle\phi|\,(M\,|\psi\rangle) = (\,(\langle\phi|\,M)\,|\psi\rangle = \langle\phi|\,M\,|\psi\rangle
\tag{3.24}
$$

The rightmost version might look like something new, but it's just telling us to multiply together three matrices: the row matrix $\langle\phi|$, the square matrix M, and the column matrix $|\psi\rangle$.

Now let's include a vital condition. Remember that Postulate 1 tells us that all quantum states have unit norm, or a magnitude of 1. Since the inner product of any ket with itself gives us the squared magnitude of that ket, this inner product will always be 1, as shown in Equation 3.25.

$$
\langle\psi|\psi\rangle = 1, \quad \text{for all } |\psi\rangle
\tag{3.25}
$$

So if we start with any quantum state $|\psi\rangle$ (which must have unit norm), and we transform it with some operator M to make $M|\psi\rangle$, that output is a new quantum state, which also must have unit norm.

Thus, we can only use operators that *preserve unit magnitude*. That is, when the input has a magnitude of 1, so does the output. We need that condition because nature doesn't let us create quantum states that don't have unit norm! If you tried to apply an operator that didn't obey this rule on actual quantum computing hardware, you might not even be able to do it. If you could somehow perform the operation, things would go wrong in some way, depending on the particular technology.

It would be nice to have a simple rule that characterized such operators. Of course, I only suggested that because there is such a rule! Let's find it.

Since the output of an operator must have unit norm, the inner product of that output with itself must be 1. Suppose we have a candidate operator U. Applying it to some arbitrary state $|\psi\rangle$ would give us $U|\psi\rangle$. Let's write the inner product of this result with itself, starting in Equation 3.26, where I've expanded the inner product into the product of two matrices.

$$\left\langle U|\psi\rangle \,\big|\, U|\psi\rangle \right\rangle = \left(U|\psi\rangle\right)^{\dagger}\left(U|\psi\rangle\right) \tag{3.26}$$

The trick to simplifying the right-hand side is to find a different way to write $(U|\psi\rangle)^{\dagger}$. Let's pause for a moment to work that out. Then we'll substitute it into Equation 3.26 and keep rolling on our quest to characterize those operators that we can use in quantum computing.

I think the easiest approach is to write out the matrices and observe that they have a simpler form. The steps are shown in Equation 3.27.

$$
\begin{aligned}
\left(U|\phi\rangle\right)^{\dagger} &= \left(\begin{bmatrix} v_{00} & v_{01} \\ v_{10} & v_{11} \end{bmatrix}\begin{bmatrix} \phi_0 \\ \phi_1 \end{bmatrix}\right)^{\dagger} && \text{Expand the matrices} \\[2mm]
&= \begin{bmatrix} v_{00}\phi_0 + v_{01}\phi_1 \\ v_{10}\phi_0 + v_{11}\phi_1 \end{bmatrix}^{\dagger} && \text{Multiply the matrices} \\[2mm]
&= \begin{bmatrix} \overline{v_{00}\phi_0 + v_{01}\phi_1} & \overline{v_{10}\phi_0 + v_{11}\phi_1} \end{bmatrix} && \text{Form the adjoint} \\[2mm]
&= \begin{bmatrix} \overline{v_{00}}\,\overline{\phi_0} + \overline{v_{01}}\,\overline{\phi_1} & \overline{v_{10}}\,\overline{\phi_0} + \overline{v_{11}}\,\overline{\phi_1} \end{bmatrix} && \text{Because } \overline{\alpha + \beta} = \overline{\alpha}+\overline{\beta} \\[2mm]
&= \begin{bmatrix} \overline{\phi_0} & \overline{\phi_1} \end{bmatrix}\begin{bmatrix} \overline{v_{00}} & \overline{v_{10}} \\ \overline{v_{01}} & \overline{v_{11}} \end{bmatrix} && \text{Write as a matrix multiplication} \\[2mm]
&= \langle\phi|\,U^{\dagger} && \text{Simplify}
\end{aligned}
\tag{3.27}
$$

I've summarized what we just learned in Equation 3.28.

$$\left(U|\phi\rangle\right)^{\dagger} = \langle\phi|\,U^{\dagger} \tag{3.28}$$

We can use this to simplify Equation 3.26, as in Equation 3.29.

$$\left(U\,|\psi\rangle\right)^{\dagger}\left(U\,|\psi\rangle\right) = \left(\langle\psi|\,U^{\dagger}\right)\left(U\,|\psi\rangle\right) \quad \text{Eq. 3.26 with Eq. 3.28}$$

$$= \langle\psi|\,U^{\dagger}\,U\,|\psi\rangle \quad\quad\quad \text{Remove parentheses} \quad\quad (3.29)$$

$$= \langle\psi|\,(U^{\dagger}\,U)\,|\psi\rangle \quad\quad \text{Gather middle two matrices}$$

Phew, this is much more manageable.

Now if it turned out that $U^{\dagger}U = I$, then that term would go away, leaving us with just $\langle\psi|\psi\rangle$, which we know is 1.

And *that* is our simple test for permissible operators! If it so happens that $U^{\dagger}U = I$, then we can simplify Equation 3.29 as in Equation 3.30.

$$\langle\psi|\,(U^{\dagger}\,U)\,|\psi\rangle = \langle\psi|\,I\,|\psi\rangle \quad \text{Apply } U^{\dagger}U = I$$

$$= \langle\psi|\psi\rangle \quad\quad\quad \text{Remove } I \quad\quad\quad\quad (3.30)$$

That tells us that if U obeys the property $U^{\dagger}U = I$, then U is a permissible operator in quantum computing. In fact, because unit norms must be maintained in any program, the *only* operators we can use are those obeying this rule.

Operators that obey this rule are called *unitary operators*. Think of "unitary" as referring to the unit norm of the quantum states, which is what these operators preserve: A unit norm vector goes in, and a unit norm vector comes out. This defining property of unitary matrices is summarized in Equation 3.31 for any arbitrary operator M.

$$\text{If } MM^{\dagger} = M^{\dagger}M = I, \text{ then } M \text{ is unitary.} \quad\quad (3.31)$$

The condition holds either way you choose to multiply the matrices [281].

Notice that both forms of Equation 3.31 multiply a matrix M with another matrix to give us the identity. Therefore, that other matrix is the inverse matrix we met in Equation 3.20. In symbols, if M is unitary, then $M^{\dagger} = M^{-1}$. I've summarized this in Equation 3.32.

$$\text{If } M^{-1} = M^{\dagger}, \text{ then } M \text{ is unitary.} \quad\quad (3.32)$$

Equations 3.31 and 3.32 say the same thing in different ways, so we can use either one to test if an operator is unitary.

Be careful not to conclude that this means that unitary operators are their own inverses. This slip is easy to make because some unitary operators indeed *are* their own inverses, like the operators I, X, and H that we saw earlier. But many unitary operators are not their own inverses. To know for sure, we have to check each operator by computing its inverse and comparing that to the original.

In this book, every operator in a quantum program must transform an input with unit norm into an output with unit norm.

Unitary matrices have the property of being *reversible*, meaning that if we use a unitary matrix to transform a state, it's always possible to undo that operation and recover the original state. For example, if we transform some state with a unitary matrix U to get $U\,|\psi\rangle$, we can recover $|\psi\rangle$ with the adjoint of U, since $U^{\dagger}U\,|\psi\rangle = I\,|\psi\rangle = |\psi\rangle$.

Naming a Matrix Element

Expressions of the form $\langle\phi|\,M\,|\psi\rangle$ have a special case that can sometimes make complicated expressions easier to understand. Suppose that r and c are integers. We can use them to form the bra $\langle r|$ and ket $|c\rangle$, respectively. Then, $\langle r|M|c\rangle$ is another name for the element of M at row r and column c.

Let's see an example of this. I've written out $\langle r|M|c\rangle$ for states with four terms in Equation 3.33, using $r = 2$ and $c = 3$. I first found $\langle 2|\,M$ and then formed $\langle 2|M|3\rangle$. Because matrix multiplication is associative, we'd get the same results by first finding $M\,|3\rangle$.

$$\langle 2|M|3\rangle = \begin{bmatrix} 0 & 0 & 1 & 0 \end{bmatrix} \begin{bmatrix} \mu_{00} & \mu_{01} & \mu_{02} & \mu_{03} \\ \mu_{10} & \mu_{11} & \mu_{12} & \mu_{13} \\ \mu_{20} & \mu_{21} & \mu_{22} & \mu_{23} \\ \mu_{30} & \mu_{31} & \mu_{32} & \mu_{33} \end{bmatrix} \begin{bmatrix} 0 \\ 0 \\ 0 \\ 1 \end{bmatrix} \quad \text{Expand the matrices}$$

(3.33)

$$= \begin{bmatrix} \mu_{20} & \mu_{21} & \mu_{22} & \mu_{23} \end{bmatrix} \begin{bmatrix} 0 \\ 0 \\ 0 \\ 1 \end{bmatrix} = \mu_{23} \quad \text{Multiply } \langle 2|\,M \text{ with } |3\rangle$$

We can summarize this in general as in Equation 3.34.

$$\langle r|\,M\,|c\rangle = \mu_{r,c} \tag{3.34}$$

Sometimes we'll form expressions like the one on the left in the course of doing algebra, so it's nice to be able to replace that big chunk with the explicit matrix element on the right.

Revisiting *I*, *X*, and *H*

The three operators we saw in this chapter are all their own transpose. We can see this by noting that they're each symmetrical with respect to the main diagonal. This means that they're each their own conjugate, since they're composed of only real numbers.

Even better, each is its own inverse. Just multiply the matrix for I, X, or H with itself, and you'll get back the identity matrix every time.

Best of all, because each matrix is its own inverse and adjoint, that means the inverse and adjoint are equal, so each matrix is unitary. If they weren't, they wouldn't be of any use for manipulating quantum states, and I wouldn't have discussed them in the first place!

Putting It All Together

We can now return to Postulate 2 and interpret it. Here's the postulate again, for reference:

Postulate 2   Quantum states are changed by unitary operators.

This postulate flows directly from Postulate 1, which tells us that quantum states must always have a magnitude of 1, and our study of linear operators and their matrices. We use the label "unitary" to refer to those operators that preserve the unit magnitude (that is, a magnitude of 1) of their inputs.

Because we need the output of every operator to be a state that represents a physical quantum object, those outputs need to have a magnitude of 1, like their inputs, and thus those operators must preserve the magnitude of their inputs. Unitary operators are the ones that have this property.

There are an infinite number of unitary operators. We saw three of them in this chapter, and we'll meet more in coming chapters.

It's nice that there's a short test that we can apply to any matrix to determine if it's unitary. We've just seen two such tests, Equations 3.31 and 3.32, which tell us that any operator A is unitary if its inverse is its adjoint, or $A^{-1} = A^{\dagger}$. Any matrix that satisfies this rule may be used for quantum computation, and any matrix that does not satisfy this rule may not be used for quantum computation.

Summary

In this chapter, we looked at how to change quantum states, the process at the heart of quantum computing. These changes are performed by operators, which in a given basis (such as the computational basis that we're using) can be represented as specific square matrices.

We met the identity operator I, which does nothing to its input; the X (or *NOT*) operator, which exchanges the two complex numbers in its input; and the Hadamard operator H, which modifies both of the complex numbers in its input.

We saw that quantum operations must preserve the norm of their inputs. The square matrices that have this property are those whose adjoint (the transpose of their conjugated elements) is the same as their inverse. We call these unitary matrices. For any such matrix U, we can be sure that $U^{\dagger} = U^{-1}$.

We looked at the notation $\langle \phi | M | \psi \rangle$ and saw that it's three matrices multiplied together. When $|\phi\rangle$ and $|\psi\rangle$ are computational basis states, then they extract a single element from the matrix M.

Now we know how to describe a quantum state and how to change it. We're well on our way to writing quantum algorithms.

4

WORKING WITH QUBITS

You see it all, don't you? You can see how everything is just a random rearrangement of particles in a vibrating superposition.
—Daniel Kwan and Daniel Scheinert, *Everything Everywhere All at Once*, 2022 [118]

In the previous two chapters, we've seen what quantum states look like and how we can modify them with quantum operators. We can now put these two ideas together. In this chapter, we'll assemble a few operators to make our first quantum algorithm!

Generally, a quantum algorithm has three stages. First, we initialize one or more qubits to a known state. Second, we process or modify those qubits with operators. Third, we measure the results.

The art of quantum programming is in the second step, where we design a sequence of operators that each modify the qubit states in such a way that our final measurement will give us a useful answer. We often do this by exploiting the unconventional properties of quantum states and operators.

Hello, World!

Probably the most famous computer program has no official name but is widely known as *Hello, World!* (with or without the exclamation point) [266].

It became so well known because it was the first program in the first book on the C programming language, which has influenced untold numbers of programmers (including me) since it appeared in 1978 [112].

It's not much of a program, since all it does is print Hello, World!, like a child opening their eyes for the first time and beholding the world around them.

The point of this little program was that if you could get it running, you'd have solved many essential practical steps in programming, from logging in to the computer and using a text editor to compiling the program and getting the computer to run it. If you could get the computer to print out Hello, World!, then you could build from there and get it to do anything.

There isn't a consensus yet, but I think that the quantum computing version of *Hello, World!* is the algorithm shown in graphical form in Figure 4-1.

Figure 4-1: My vote for the quantum computing version of Hello, World!

Starting at the left in this figure, we have a qubit that's been initialized to the state $|0\rangle$. Moving to the right along a solid line, that qubit enters a device called a *meter* (indicated by the triangle) that measures the qubit. The double line coming out of the meter represents a classical bit, which I've called b. The meter is set up so that if the state $|0\rangle$ comes in, it produces a signal that we agree to interpret as a classical bit with the value 0. Similarly, if the meter's input is the state $|1\rangle$, its output is a bit with value 1. In this figure, the input of the meter is always $|0\rangle$, so its output will always be 0.

Instead of using a triangle to represent the meter, as in Figure 4-1, many people use a simplified version of a sound level meter, like one you might find on a piece of audio equipment (you can see the different icons in the Appendix).

Introducing Hello, XWorld!

Let's make our *Hello, World!* algorithm in Figure 4-1 a little more interesting by getting it to do some quantum computation. We'll make an addition between the initialized qubit and the output meter. The new element is a box with the letter X inside. This new algorithm, which I'll call *Hello, XWorld!*, is shown in Figure 4-2.

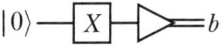

Figure 4-2: The Hello, XWorld! algorithm

This new box is a *quantum gate*, or *qugate* (pronounced CUE-gate). Many people refer to this simply as a "gate," but I prefer the term qugate; not only is it consistent with qubit, but it also helps us distinguish quantum gates from the logic elements used in classical logic diagrams, which have been called gates for many decades.

The X inside the box tells us that this qugate applies the X operator that we met in Chapter 3 and defined in Equation 3.7. Its input arrives via the line on the left side of the box. Its output is represented by the line coming out of the right side of the box.

Let's break down what's happening in Figure 4-2 step by step. This is how we usually analyze quantum algorithms, whether simple or complicated.

To help, I'll add some labels to the drawing in Figure 4-2 so I can talk about the value of the qubit at different stops along its journey from the left side to the right. I've labeled these stops with kets $|\psi_0\rangle$ and $|\psi_1\rangle$ in Figure 4-3. These labels, and the vertical dashed lines associated with them, aren't part of the algorithm. They're just there to help us unambiguously discuss the qubit's value at the marked locations.

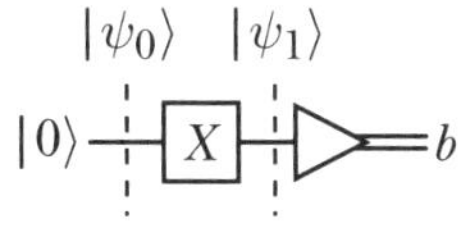

Figure 4-3: A labeled version of Hello, XWorld! from Figure 4-2

The algorithm begins at the far left, with a single qubit in state $|0\rangle$. I've labeled that as $|\psi_0\rangle$, so $|\psi_0\rangle = |0\rangle$.

Now the qubit enters the X qugate. The output of the qugate is marked $|\psi_1\rangle$, so we can write this as $|\psi_1\rangle = X|\psi_0\rangle$.

If we want to know the components of $|\psi_1\rangle$, we can write out $|\psi_0\rangle$ and X in matrix form, so $|\psi_1\rangle$ is the result of multiplying those matrices. I've written that out in Equation 4.1.

$$|\psi_1\rangle = X|\psi_0\rangle = X|0\rangle = \begin{bmatrix} 0 & 1 \\ 1 & 0 \end{bmatrix} \begin{bmatrix} 1 \\ 0 \end{bmatrix} = \begin{bmatrix} 0 \\ 1 \end{bmatrix} = |1\rangle \qquad (4.1)$$

Finally, the output of $|1\rangle$ goes into the meter. When the meter receives $|1\rangle$, it always produces a classical bit with value 1. I've called this output b, so in this algorithm, the classical output b always has a value of 1.

Superpositions

We know that a quantum state is described by a two-element complex vector. The complex elements of this vector, called *amplitudes*, are the complex coefficients α and β on the basis states $|0\rangle$ and $|1\rangle$, as shown in Equation 4.2.

$$|\psi\rangle = \alpha|0\rangle + \beta|1\rangle, \qquad \text{where } |\alpha|^2 + |\beta|^2 = 1 \qquad (4.2)$$

This kind of state, which involves more than one ket, is called a *superposition*.

When α is 1 and β is 0, the superposition degenerates to just the state $|0\rangle$. Similarly, when α is 0 and β is 1, the superposition is just the state $|1\rangle$. The more interesting case is when both α and β are not 0. Then, the superposition is some combination of the states $|0\rangle$ and $|1\rangle$. What does the meter show if we measure such a superposition?

I'll state the result informally for now (we'll come back to this topic in more detail in Chapter 6). When we measure a state in superposition, the superposition disappears, and the state of the quantum object is replaced by exactly one of the basis states that make it up (in this book I always use the computational basis, so a qubit will always become either $|0\rangle$ or $|1\rangle$).

When we measure the single qubit from Equation 4.2, it turns into either $|0\rangle$ or $|1\rangle$, producing a meter output of either 0 or 1, respectively. This replacement is an unpredictable process in which each state has a probability of being "selected" given by the squared magnitude of its amplitude. In symbols, when we measure $|\psi\rangle$, the probability of it becoming $|0\rangle$ and the meter showing 0 is $|\alpha|^2$, and the probability of it becoming $|1\rangle$ and the meter showing 1 is $|\beta|^2$. For any given measurement of a state in superposition, we cannot say for sure whether it will output a 0 or a 1. All we know for sure are these probabilities of measuring a 0 or a 1.

Suppose we have 100 independent qubits that have all been set to the same state, $|\psi\rangle = \alpha |0\rangle + \beta |1\rangle$. If we measure them all, about $100|\alpha|^2$ times the meter will report 0, and the remaining $100|\beta|^2$ measurements will be 1. If we measure those same qubits again any number of times, they will always return the same bits, since the process of measuring them has caused them to become either $|0\rangle$ or $|1\rangle$.

This tells us why we have the condition $|\alpha|^2 + |\beta|^2 = 1$. Since each measurement has only two possible results, and we always get one of them, their probabilities must add up to 1. We'll look more closely at measurements in Chapter 6.

Let's look at this in action. I'll create a quantum state that's an equal combination of $|0\rangle$ and $|1\rangle$ and then measure it. The circuit is shown in Figure 4-4.

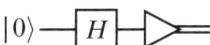

Figure 4-4: Computing and then measuring $H |0\rangle$

You can think of this circuit as a quantum coin flipper. Each time we run it, we'll get back either 0 or 1, with equal probability. We can never say which bit we'll observe, but we can say that over time we'll get back about the same number of 0s and 1s.

The heart of this process is the Hadamard qugate, written H. The beauty of the H qugate is that it can *create superpositions* from states that are not in superposition. This is an essential step in almost every quantum algorithm.

Let's look more closely at the H qugate to see how it performs this incredible, and incredibly useful, feat.

Properties of H

Let's start our discussion of the Hadamard qugate by looking at what it does to some important quantum states.

We can work out what H does to $|0\rangle$ and $|1\rangle$ by explicitly multiplying the matrix and each ket, as in Equation 4.3. I'll use the form of the H matrix given in Equation 3.12. Remember, $\vee$ is a real number with value $1/\sqrt{2}$.

$$H|0\rangle = \vee \begin{bmatrix} 1 & 1 \\ 1 & -1 \end{bmatrix} \begin{bmatrix} 1 \\ 0 \end{bmatrix} = \vee \begin{bmatrix} 1 \\ 1 \end{bmatrix} = \begin{bmatrix} \vee \\ \vee \end{bmatrix}$$

(4.3)

$$H|1\rangle = \vee \begin{bmatrix} 1 & 1 \\ 1 & -1 \end{bmatrix} \begin{bmatrix} 0 \\ 1 \end{bmatrix} = \vee \begin{bmatrix} 1 \\ -1 \end{bmatrix} = \begin{bmatrix} \vee \\ -\vee \end{bmatrix}$$

These two output states have standard shorthand names: $|+\rangle$ and $|-\rangle$, respectively. They're defined in Equation 4.4 as $H|0\rangle$ and $H|1\rangle$.

$$|+\rangle \overset{\Delta}{=} H|0\rangle = \vee \begin{bmatrix} 1 \\ 1 \end{bmatrix} = \begin{bmatrix} \vee \\ \vee \end{bmatrix} = \vee(|0\rangle + |1\rangle)$$

(4.4)

$$|-\rangle \overset{\Delta}{=} H|1\rangle = \vee \begin{bmatrix} 1 \\ -1 \end{bmatrix} = \begin{bmatrix} \vee \\ -\vee \end{bmatrix} = \vee(|0\rangle - |1\rangle)$$

Equation 4.4 shows a few different equivalent ways to write $|+\rangle$ and $|-\rangle$.

We'll use $|+\rangle$ and $|-\rangle$ frequently throughout this book, so it's worth taking a moment to memorize these definitions (remember that all the notation in this book is collected in the Appendix).

We know from "Revisiting I, X, and H" in Chapter 3 that H is its own inverse, so we'd expect that $HH|0\rangle = H|+\rangle$ should be $|0\rangle$, and similarly $HH|1\rangle = H|-\rangle$ should be $|1\rangle$. A little matrix multiplication confirms those expectations, as shown in Equation 4.5.

$$H|+\rangle = \vee \begin{bmatrix} 1 & 1 \\ 1 & -1 \end{bmatrix} \begin{bmatrix} \vee \\ \vee \end{bmatrix} = \vee \begin{bmatrix} 2\vee \\ 0 \end{bmatrix} = \begin{bmatrix} 2\vee^2 \\ 0 \end{bmatrix} = \begin{bmatrix} 2\left(\frac{1}{\sqrt{2}}\right)^2 \\ 0 \end{bmatrix} = \begin{bmatrix} 1 \\ 0 \end{bmatrix} = |0\rangle$$

(4.5)

$$H|-\rangle = \vee \begin{bmatrix} 1 & 1 \\ 1 & -1 \end{bmatrix} \begin{bmatrix} \vee \\ -\vee \end{bmatrix} = \vee \begin{bmatrix} 0 \\ 2\vee \end{bmatrix} = \begin{bmatrix} 0 \\ 2\vee^2 \end{bmatrix} = \begin{bmatrix} 0 \\ 2\left(\frac{1}{\sqrt{2}}\right)^2 \end{bmatrix} = \begin{bmatrix} 0 \\ 1 \end{bmatrix} = |1\rangle$$

The four results we've just seen appear frequently in quantum computing and in this book. I've collected them all in Equation 4.6.

$$H\left|0\right\rangle = \left|+\right\rangle$$
$$H\left|1\right\rangle = \left|-\right\rangle$$
$$H\left|+\right\rangle = \left|0\right\rangle \tag{4.6}$$
$$H\left|-\right\rangle = \left|1\right\rangle$$

Equation 4.6 is a nice demonstration that H is its own inverse. In symbols, $H^2 = I$, so $HH\left|\psi\right\rangle = \left|\psi\right\rangle$ for any $\left|\psi\right\rangle$.

Thanks to H being linear, if we know what H does to the basis states $\left|0\right\rangle$ and $\left|1\right\rangle$, then we know what it does to all states, since they're just sums of scaled versions of these basis states.

Let's demonstrate this using the linearity properties L1 and L2. We can write the operation of H on an arbitrary ket $\left|\psi\right\rangle = \alpha\left|0\right\rangle + \beta\left|1\right\rangle$, shown in Equation 4.7.

$$
\begin{aligned}
H\left|\psi\right\rangle &= H\big(\alpha\left|0\right\rangle + \beta\left|1\right\rangle\big) && \text{Expand } \left|\psi\right\rangle \\
&= H(\alpha\left|0\right\rangle) + H(\beta\left|1\right\rangle) && \text{Use linearity property L2} \\
&= \alpha H\left|0\right\rangle + \beta H\left|1\right\rangle && \text{Use linearity property L1} \\
&= \alpha\left|+\right\rangle + \beta\left|-\right\rangle && \text{From definitions in Eq. 4.6}
\end{aligned}
\tag{4.7}
$$

The series of steps shown in Equation 4.7 are a useful guide for understanding how any qugate works. Let's see why.

Qugates and Basis States

Equation 4.7 showed us that the operation of H on any arbitrary state $\left|\psi\right\rangle$ was the effect of H on each of the basis states $\left|0\right\rangle$ and $\left|1\right\rangle$, which is given by the final line.

In the rest of this book, we'll often look at how qugates manipulate quantum states by working out their effects on the computational bases $\left|0\right\rangle$ and $\left|1\right\rangle$. Because all qugates are linear, we can always follow the steps of Equation 4.7 with a new qugate in place of H, and we'll get back the effect of that qugate on any state $\left|\psi\right\rangle$.

So in general, to find the result of applying an arbitrary qugate A to an arbitrary state $\left|\psi\right\rangle$, the third line of Equation 4.7 gives us Equation 4.8.

$$A\left|\psi\right\rangle = \alpha A\left|0\right\rangle + \beta A\left|1\right\rangle \tag{4.8}$$

So, if we know $A\left|0\right\rangle$ and $A\left|1\right\rangle$, we know exactly what A does to any arbitrary ket $\left|\psi\right\rangle$.

Initializing with H

Let's return to the Hadamard qugate. A remarkable property of H is that it creates a superposition state out of either $\left|0\right\rangle$ or $\left|1\right\rangle$. Creating superpositions from basis states is of tremendous practical value, because quantum computing hardware usually encourages (or even requires) us to initialize qubits to $|0\rangle$. Running those qubits through an H qugate turns each $|0\rangle$ into a $|+\rangle$, which is an equal superposition of $|0\rangle$ and $|1\rangle$. Putting each of our input qubits into superposition is the key step that enables many of the other useful properties of quantum computing.

As a result, many quantum algorithms begin with a bunch of qubits initialized to $|0\rangle$, which immediately go into H qugates. Then we do some computing and finally measure the qubits. Figure 4-5 shows this common architecture using k qubits.

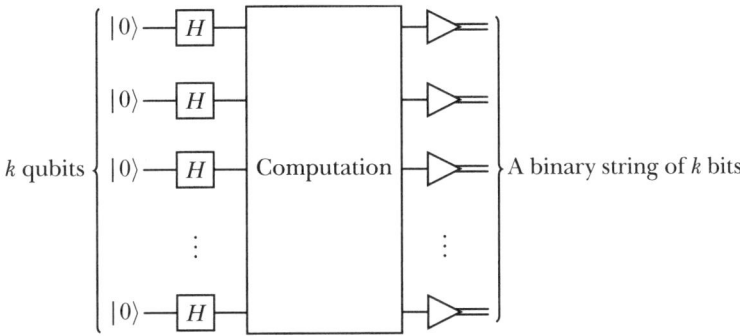

Figure 4-5: Sending k qubits, each initialized as $|0\rangle$, through an H qugate, performing a computation on those qubits, and then measuring them

Each qubit coming out of its H qugate in Figure 4-5 is in the state $|+\rangle$, or an equal superposition of $|0\rangle$ and $|1\rangle$. Suppose for the moment that the box marked *Computation* does nothing, and we immediately measure these qubits. Each meter will then produce either a 0 or a 1, with equal probability. So, the output of the algorithm is a binary string of k bits, though we can't predict the specific values of those bits. In fact, each time we run this algorithm, we'll usually get back a completely different string of bits. Interpreting these bits as a k-bit binary number, we could get outputs from 0 up to and including $2^k - 1$.

Many quantum algorithms boil down to replacing that big box in Figure 4-5 with a sequence of quantum gates. The goal is to transform the qubits so that, when they're measured, the outputs of the meters are a binary number that solves (or helps us to solve) our problem.

Interference

Let's look at another important operation that we use frequently in quantum algorithms. When H processes an input state in superposition, it can change the state's coefficients. This moment is when those coefficients can *interact*. This interaction is called *interference*.

Interference is one of the key phenomena that distinguish quantum computing from classical computing. Typically, classical bits don't interact unless we tell them to. And even then, we can't nudge a 0 a little bit toward a 1, or vice versa.

But qubits do interact, or interfere, with one another and can change one another in small or large ways. Interference can change the amplitudes of a qubit, which in turn changes the probabilities of what we measure at the final output. By controlling these interactions, we can change the states of our qubits to increase the probability that the final value we measure from our algorithm will be useful to us.

Let's see interference in action with another small algorithm. As usual, I'll start with one qubit in the initial state $|0\rangle$. Then I'll apply two H qugates in a row, and wrap up with a meter. Figure 4-6 shows the algorithm.

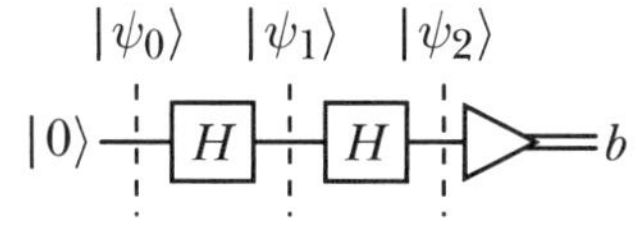

Figure 4-6: A small quantum algorithm of two H qugates

We know that $H|0\rangle = |+\rangle$, so I'll pick up the analysis at the location marked $|\psi_1\rangle$, where the qubit has the value $|+\rangle$. The analysis itself is in Equation 4.9. These steps start out using braket notation but then switch to explicit matrices so we can more clearly see the interaction between the amplitudes.

$$
\begin{aligned}
H|+\rangle &= H\big(\vee\,(|0\rangle + |1\rangle)\big) && \text{Expand } |+\rangle \text{ using Eq. 4.4}\\
&= H(\vee\,|0\rangle + \vee\,|1\rangle) && \text{Distribute the number } \vee\\
&= \vee H\,|0\rangle + \vee H\,|1\rangle && \text{Use L2 to distribute } H\\
&= \vee\,|+\rangle + \vee\,|-\rangle && \text{Use shorthands for } H\,|0\rangle \text{ and } H\,|1\rangle\\[4pt]
&= \vee\begin{bmatrix}\vee\\\vee\end{bmatrix} + \vee\begin{bmatrix}\vee\\-\vee\end{bmatrix} && \text{Write matrices for } |+\rangle \text{ and } |-\rangle\\[4pt]
&= \begin{bmatrix}\vee^2\\\vee^2\end{bmatrix} + \begin{bmatrix}\vee^2\\-\vee^2\end{bmatrix} && \text{Multiply each matrix by } \vee \qquad (4.9)\\[4pt]
&= \begin{bmatrix}\vee^2 + \vee^2\\\vee^2 - \vee^2\end{bmatrix} = \begin{bmatrix}2\vee^2\\0\end{bmatrix} && (\star)\ \text{Interference happens!}\\[4pt]
&= \begin{bmatrix}2(1/2)\\0\end{bmatrix} && \text{Use } \vee^2 = 1/2\\[4pt]
&= \begin{bmatrix}1\\0\end{bmatrix} = |0\rangle && \text{The result is } H\,|+\rangle = |0\rangle
\end{aligned}
$$

The magic is at the line marked with a star. That's interference! This is where the coefficients on the $|1\rangle$ state in the superposition cancel one another out. The result is that the amplitude of the $|1\rangle$ state goes to 0 and thus disappears from the superposition. We call this *destructive interference*.

At the same time, the coefficients for $|0\rangle$ add together to produce an amplitude of 1. This effect of increasing the amplitude is called *constructive interference*.

Interference is a key tool in quantum computing, and we'll see it many more times.

Summary

In this chapter, we met our first quantum algorithms and saw how to follow the state of a qubit as it is operated upon by quantum gates, or qugates.

We began with the quantum version of the classic, minimal *Hello, World!* program, and then we added an *X* qugate so that it actually computed something.

The Hadamard qugate, *H*, is a key tool in quantum algorithms, because it can both create and destroy superpositions. We saw how it creates the superpositions $|+\rangle$ and $|-\rangle$ from inputs $|0\rangle$ and $|1\rangle$, respectively. Applying *H* to those superpositions destroys them, turning $|+\rangle$ to $|0\rangle$ and $|-\rangle$ to $|1\rangle$.

We also looked at the phenomenon of interference, where amplitudes combine to produce a smaller or larger value for one or more basis states.

With these basics under our belt, we're now ready to move past single qubits and see how to build more powerful algorithms that work with multiple qubits.

5

SYSTEMS

True stability results when presumed order and presumed disorder are balanced. A truly stable system expects the unexpected, is prepared to be disrupted, waits to be transformed.
—Tom Robbins, *Even Cowgirls Get the Blues*, 1976 [173]

One does not discover new lands without consenting to lose sight of the shore for a very long time.
—André Gide, *The Counterfeiters: A Novel*, 1927 [72]

 A classical computer that uses only a single bit can't do much. The same is true for a quantum computer with a single qubit. To do interesting work, we'll need lots of qubits.

To represent a group of qubits, in this chapter we'll use a new mathematical operator that builds a collection of qubits. That same operator will also let us build up collections of qugates. We'll call each of these collections a *system*. Together, these systems will enable us to work with groups of qubits and operators, which is essential to understanding existing quantum algorithms and designing new ones.

Postulate 3

In this chapter we'll pull apart and then reassemble another of the postulates of quantum mechanics. Here's a version that's appropriate for our use in this chapter [146, p. 94]:

> **Postulate 3** A system of multiple quantum objects can be represented by the tensor product of those objects.

This postulate introduces a new term, *tensor product*. This is a mathematical operator that lets us combine qubits or qugates. This chapter is about understanding this mathematical tool and how we use it to build systems of multiple qubits and qugates.

We'll first look at how the tensor product works to build these collections.

Then we'll see the style of drawings that are typically used to represent quantum algorithms (rather than, say, a text listing). The drawings are great because they offer us a useful overview of how an algorithm processes the qubits that flow through it.

We'll then bring in the tensor product to help us simplify and analyze these drawings and develop their corresponding algebraic representations. We'll see in Part II that it's these algebraic representations that we usually depend on to analyze and understand how a quantum algorithm works.

Finally, we'll meet a new kind of quantum gate called a controlled qugate, and we'll discuss the most important one of these, called the *CX* qugate.

Combining Quantum States

If we build a quantum computer that contains multiple qubits, it would be useful to come up with one unified description that describes the collection of qubits rather than always having to list them out one by one. How should we build such a description?

Let's start with an analogy. Suppose that a married couple lives up the street from you. You might refer to them as an aggregate entity, saying something like "That couple has two children." By referring to them this way, you're treating the couple as a single, unified object. But you can always distinguish the two people as individuals, perhaps saying "Niko got a nice haircut today."

However, there are situations where combined objects cannot be separated into their components. For example, suppose you order a latte at your favorite coffee shop. This is a combination of espresso and milk. Once they've been mixed together, you can't separate the milk from the coffee. They're now a single, indivisible combined unit.

When we mathematically combine the states of two qubits, sometimes we can treat them like the couple up the street and refer to them either collectively or individually. For example, we might operate on a combined state with some qugate, producing a new combined state. Or we might prefer to separate the combined state into its individual qubits, operate on those qubits, and then assemble them back into a combination. We're free to choose whichever approach is more convenient or informative.

But we don't always have this choice. Sometimes combined quantum states are like the milk and coffee in a latte and cannot be separated. In that case, we *must* treat them as a single object.

If we can't assign individual values to the two nonseparable qubits, what's going on inside the quantum computer? We know that there are two distinct physical entities that are serving as our qubits. So we could physically carve up the computer into two pieces and take one qubit to the north pole and the other one to the south pole. Surely those two qubits, so widely separated and physically distinct, each have their own state, regardless of whether we choose to write them in a merged way. That's just common sense.

And in this case, common sense is just plain wrong. Our intuition for the behavior of nature at the sizes and time scales we're used to just isn't reliable in the quantum realm. Sometimes, when we have two separate objects, they don't have separate states. They are two objects in one shared state, not merely mixed together like milk and coffee but inextricably linked, even though they're physically distinct, or even far apart. Knowing about either one tells us about the other. However unlikely this seems, it has been experimentally verified countless times, proving that this phenomenon really does exist. It's an important practical building block for many quantum algorithms.

Let's dig into the details of the tensor product, which will enable us to understand how collections of qubits and qugates are formed.

The Tensor Product

Let's start by looking at how we might combine two states, $|\psi\rangle$ and $|\phi\rangle$.

One approach would be to list one after the other, producing a sequence of four complex numbers (two each from each state). While reasonable, that's not going to work in the long run, because that operation won't generalize to usefully combine the matrices that describe operators. We'd like one tool that lets us combine both quantum states and quantum operators. Happily, the tensor product is just that tool.

The tensor product combines states by multiplying together each element of $|\psi\rangle$ with each element of $|\phi\rangle$. We could visualize the four terms created by this process in the form of a two-by-two matrix, as shown in Figure 5-1. In part (a) I've written out the elements, and in part (b) I've shown a graphical version using shapes and colors.

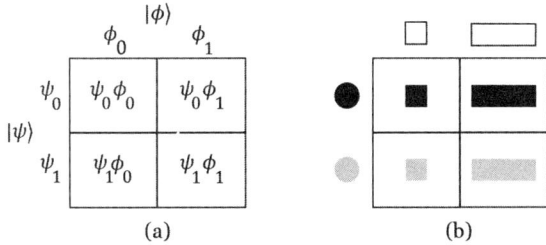

Figure 5-1: Combining states. (a) The elements of $|\psi\rangle$ and $|\phi\rangle$. (b) A graphical interpretation using shapes and colors.

If we agreed on how to sequence the elements of the left-hand grid, we'd have an unambiguous list of four elements that describes both of its component states at once. Unfortunately, like just making a list, this grid won't generalize nicely to handling operators.

A third choice would be to use the combining process called the tensor product. This operation not only generalizes the grid of Figure 5-1, but also will let us make combinations of operators.

The tensor product will be a central tool in this book. It will greatly simplify all of our work with quantum algorithms, from analyzing existing programs to writing new ones. Let's dig in!

Most of us are used to building up collections of things by simply listing one after another. Mathematically, we call this building a *Cartesian product*, and we represent it with the $\times$ operator. In this context, $\times$ doesn't mean multiplication but rather is a kind of glue to assemble a list.

We saw this use of $\times$ in Chapter 2 when we created lists of numbers. For example, a point in traditional 3D space is given as a list of three real numbers, (x, y, z), which we can say is described by the structure, or format, $\mathbb{R} \times \mathbb{R} \times \mathbb{R}$, or $\mathbb{R}^3$.

Let's look at another way to build a collection of objects in lists. We'll create a new list that contains the product of every pairing of an object from the first list with an object from the second.

For example, consider two lists named A and B. List A contains $\dim(A)$ elements, and list B contains $\dim(B)$ elements. Let's fill these up with shapes, as in Figure 5-2. I've drawn these as column vectors because we'll ultimately use these techniques with quantum states, so we might as well start with their column matrix structures. In Figure 5-2, list A has three elements and list B has two.

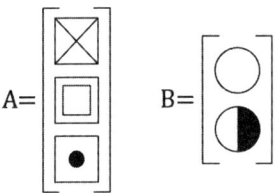

Figure 5-2: Two column matrices, A and B, each composed of shapes

When our lists are made of numbers (as they usually will be), we'll combine pairs of elements by multiplying them. Thus, I'll say we form their *product*. When we're using pictures for elements, as in Figure 5-2, let's say that their product (or the result of multiplying them) is a new picture with the two pictures side by side, with the first picture on the left and the second on the right. Note that *order matters* when we form this pairing.

I'll describe the process of forming the tensor product first, so you'll know what it does. That process is a little complicated, so I'll then show you a shortcut that everyone uses.

We start with an empty list, which I'll name C. The process begins by selecting the first element of A (named a_0) and the first element of B (named b_0). We multiply these two elements together to produce $a_0 b_0$, and we place this result into C. At this point, C is a column matrix that has one element: $a_0 b_0$. Because B has more than one element, we'll hang on to this element from A and now get the next element of B and form the new product $a_0 b_1$. We tack that onto the bottom of C, so it's now a column matrix with two elements: $a_0 b_0$ followed by $a_0 b_1$. We keep doing this until we've paired a_0 with every element from B. Then we take the next element of A and repeat the process, forming $a_1 b_0$, then $a_1 b_1$, and so on. We keep this up until we've paired every element of A with every element of B. For our A and B in Figure 5-2, the result is shown in Figure 5-3.

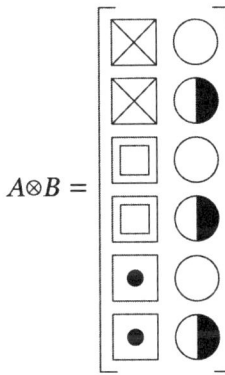

$A \otimes B =$

Figure 5-3: The tensor product $A \otimes B$

We call the result the *tensor product* of A and B. We write this as $A \otimes B$, pronounced "A tensor B."

Let's write this process a little more formally, as a series of steps:

1. Create a new column matrix, C, with no elements.

2. Set the integer $v \leftarrow 0$.

3. Set the integer $w \leftarrow 0$.

4. Create the product $a_v b_w$ and append it to the bottom of C.

5. If $w < \dim(B) - 1$, then set $w \leftarrow w + 1$ and go to step 4.

6. If $v < \dim(A) - 1$, then set $v \leftarrow v + 1$ and go to step 3.

7. Return C, now referred to as $A \otimes B$.

Programmers will recognize this process as a pair of nested loops.

In practice, we rarely think of the tensor product operation in terms of these steps. We can get the same result with less conceptual overhead by following the shortcut I promised you, illustrated by the three graphical steps shown in Figure 5-4. First, we create the overall structure of $A \otimes B$ by making a new column matrix that has every entry of the first list, A, written individually, followed by the symbol B representing the second list. Second, we replace each instance of B with an explicit column matrix of its elements. Finally, we multiply each element of A with each of the elements in B to its right, and we drop the brackets that we drew around the B entries. This technique always gives us the same result as the stepwise process. With some experience, you can often do one or more of these steps in your head.

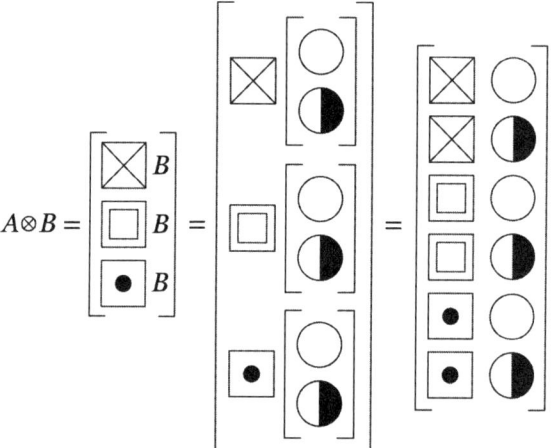

Figure 5-4: Picturing the steps of creating the tensor product $A \otimes B$

Both the formal approach and the shortcut produce the same results in the same order. This is important, because when we use tensor products to represent systems of qubits and qugates, keeping everything in a consistent order is vital to getting the results we want.

The tensor product is not commutative, so in general, $A \otimes B \neq B \otimes A$. To illustrate this, I've drawn $B \otimes A$ in Figure 5-5.

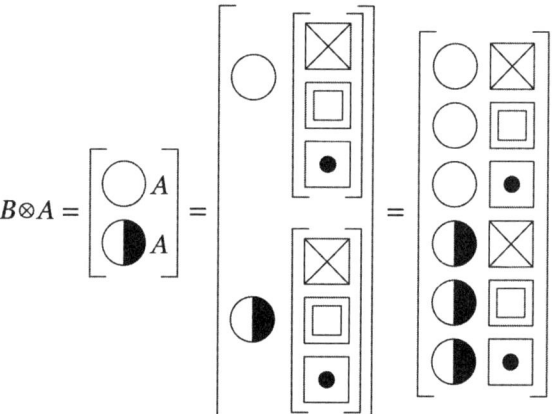

Figure 5-5: Creating the tensor product $B \otimes A$

Figure 5-6 shows $A \otimes B$ from Figure 5-4 alongside $B \otimes A$ from Figure 5-5. As is usually the case, they are not the same! Even if we ignore the ordering of the two pictures in each element, the top to bottom sequences are different.

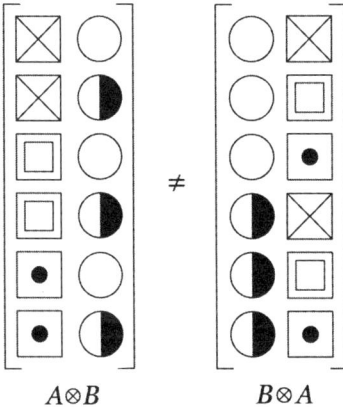

$A \otimes B \qquad B \otimes A$

Figure 5-6: Tensor products are usually not commutative, as shown here for $A \otimes B$ and $B \otimes A$.

How many elements will $A \otimes B$ have? We know that A has dim(A) elements and B has dim(B) elements, and since every element of A is eventually paired up with every element of B, the tensor product $A \otimes B$ will have dim(A) $\times$ dim(B) elements (where we're using $\times$ for multiplication). That is, the number of elements in the new list is the *product* of the number of elements in the input lists. In the figures we just saw, dim(A) = 3 and dim(B) = 2, so dim($A \otimes B$) = 3×2 = 6.

This is all we're going to need from the tensor product, though it's just the tip of a mathematically deep and powerful idea. There's much more to learn about the tensor product if you're intrigued [24] [41] [79] [114] [115].

Product States

> *From this viewpoint [of tensor products], the situation with quantum systems is extremely paradoxical. . . . One might wonder where nature finds the extra storage space when we put these two subsystems together.*
>
> —Umesh Vazirani, *Hilbert Spaces, Tensor Products, Teleportation,*
> 2021 [224]

When we use the tensor product to combine quantum states, we call the result a *tensor product state*, or just a *product state* for short. In this section, we'll look at the structure of these product states.

We'll begin by going back to Postulate 3, which tells us that we can create systems of multiple qubits using the tensor product. Let's try this now.

I'll tensor together the states $|\psi\rangle$ and $|\phi\rangle$. Each is a vector composed of two complex numbers. I've written them out in Equation 5.1.

$$|\psi\rangle = \psi_0 \, |0\rangle + \psi_1 \, |1\rangle = \begin{bmatrix} \psi_0 \\ \psi_1 \end{bmatrix}, \quad |\phi\rangle = \phi_0 \, |0\rangle + \phi_1 \, |1\rangle = \begin{bmatrix} \phi_0 \\ \phi_1 \end{bmatrix} \quad (5.1)$$

Following the steps in Figure 5-4, I'll first write a new state that has each element of $|\psi\rangle$ listed explicitly, followed by the ket $|\phi\rangle$. Then I'll replace each $|\phi\rangle$ by its elements. We then multiply the elements and remove the inner square brackets, which gives us a new four-element vector $|\psi\rangle \otimes |\phi\rangle$, as shown in Equation 5.2. Each element of $|\psi\rangle \otimes |\phi\rangle$ is the product of two complex numbers, one each from $|\psi\rangle$ and $|\phi\rangle$.

$$|\psi\rangle \otimes |\phi\rangle = \begin{bmatrix} \psi_0 \, |\phi\rangle \\ \psi_1 \, |\phi\rangle \end{bmatrix} = \begin{bmatrix} \psi_0 \begin{bmatrix} \phi_0 \\ \phi_1 \end{bmatrix} \\ \psi_1 \begin{bmatrix} \phi_0 \\ \phi_1 \end{bmatrix} \end{bmatrix} = \begin{bmatrix} \psi_0 \times \phi_0 \\ \psi_0 \times \phi_1 \\ \psi_1 \times \phi_0 \\ \psi_1 \times \phi_1 \end{bmatrix} = \begin{bmatrix} \psi_0 \phi_0 \\ \psi_0 \phi_1 \\ \psi_1 \phi_0 \\ \psi_1 \phi_1 \end{bmatrix} \quad (5.2)$$

What is this new four-element column matrix we've just made? It's a single mathematical object that describes both qubits, taken together.

Note that Equation 5.2 represents a single quantum state. Just as the expression $|\psi\rangle = \psi_0 \, |0\rangle + \psi_1 \, |1\rangle$ represents a single quantum bit, the four-element column matrix $|\psi\rangle \otimes |\phi\rangle$ represents a system of two quantum bits. It's a single state built from the two qubits. With larger operators, we can manipulate this single state just as we manipulated single-qubit states and thereby change both qubits in one operation.

As discussed in Chapter 2, in regular algebra, we often drop the multiplication sign between objects and leave it implied. For example, the product $a \times b$ of two numbers a and b is often written ab. In the same way, rather than always writing the tensor product of two states as $|\psi\rangle \otimes |\phi\rangle$, we can write them side by side, like $|\psi\rangle \, |\phi\rangle$, where the tensor product sign is implied. We'll write the tensor product of two states so often that we frequently compress expressions like $|\psi\rangle \, |\phi\rangle$ even further and drop the middle symbols, writing just $|\psi\phi\rangle$.

In short, $|\psi\rangle \otimes |\phi\rangle$, $|\psi\rangle \, |\phi\rangle$, $|\psi\phi\rangle$, and the four-element column vector at the right of Equation 5.2 are all equivalent ways to write the quantum state that describes a system built from the qubits $|\psi\rangle$ and $|\phi\rangle$.

The elements of the final matrix in Equation 5.2 appear in this somewhat strange form because it will dovetail perfectly with the corresponding form we'll get from creating the tensor products of operators. If you square each element of $|\psi\rangle \otimes |\phi\rangle$ and add up the results, you'll be relieved to find that they add up to 1, so this is a valid quantum state.

The *key thing to remember* as you get used to tensor products is that the tensor product $|\psi\rangle \otimes |\phi\rangle$ is *not* the two coefficients of $|\psi\rangle$ in the top two positions and those of $|\phi\rangle$ under them in the lower two positions.

This can be a tempting mental picture, but it leads only to confusion and sorrow. The vector $|\psi\rangle \otimes |\phi\rangle$ contains the *combinations*, or *interactions*, of the qubit coefficients, not the isolated coefficients. The elements of $|\psi\rangle \otimes |\phi\rangle$ are *four entirely new values* created from the four possible ways to multiply an element from the first state with an element from the second, written in a specific order.

Exploring Product States

Let's start exploring product states.

The left side of Figure 5-7 shows two qubits that both start out initialized to $|0\rangle$. Our goal is to write the two-qubit state marked $|\psi_0\rangle$ as a single mathematical object. The steps are shown on the right.

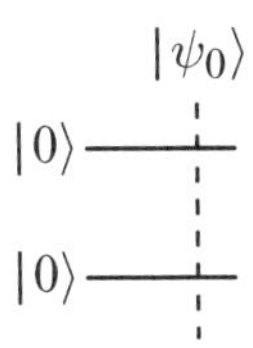

$$|\psi_0\rangle = |0\rangle \otimes |0\rangle = \begin{bmatrix} 1 \, |0\rangle \\ 0 \, |0\rangle \end{bmatrix} = \begin{bmatrix} 1 \begin{bmatrix} 1 \\ 0 \end{bmatrix} \\ 0 \begin{bmatrix} 1 \\ 0 \end{bmatrix} \end{bmatrix} = \begin{bmatrix} 1 \times 1 \\ 1 \times 0 \\ 0 \times 1 \\ 0 \times 0 \end{bmatrix} = \begin{bmatrix} 1 \\ 0 \\ 0 \\ 0 \end{bmatrix}$$

Figure 5-7: Writing a two-qubit system as a single state

We've combined two state vectors of two elements each into a single system state vector of four elements. We'll see that with each additional qubit we include in our system, the dimensionality of the system vector *doubles*. That means a system of three qubits (represented by a vector of $2^3 = 8$ elements) is as big as we're ever going to want to actually write down on the page (and even that's almost too big for comfort). So, to avoid drawing huge column matrices, it seems that we need a new, compact notation for representing product states of multiple qubits.

And . . . drum roll please . . . we don't need a new notation! You may recall that in Chapter 2 I explained that the labels $|0\rangle$ and $|1\rangle$ referred to vectors where every element had a value of 0 except for a single 1 at index 0 and 1, respectively. That probably sounded somewhat strange at the time, as "every element" referred to only one element. But now we can see the purpose behind that convention. The rightmost vector in Figure 5-7 follows the same pattern as $|0\rangle$, because it's 1 at index 0 and 0 everywhere else. So, the rightmost vector of Figure 5-7 is written $|0\rangle$.

Now we have a new problem, because we already used the label $|0\rangle$ for the two-element states we've been using until now. We usually deal with this problem by ignoring it. When we write the product state of two qubits, we know that the result *must* be a vector with four elements, so in this context, $|0\rangle$ is unambiguously a vector of four elements, three of which are 0, with a single 1 at index 0. We usually rely on context to tell us how many elements are in a vector we write as $|0\rangle$.

When the context isn't clear, or we otherwise want to emphasize the dimensionality of a state, we write the dimensionality with a subscript. If there's any chance of confusion, we can write the final state of Figure 5-7 as $|0\rangle_4$. Because adding subscripts to every state makes expressions more cluttered and harder to read (and also prevents us from using that subscript slot for anything else), I'll follow the usual convention in this book and mostly rely on context, except when there's any ambiguity.

Just as $|0\rangle$ refers to a state with a 1 at index 0 and $|1\rangle$ refers to a state with a 1 at index 1, putting *any* integer inside a ket refers to a state that is 0 everywhere except for a single 1 at that integer's location. If the ket has d elements, then the integer needs to be in the range $[d]$. So, $|3\rangle$ refers to a state of at least four elements that are all 0 except for a 1 at index 3. Four such four-element states are shown in Equation 5.3.

$$|0\rangle = \begin{bmatrix} 1 \\ 0 \\ 0 \\ 0 \end{bmatrix}, \quad |1\rangle = \begin{bmatrix} 0 \\ 1 \\ 0 \\ 0 \end{bmatrix}, \quad |2\rangle = \begin{bmatrix} 0 \\ 0 \\ 1 \\ 0 \end{bmatrix}, \quad |3\rangle = \begin{bmatrix} 0 \\ 0 \\ 0 \\ 1 \end{bmatrix} \quad (5.3)$$

For the rest of this book, $|k\rangle$ refers to a state of at least $k + 1$ elements that is 0 everywhere except for a 1 at index k. Almost always, the dimensionality of $|k\rangle$ will be a power of 2 that needs to be inferred from context.

These states, written $|k\rangle$, are special. *They are the computational basis states!* We will use them for every system of qubits in this book, including those we've seen so far with only a single qubit.

Any four-element state can be written as a sum of the four basis states in Equation 5.3, each scaled by a number. Equation 5.4 shows the mechanics for an arbitrary four-element state.

$$\begin{bmatrix} \alpha \\ \beta \\ \gamma \\ \delta \end{bmatrix} = \alpha \begin{bmatrix} 1 \\ 0 \\ 0 \\ 0 \end{bmatrix} + \beta \begin{bmatrix} 0 \\ 1 \\ 0 \\ 0 \end{bmatrix} + \gamma \begin{bmatrix} 0 \\ 0 \\ 1 \\ 0 \end{bmatrix} + \delta \begin{bmatrix} 0 \\ 0 \\ 0 \\ 1 \end{bmatrix} = \alpha |0\rangle + \beta |1\rangle + \gamma |2\rangle + \delta |3\rangle \quad (5.4)$$

We know that we can write the system $|0\rangle \otimes |0\rangle$ more compactly as $|00\rangle$. This is not the same state as $|0\rangle$! When we have multiple numbers inside a ket, we're not multiplying them, so $|00\rangle \neq |0\rangle$. The former, $|00\rangle$, is a four-element product state of two kets, $|0\rangle \otimes |0\rangle$, while the latter, $|0\rangle$, is a two-element state. When you see multiple digits inside a ket, remember that they refer to the tensor product of their individual states.

This notational convention is summarized in Equation 5.5 for two arbitrary states. They don't need to be of the same dimensionality (that is, $|\psi\rangle$ and $|\phi\rangle$ may have different numbers of elements, as we saw in Figure 5-4).

$$|\psi\rangle \otimes |\phi\rangle = |\psi\rangle |\phi\rangle = |\psi\phi\rangle \qquad (5.5)$$

Let's look at an algorithm, and the product state that describes it. In Figure 5-8, I've inserted an X qugate on the upper qubit, so its state at the position marked $|\psi_0\rangle$ is $|1\rangle$.

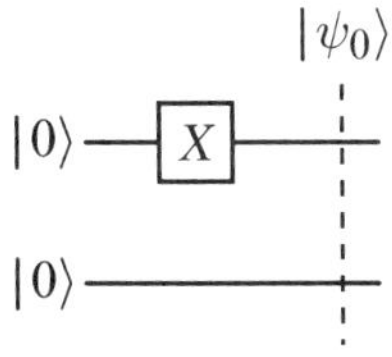

$$|\psi_0\rangle = |1\rangle \otimes |0\rangle = \begin{bmatrix} 0\,|0\rangle \\ 1\,|0\rangle \end{bmatrix} = \begin{bmatrix} 0\begin{bmatrix}1\\0\end{bmatrix} \\ 1\begin{bmatrix}1\\0\end{bmatrix} \end{bmatrix} = \begin{bmatrix} 0\times1 \\ 0\times0 \\ 1\times1 \\ 1\times0 \end{bmatrix} = \begin{bmatrix} 0\\0\\1\\0 \end{bmatrix} = |2\rangle$$

Figure 5-8: Finding the elements of the two-qubit system $|10\rangle$

Using our new convention, we can write the final state $|\psi_0\rangle$ in Figure 5-8 as $|2\rangle$, or $|2\rangle_4$ if we want to clarify that the vector has four elements.

Figure 5-8 illustrates another convention in this book: When I tensor together a system of qubits, I write them from *left to right*, corresponding to reading their diagram version from *top to bottom*. Many, but not all, authors use this convention. If you're reading someone else's work and the equations don't seem to match the diagrams, it's worth checking to see if they're tensoring the qubits in a diagram from top to bottom (as I do in this book) or from bottom to top.

Let's take on a more challenging diagram. Building on Figure 5-8, I'll insert an H qugate on the lower qubit, so at $|\psi_0\rangle$ that qubit has the state $H|0\rangle = |+\rangle$, as shown in Figure 5-9.

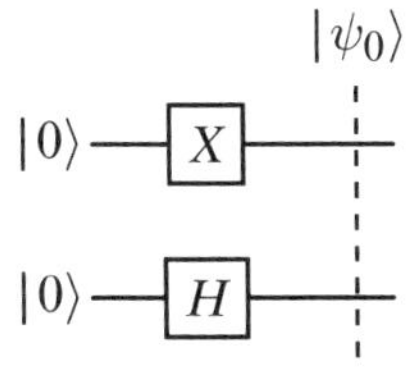

$$|\psi_0\rangle = |1\rangle \otimes |+\rangle = \begin{bmatrix} 0\,|+\rangle \\ 1\,|+\rangle \end{bmatrix} = \begin{bmatrix} 0\begin{bmatrix}\vee\\\vee\end{bmatrix} \\ 1\begin{bmatrix}\vee\\\vee\end{bmatrix} \end{bmatrix} = \begin{bmatrix} 0\times\vee \\ 0\times\vee \\ 1\times\vee \\ 1\times\vee \end{bmatrix} = \begin{bmatrix} 0\\0\\\vee\\\vee \end{bmatrix}$$

Figure 5-9: Writing the system $|1\rangle \otimes |+\rangle$

That's an interesting result! Unlike the states we saw previously, we can't write this in the form $|k\rangle$, because the elements aren't all 0 with a single 1. But, like every other quantum state, we can write it as a sum of scaled basis states. Equation 5.6 shows four equivalent ways to write the final state of Figure 5-9. All of these expressions describe the quantum state of the system $|\psi_0\rangle$ in Figure 5-9.

$$\begin{bmatrix} 0 \\ 0 \\ \vee \\ \vee \end{bmatrix} = \begin{bmatrix} 0 \\ 0 \\ \vee \\ 0 \end{bmatrix} + \begin{bmatrix} 0 \\ 0 \\ 0 \\ \vee \end{bmatrix} = \vee \begin{bmatrix} 0 \\ 0 \\ 1 \\ 0 \end{bmatrix} + \vee \begin{bmatrix} 0 \\ 0 \\ 0 \\ 1 \end{bmatrix} = \vee |2\rangle + \vee |3\rangle = \vee(|2\rangle + |3\rangle) \quad (5.6)$$

As Figure 5-9 shows, we can also write $|\psi_0\rangle$ as the product state $|1\rangle \otimes |+\rangle$, or $|1+\rangle$, explicitly identifying the qubits involved. The version $\vee(|2\rangle + |3\rangle)$ describes the system in another way, by multiplying out the tensor product $|1\rangle \otimes |+\rangle$. All of these descriptions of $|\psi_0\rangle$ are fine, and we're free to choose the one that's most helpful or informative at any time.

Let's look at one last system, where I'll pass each of two qubits (both initially $|0\rangle$) through their own H qugates. As Figure 5-10 shows, this means both qubits will be in the state $|+\rangle$ when they get to the position marked $|\psi_0\rangle$. The final system is a four-element column vector where each entry is $1/2$.

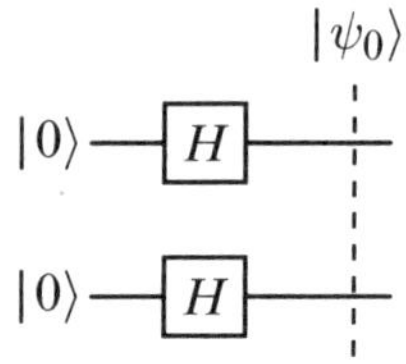

$$|\psi_0\rangle = |+\rangle \otimes |+\rangle = \begin{bmatrix} \vee\,|+\rangle \\ \vee\,|+\rangle \end{bmatrix} = \begin{bmatrix} \vee \begin{bmatrix} \vee \\ \vee \end{bmatrix} \\ \vee \begin{bmatrix} \vee \\ \vee \end{bmatrix} \end{bmatrix} = \begin{bmatrix} \vee^2 \\ \vee^2 \\ \vee^2 \\ \vee^2 \end{bmatrix} = \frac{1}{2} \begin{bmatrix} 1 \\ 1 \\ 1 \\ 1 \end{bmatrix}$$

Figure 5-10: Writing the system $|+\rangle \otimes |+\rangle$; remember that $\vee^2 = (1/\sqrt{2})^2 = 1/2$

This is an important result! Let's dig into why.

First, let's make sure it's a valid quantum state. As we know, every physically realizable state (that is, not merely a vector of complex numbers but a vector that can describe an actual quantum object) must have a magnitude of 1. This tensor product describes a state made of two qubits, but it's still a single quantum state and thus must have a magnitude of 1. Each term in the final state of Figure 5-10 has an amplitude of $1/2$. Squaring that gives us a probability of $(1/2)^2 = 1/4$. There are four terms in the state, and $4 \times (1/4) = 1$, so yes, this is a valid quantum state.

Recall that I said that Figure 4-5 was a common way to start a quantum algorithm. Now we can better understand why that's the case. Figure 5-10 is a two-qubit version of the left side of Figure 4-5. Equation 5.7 expands the final state of Figure 5-10 in terms of the four basis states for a four-element ket.

$$|\psi_0\rangle = \frac{1}{2}\begin{bmatrix}1\\1\\1\\1\end{bmatrix} = \frac{1}{2}\left(\begin{bmatrix}1\\0\\0\\0\end{bmatrix} + \begin{bmatrix}0\\1\\0\\0\end{bmatrix} + \begin{bmatrix}0\\0\\1\\0\end{bmatrix} + \begin{bmatrix}0\\0\\0\\1\end{bmatrix}\right) \tag{5.7}$$

$$= \frac{1}{2}\left(|0\rangle + |1\rangle + |2\rangle + |3\rangle\right)$$

We can write this result more compactly as shown in Equation 5.8.

$$|\psi_0\rangle = \sum_{k\in[4]}\frac{1}{2}|k\rangle = \frac{1}{2}\sum_{k\in[4]}|k\rangle = \frac{1}{2}\sum_{k\in\mathbb{B}^2}|k\rangle \tag{5.8}$$

In the rightmost version of Equation 5.8, I used the set $\mathbb{B}^2$ as the range of values for k. Recall that this set contains all of the bitstrings of two elements. These four bitstrings, 00, 01, 10, and 11, correspond to the digits 0, 1, 2, and 3 in the sequence [4].

Each basis vector $|k\rangle$ has a magnitude of 1 (that is, each vector is normalized). Recall that the braket (or inner product) of a vector with itself, $\langle\psi|\psi\rangle$, gives us the squared magnitude of $|\psi\rangle$, so for every basis state $|k\rangle$, we have $\langle k|k\rangle = 1$.

Each basis vector is also orthogonal to all the others. Because the inner product is a generalization of the dot product, the inner product of two different vectors, $\langle\phi|\psi\rangle$, depends on the angle between them. When both vectors have nonzero magnitudes but their inner product is 0, it means that the cosine of the angle between them is 0, or they are at right angles to one another. So, two basis vectors $|k\rangle$ and $|m\rangle$ are orthogonal if $\langle k|m\rangle = 0$.

We can summarize both of these conditions in the expression given in Equation 5.9, using the Kronecker delta we defined in Equation 2.72. This is a compact way of stating that our set of basis vectors is orthonormal.

$$\langle k|m\rangle = \delta_{k,m} \tag{5.9}$$

We used this relationship on line 5 of Equation 2.74, and we'll use it again in this book to simplify equations into forms that will be easier for us to manipulate and understand.

The tensor product preserves the norm of its inputs. That is, given two states that each have a magnitude of 1 (as they must), their tensor product also has a magnitude of 1. Proving this directly for states of many qubits involves a lot of bookkeeping. If you're inclined to check, try first expanding everything out and then grouping the results.

More Qubits

Two qubits are fine, but how about three?

Let's roll up our sleeves and work out the state vector for a system of three qubits. As I said before, a state vector for such a system has $2^3 = 8$ elements, and that's about as complicated as I like to go when writing out all of these components. Equation 5.10 shows the steps.

$$|\psi\rangle \otimes |\phi\rangle \otimes |\tau\rangle = \left(|\psi\rangle \otimes |\phi\rangle \right) \otimes |\tau\rangle$$

$$= \begin{bmatrix} \psi_0\phi_0 \\ \psi_0\phi_1 \\ \psi_1\phi_0 \\ \psi_1\phi_1 \end{bmatrix} \otimes |\tau\rangle = \begin{bmatrix} \psi_0\phi_0 \, |\tau\rangle \\ \psi_0\phi_1 \, |\tau\rangle \\ \psi_1\phi_0 \, |\tau\rangle \\ \psi_1\phi_1 \, |\tau\rangle \end{bmatrix}$$

$$= \begin{bmatrix} \psi_0\phi_0 \begin{bmatrix} \tau_0 \\ \tau_1 \end{bmatrix} \\ \psi_0\phi_1 \begin{bmatrix} \tau_0 \\ \tau_1 \end{bmatrix} \\ \psi_1\phi_0 \begin{bmatrix} \tau_0 \\ \tau_1 \end{bmatrix} \\ \psi_1\phi_1 \begin{bmatrix} \tau_0 \\ \tau_1 \end{bmatrix} \end{bmatrix} = \begin{bmatrix} \psi_0\phi_0\tau_0 \\ \psi_0\phi_0\tau_1 \\ \psi_0\phi_1\tau_0 \\ \psi_0\phi_1\tau_1 \\ \psi_1\phi_0\tau_0 \\ \psi_1\phi_0\tau_1 \\ \psi_1\phi_1\tau_0 \\ \psi_1\phi_1\tau_1 \end{bmatrix} \qquad (5.10)$$

Because the tensor product is associative, we're free to group the operations in any order, as long as we never change the order of the states. In the first line, I chose to begin by tensoring $|\psi\rangle$ and $|\phi\rangle$, but you could also first tensor $|\phi\rangle$ and $|\tau\rangle$, then tensor $|\psi\rangle$ with that (a list of many more properties of the tensor product appears in the Appendix).

There are eight coefficients in the state at the end of Equation 5.10. Let's call this state $|\lambda\rangle$. We can name the elements λ from top to bottom, so $\lambda_0 = \psi_0\phi_0\tau_0$, $\lambda_1 = \psi_0\phi_0\tau_1$, and so on. Then we can write the conclusion of Equation 5.10 more succinctly, as shown in Equation 5.11, assuming that we know the value of each λ_k.

$$|\psi\rangle \otimes |\phi\rangle \otimes |\tau\rangle = \sum_{k\in[8]} \lambda_k \, |k\rangle, \quad \text{where} \quad \sum_{k\in[8]} |\lambda_k|^2 = 1 \qquad (5.11)$$

We can write Equation 5.11 as an explicit sum of the basis states in tableau form, as shown in Equation 5.12.

$$\psi_0\phi_0\tau_0 \begin{bmatrix} 1 \\ 0 \\ 0 \\ 0 \\ 0 \\ 0 \\ 0 \\ 0 \end{bmatrix} + \psi_0\phi_0\tau_1 \begin{bmatrix} 0 \\ 1 \\ 0 \\ 0 \\ 0 \\ 0 \\ 0 \\ 0 \end{bmatrix} + \psi_0\phi_1\tau_0 \begin{bmatrix} 0 \\ 0 \\ 1 \\ 0 \\ 0 \\ 0 \\ 0 \\ 0 \end{bmatrix} + \psi_0\phi_1\tau_1 \begin{bmatrix} 0 \\ 0 \\ 0 \\ 1 \\ 0 \\ 0 \\ 0 \\ 0 \end{bmatrix} +$$

$$\psi_1\phi_0\tau_0 \begin{bmatrix} 0 \\ 0 \\ 0 \\ 0 \\ 1 \\ 0 \\ 0 \\ 0 \end{bmatrix} + \psi_1\phi_0\tau_1 \begin{bmatrix} 0 \\ 0 \\ 0 \\ 0 \\ 0 \\ 1 \\ 0 \\ 0 \end{bmatrix} + \psi_1\phi_1\tau_0 \begin{bmatrix} 0 \\ 0 \\ 0 \\ 0 \\ 0 \\ 0 \\ 1 \\ 0 \end{bmatrix} + \psi_1\phi_1\tau_1 \begin{bmatrix} 0 \\ 0 \\ 0 \\ 0 \\ 0 \\ 0 \\ 0 \\ 1 \end{bmatrix} \qquad (5.12)$$

$$= \psi_0\phi_0\tau_0 \left|0\right\rangle + \psi_0\phi_0\tau_1 \left|1\right\rangle + \psi_0\phi_1\tau_0 \left|2\right\rangle + \psi_0\phi_1\tau_1 \left|3\right\rangle +$$
$$\psi_1\phi_0\tau_0 \left|4\right\rangle + \psi_1\phi_0\tau_1 \left|5\right\rangle + \psi_1\phi_1\tau_0 \left|6\right\rangle + \psi_1\phi_1\tau_1 \left|7\right\rangle$$

At this point, you might be wondering why we're combining qubits with the tensor product rather than in some other way. The answer is both simple and profound: This matches experimental observations. In Chapter 6, we'll see that when we model groups of qubits and operators using the tensor product, the outputs that come from that description match the measured outputs produced by actual hardware. That's why Postulate 3 says that collections of qubits (and, later in this chapter, qugates) are created with tensor products.

Like Newton's laws, or Einstein's theory of general relativity, we use the mathematical tools that match the reality we observe. Exactly why the universe works this way, and not some other way, nobody can say with certainty. What we can say is that combining quantum objects with the tensor product describes and correctly predicts the world we live in, to the limits of our ability to observe and measure it.

Quantum Algorithm Diagrams

We often draw quantum algorithms in picture form, at least while discussing them at a high level. This is not at all how we write conventional programs.

The majority of today's conventional programs are created in text form, which is then compiled, or transformed, into lower-level operations that hardware can execute. We also sometimes create classical programs with diagrams and graphics, though there's usually some text in there as well. These graphical approaches, used in languages like Scratch [190] and LEGO MINDSTORMS [121], are often appealing for writing casual programs and for introducing people to the craft of programming. They're also used in specialist software when a primary task is controlling the flow of information, such as in the Houdini computer graphics system [197] or the Max system for music and sound design [42].

In the same way, quantum algorithms can be written in multiple ways, including text, equations, and graphics. All three are equivalent in power, and we can turn any description into any other. Currently, equations and graphics are the most common forms, with graphics dominating when the algorithm is not too complicated or when discussing a particular part of the algorithm that has a nice visual interpretation. In this book, I'll usually use diagrams to present the structure of an algorithm and math to discuss how it processes qubits to achieve its goal.

There is a standard visual language for drawing quantum algorithms, which we've already seen in small examples. The big idea is that we draw qubits starting on the left, and the algorithm works by applying a series of qugates to those qubits as we read from left to right. You can think of this horizontal flow as a timeline, where the qubits begin with initial values at the left, are transformed over time as they move to the right, and then are (usually) measured at the end, at the far right. This structure is somewhat like the schematics we use to represent electronic circuits, so these quantum diagrams are also sometimes called *circuit diagrams*, or just *circuits*. I'll use these terms interchangeably in this book.

In quantum computing, our operations on qubits are *discrete* rather than *continuous*. A continuous process is one that has a smoothly changing description, like the height of a balloon as it flies away into the clouds. In contrast, a discrete process is described by a series of individual jumps, like an old-fashioned digital clock whose display only switches after each second, rather than changing smoothly. We often think of the real numbers as a model for continuous changes and the integers as a model for discrete changes.

In quantum computing, qubits are like those digital clocks. They only change when they are being processed by a qugate. Between qugates the qubit is unchanged, retaining its state until it reaches another qugate that might change it (this is an idealization, but hardware designers are making it ever closer to reality).

Let's see this in action. Figure 5-11 shows a generic quantum algorithm that I'll use to discuss quantum circuits, or diagrams. The qugates are arbitrarily named *A*, *B*, and *C* so that we can discuss them and don't refer to any specific operators.

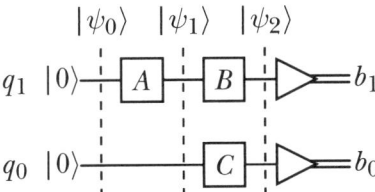

Figure 5-11: A simple quantum circuit
showing most of the usual elements

Every qubit is represented by a single unbroken horizontal line. I'll often name the qubits so I can refer to them. In Figure 5-11, the qubits are named q_1 and q_0. My convention is to name the qubits starting with 0 at the bottom and adding 1 to each line as we go upward. Note that these names aren't kets, because they're labels, not states. Assigning the labels from bottom to top means that when we combine final measurements in our usual order from top to bottom, we'll get back a binary number in the form we're used to from classical computing. In this example, we'd get the binary number $b_1 b_0$, with the most significant bit, b_1, at the left, where it belongs. This convention will prove to be convenient in many algorithms that produce a binary number as output.

To the right of each qubit's name in this figure is its initial value. This is almost always $|0\rangle$, as that's often a convenient state for the hardware to create. When an initial state is not explicitly provided, it's usually assumed to be $|0\rangle$ (that is, the two-element state $|0\rangle_2$).

People often treat a diagram, or circuit, as being the same as the algorithm itself, just as we often treat a piece of classical program text as being the same as the program that's run on the hardware. Different implementations of a classical or quantum algorithm can have different performance characteristics, so when that matters, it's important to distinguish the abstract description from any specific realization of it.

We follow each qubit rightward from its label and starting value at the left, through the qugates that modify it. We call the qubit's changing values its *evolution*, and we say that the algorithm sequentially *evolves* the qubit through a sequence of states.

The most common thing to see on a line is a qugate. A single-input qugate appears as a box with the qugate's name inside. People try to keep these names short so the boxes don't grow too big.

On the top line of Figure 5-11, qubit q_1 first goes into an A qugate. Recall that here I'm using letters such as A and B as generic qugate names. They could stand for I, X, H, or any other specific qugate. The output of the A qugate goes into B. The triangular meter that comes next indicates *measurement*. Measurement is a big topic, and it's the focus of Chapter 6. For now, we'll stick with the description given in Chapter 3 that measuring a qubit will produce one of two possible observations, named 0 and 1, like classical bits. We draw the output of the meter with two parallel lines to indicate that they carry a classical bit, rather than a single line, which carries a quantum bit. I've labeled that output bit b_1.

Similarly, qubit q_0 starts out in state $|0\rangle$, is modified by the single-qubit qugate C, and finally is measured, producing output b_0.

I drew C directly under B. Generally speaking, we like the qugates to line up vertically, because that usually makes the circuit easier to analyze and discuss. In this case, I could have drawn C under A, if that made more conceptual sense for this circuit.

As mentioned in Chapter 4, it's often useful to write down the values of the qubits at various moments along their evolution. We mark each of these particular moments of interest with a vertical dashed line, named with a ket above or below. In this example, $|\psi_0\rangle$, $|\psi_1\rangle$, and $|\psi_2\rangle$ refer to different four-element product states $|q_1\rangle \otimes |q_0\rangle$, using the values of the qubits named q_1 and q_0 at the moment marked by the associated dashed vertical line. You'll rarely see subscripts on such states to identify their dimensionality, since their dimensionality can be inferred from the diagram.

These dashed vertical lines reinforce the idea that computation proceeds in discrete steps. In short, we start with initialized qubits. Then we apply qugates to them. Then we apply more qugates to them. We keep doing that until we're done and we measure the qubits. That's a quantum algorithm.

Often we pair quantum algorithms with classical programs, so each type of computer can do the sorts of tasks it's best at. For example, a classical computer might itself construct the quantum algorithm based on the nature of the input data, or it might preprocess the data in some way, or it might postprocess the measurements from the quantum circuit to give us more useful results. We call these classical–quantum collaborations *hybrid algorithms*.

Quantum circuits are programs. Some online quantum computers allow you to create your program using their web-based drawing tools that enable you to interactively place qugates, meters, and other symbols. Then you submit this drawing for execution. Classical programmers often refer to the text of their programs as *code*. That language works just as well for quantum programs, whose text representation can be extracted from a diagram.

Quantum circuit diagrams follow a few rules.

Generally speaking, we don't have any explicit storage of variables as we think of them in classical computing. The variables in our computation are the values of the qubits. This means that there are no explicit higher-level data structures.

A quantum diagram is called a *directed acyclic graph (DAG)*. It's *directed* because the flow is generally left to right, *acyclic* because there are no cycles (or loops), and a *graph* because it's made up of lines and nodes.

Today's quantum computers and programming environments lack many of the conveniences of modern classical programming, such as subroutines, functions, data structures, classes, and loops. These might come someday, as the field matures. Today's systems don't even provide mathematical operations such as addition and multiplication on arbitrary integers or floating-point numbers, though libraries of circuits for these operations are starting to emerge [64] [86].

A keystone of conventional programming is the ability to *test* a variable and then perhaps *branch* to another piece of code. We call this an *if-then* construction (or *if-then-else*). In quantum algorithms, we have only one rudimentary kind of test, which essentially either enables or disables one or more qugates that operate on other qubits. This test itself is a qugate, and we'll meet it in "The *CX* Qugate" on page 141.

Finally, lines carrying qubits cannot split.

Wait, no, that can't be right! Splitting a line in two would correspond to copying the qubit. Copying bits is a foundational tool of conventional computers. Copying variables lets us say things like a = b + 1, which requires making a copy of b that we then add 1 to and store in a. Without such assignment statements, most of the programs written today would become useless. Surely we can perform the equivalent of num_apples = num_oranges in a quantum computer?

Nope! Nature forbids copying qubits, and we'll see exactly why in "The No-Cloning Theorem" on page 138. That is going to have a profound effect on how we think when we write quantum programs.

Systems of Qugates

Just as we can combine multiple qubits into a single mathematical object, we can also combine multiple qugates. But qugates are a little more interesting because we combine them in two different ways, depending on whether they're acting *serially* (one after another on the same qubits) or *in parallel* (acting simultaneously but independently on different qubits).

Let's take the serial case first.

Horizontal Systems of Qugates

Consider the fragment of a quantum algorithm in Figure 5-12(a). Think of this as a piece extracted from some larger algorithm, so there aren't any initializations or measurements.

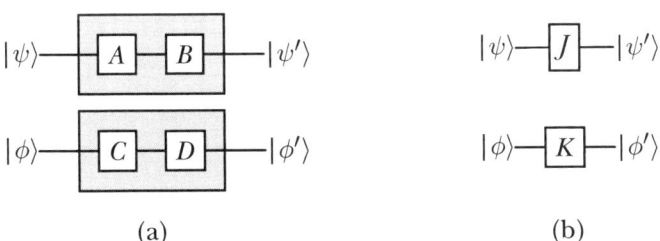

(a) (b)

Figure 5-12: (a) Treating A and B as a single system, and the same for C and D. (b) Using one qugate for each system, where J = BA and K = DC.

In this diagram, we have two qubits, each following its own independent path through the algorithm. The upper qubit starts out in state $|\psi\rangle$, gets processed by generic qugates A and then B, and finally emerges in some new state, $|\psi'\rangle$. Similarly, the lower qubit enters as $|\phi\rangle$, is transformed by C and then D, and emerges in the new state $|\phi'\rangle$.

Qugates A and B can be considered a single horizontally aligned system, so in Figure 5-12(a) I joined them inside a shaded box. We can also think of C and D as a system, so they're in a box as well.

Let's focus on the operations A and B that affect the top qubit. A and B are aligned to one another in the *horizontal* direction. Both A and B are qugates, and we know that every qugate has an associated matrix.

The upper qubit starts as $|\psi\rangle$, becomes $A\,|\psi\rangle$, then exits as $B(A\,|\psi\rangle)$, or more simply $BA\,|\psi\rangle$ (remember that although the qugates are applied in the order A then B, we write them in the order BA because we apply algebraic operators from right to left). The key insight is to recognize that $|\psi\rangle$, A, and B are all matrices and thus can be multiplied together ($|\psi\rangle$ is a column matrix, while the others are square). Because matrix multiplication is associative, for any three matrices P, Q, and R, $(PQ)R = P(QR)$, so we can regroup the operations on $|\psi\rangle$ in the form shown in Equation 5.13.

$$B(A\,|\psi\rangle) = (BA)\,|\psi\rangle \tag{5.13}$$

We can group the qugates C and D that operate on $|\phi\rangle$ in the same way, giving us $(DC)\,|\phi\rangle$.

In the expression $(BA)\,|\psi\rangle$, it looks like the matrices B and A are multiplied together in that order, and then that result is applied to $|\psi\rangle$. And that's exactly right!

We can combine A and B, and also C and D, with ordinary matrix multiplication. We can replace each of these pairs of qugates with a single new qugate that holds the product of multiplying its corresponding matrices. Let's say $J = BA$ and $K = DC$, so we can draw the simpler diagram shown in Figure 5-12(b).

The general rule is that to combine qugates that come one after another, we just multiply their matrices. Drawing all the qugates in a diagram is useful to get a visual sense of the structure, while writing out the operations algebraically gives us a more abstract representation of the same process.

Vertical Systems of Qugates

Let's group our quantum objects in the *vertical* direction this time, as shown in Figure 5-13(a).

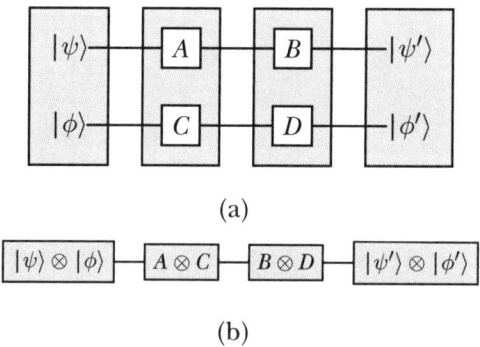

(a)

(b)

Figure 5-13: (a) Interpreting each vertical column of qubits and qugates as a single system. (b) Expressions for each system.

We know from "Product States" on page 115 that we can write the systems of qubits at the left and right ends of this diagram using the tensor product, so the shaded boxes for the input and output states can be written as $|\psi\rangle \otimes |\phi\rangle$ and $|\psi'\rangle \otimes |\phi'\rangle$, respectively.

How about the shaded boxes in the middle? *We also combine vertical qugates with the tensor product.*

To tensor two matrices, we use the very same procedure that we used to form the tensor product of states. The steps are shown in Equation 5.14. We begin by creating a new matrix that has every element of the first matrix followed by the name of the second. Then we replace every one of those names with the second matrix itself. Finally, the elements are complex numbers, so we multiply each pair together and get rid of the intermediate square brackets, which gives us our result, the tensor product of the two matrices. When our two-by-two starting matrices are tensored together, we get back a new four-by-four matrix.

$$
\begin{aligned}
A \otimes C &= \begin{bmatrix} a_{00} & a_{01} \\ a_{10} & a_{11} \end{bmatrix} \otimes C & &\text{Expand } A \text{ into its elements} \\[2em]
&= \begin{bmatrix} a_{00}C & a_{01}C \\ a_{10}C & a_{11}C \end{bmatrix} & &\text{Multiply each element with } C \\[2em]
&= \begin{bmatrix} a_{00}\begin{bmatrix} c_{00} & c_{01} \\ c_{10} & c_{11} \end{bmatrix} & a_{01}\begin{bmatrix} c_{00} & c_{01} \\ c_{10} & c_{11} \end{bmatrix} \\[2em] a_{10}\begin{bmatrix} c_{00} & c_{01} \\ c_{10} & c_{11} \end{bmatrix} & a_{11}\begin{bmatrix} c_{00} & c_{01} \\ c_{10} & c_{11} \end{bmatrix} \end{bmatrix} & &\text{Replace each } C \text{ with} \quad (5.14) \\ & & &\text{its matrix} \\[2em]
&= \begin{bmatrix} a_{00}c_{00} & a_{00}c_{01} & a_{01}c_{00} & a_{01}c_{01} \\ a_{00}c_{10} & a_{00}c_{11} & a_{01}c_{10} & a_{01}c_{11} \\ a_{10}c_{00} & a_{10}c_{01} & a_{11}c_{00} & a_{11}c_{01} \\ a_{10}c_{10} & a_{10}c_{11} & a_{11}c_{10} & a_{11}c_{11} \end{bmatrix} & &\text{Multiply the elements}
\end{aligned}
$$

Whew! The good news is that much of the time we'll be working with matrices that have lots of 0 and 1 elements, so that after some practice with this recipe, we'll often be able to write down the final four-by-four matrix by eye.

Note that I'm tensoring together the operators in top-down order in the diagram. This matches our convention of tensoring together the qubits in left-to-right order following the top-down order. This consistency is important! You can do both of them bottom-up if you prefer (a few authors do), but you have to always tensor both qubits and operators in the same direction for the algebra to properly match the diagram.

Now that we have the matrix $A \otimes C$, we have a horizontal system again. We have a single starting matrix, $|\psi\rangle \otimes |\phi\rangle$, going into a single qugate matrix, $A \otimes C$, to produce a new state matrix. Let's write that down symbolically in Equation 5.15.

$$(A \otimes C)(|\psi\rangle \otimes |\phi\rangle) \tag{5.15}$$

Does this make sense as a matrix multiplication? The matrix $(A \otimes C)$ is four by four, and the matrix $|\psi\rangle \otimes |\phi\rangle$ has four rows and one column, so Equation 5.15 obeys the rules of matrix multiplication. We get back a new matrix of one column and four rows.

Returning to Figure 5-13, the next step is to run this intermediate result into the vertical system described by B and D. Following our rule that vertical qugates combine with the tensor product, this box is the four-by-four matrix $B \otimes D$. Given Equation 5.15 as input, its output is a four-element column matrix, which we can write as the product state $|\psi'\rangle \otimes |\phi'\rangle$.

As with tensoring states, I haven't presented any reason for why tensoring vertical qugates is the right way to combine them. It's pretty great that we didn't have to come up with something new. But why should we use the tensor product to make systems from vertically aligned qugates? Again, the reason is ultimately because it matches what we observe. Our math is a tool that we use to help us understand how nature works and predict how nature will behave in new situations (like when we execute new quantum algorithms). The tensor product is the process that works. Because nobody can say exactly why, we just accept this as how it is and raise that process up to the level of the fundamental postulate at the start of this chapter. It's been over 100 years, and nobody's yet seen a single instance where using the tensor product to combine states and operators didn't match nature. We can never be sure it won't fail a half hour from now, but until someone finds a repeatable situation where the tensor product doesn't correctly describe and predict observations, we'll keep using it.

Another way to look at this is that we started with observations of quantum states and created a mathematical representation for them. That led us to define operations on those states, which in turn led us to use the tensor product to combine those operations.

In that sense, the road from states and qubits to tensor products was determined when we established that states were two-element complex column vectors. We might someday make observations that compel us to find new representations and ways to combine them. Until then, column vector kets, unitary matrix qugates, and matrix and tensor products for combining them into systems give us a tightly interlocked and mutually compatible system of mathematical manipulations that matches reality.

Horizontal and Vertical Rules

We've just used both horizontal and vertical combinations to give us two different ways to make combined systems of qugates.

Let's summarize those two rules:

Horizontal When qugates A and B are horizontal neighbors in the circuit diagram, sequentially operating on the same qubit(s), we can combine them left to right with ordinary matrix multiplication to form the system BA (note the order), which we can draw as a single qugate.

Vertical When qugates A and B are vertical neighbors in the circuit diagram, we can combine them with the tensor product to form the system $A \otimes B$ (again, note the order), which we can draw as a single qugate that takes in a multi-qubit system and produces a new multi-qubit system as output.

I've done something sneaky here without mentioning it. Until now, we've only seen qugates like I, X, and H that take a single qubit as input and produce a single qubit as output. Naturally, we call these *one-qubit qugates* or *single-qubit qugates*.

Using the tensor product, we've created a new class of qugates that take in systems of multiple qubits as input and return systems of multiple qubits as output. We call these *multi-qubit qugates*.

Later in this chapter, we'll meet a few new qugates of both varieties. These can all be combined horizontally as long as their matrices are the same size, because that's a requirement of matrix multiplication of square matrices. Because the tensor product doesn't have that restriction, we can combine any qugate vertically with any other, regardless of the sizes of their matrices.

A Circuit Analysis

Let's get some experience with our horizontal and vertical techniques by analyzing the circuit fragment in Figure 5-14. Note that this fragment is just for illustration and doesn't have any deliberate computational purpose.

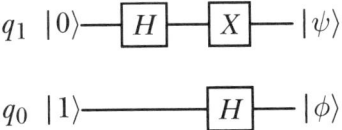

Figure 5-14: A small fragment of a circuit

In this fragment, the qubits labeled q_1 and q_0 are arriving in the states $|0\rangle$ and $|1\rangle$, respectively. I arbitrarily placed the lower H under the X, but I could have placed it anywhere, though to keep things neat the only other choice would be under the upper H. We usually pick the position that makes the most sense to us in understanding the diagram, because both versions produce the same output.

We'll find the outputs of this fragment in three ways: by algebra, by matrix elements, and by direction. That is, we'll first work with a directly algebraic representation of the drawing and simplify that to find the output.

Then we'll replace the operators in the algebraic approach with matrices and multiply them explicitly with the qugates to find the circuit's output. Finally, we'll combine the matrices of the last step with our horizontal and directional techniques to simplify them and then apply those results to the input to produce the circuit's output.

All three of these approaches will return the same result, but they use different tools to get there. Being familiar with these approaches will be helpful when we work with more complicated circuits.

Analysis by Algebra

We can find both $|\psi\rangle$ and $|\phi\rangle$ algebraically by applying the qugates to the input qubits, as shown in Equation 5.16.

$$
\begin{aligned}
|\psi\rangle &= XH|0\rangle = X(H|0\rangle) = X|+\rangle = |+\rangle \\
|\phi\rangle &= H|1\rangle = |-\rangle
\end{aligned}
\tag{5.16}
$$

For this simple fragment, the algebraic approach is short and gives us an answer right away. But when the quantum circuit we're working with becomes complicated, the algebra describing it can become correspondingly complicated.

Analysis by Matrix Elements

Let's now plug in the matrix expression for each operator and check that we get the same final results. This will show us explicitly how the operators are manipulating the states. When quantum algorithms become complex, sometimes following the matrices and watching the states that emerge can help us follow the computation.

Let's compute the output of the circuit by explicitly multiplying out the elements of the matrices we just wrote down. Equation 5.17 shows how we can find the output state $|\psi\rangle$.

$$
\begin{aligned}
|\psi\rangle &= XH|0\rangle && \text{The circuit for } |\psi\rangle \\[6pt]
&= X\left(\vee \begin{bmatrix} 1 & 1 \\ 1 & -1 \end{bmatrix} \begin{bmatrix} 1 \\ 0 \end{bmatrix}\right) && \text{Expand } H \text{ and } |0\rangle \\[6pt]
&= X\left(\vee \begin{bmatrix} 1 \\ 1 \end{bmatrix}\right) && \text{Multiply the matrices} \\[6pt]
&= \begin{bmatrix} 0 & 1 \\ 1 & 0 \end{bmatrix} \vee \begin{bmatrix} 1 \\ 1 \end{bmatrix} && \text{Expand } X \\[6pt]
&= \vee \begin{bmatrix} 1 \\ 1 \end{bmatrix} = |+\rangle && \text{Multiply the matrices to find } |\psi\rangle
\end{aligned}
\tag{5.17}
$$

Equation 5.18 shows the same process for $|\phi\rangle$.

$$|\phi\rangle = H\,|1\rangle \qquad\qquad \text{The circuit for } |1\rangle$$

$$= \vee \begin{bmatrix} 1 & 1 \\ 1 & -1 \end{bmatrix} \begin{bmatrix} 0 \\ 1 \end{bmatrix} \qquad \text{Expand } H \text{ and } |1\rangle \tag{5.18}$$

$$= \vee \begin{bmatrix} 1 \\ -1 \end{bmatrix} = |-\rangle \qquad \text{Multiply the matrices to find } |\phi\rangle$$

This is the same result as Equation 5.16.

Analysis by Direction

Calculating explicit matrices, as we just did, can provide useful insights. But it can also become tedious and error-prone when there are lots of matrices.

It's often helpful to group operators (and their matrices) into clusters. Then we can approach the whole circuit hierarchically. We can start at the bottom of the hierarchy, as in the last section, and write out every matrix. We can then go up one level of the conceptual hierarchy and combine matrices into small clusters that we understand. This lets us look at the whole circuit from a more abstract viewpoint. Repeating this process of collecting the matrices together, horizontally and vertically, gives us increasingly abstract ways to look at the overall flow of information.

To that end, let's use our techniques for merging matrices. When we use them together, I call it a *bidirectional analysis*, to let us know that we're combining qugates both horizontally and vertically.

Horizontal Analysis

Let's start with a purely horizontal approach. In Figure 5-15(a), I've marked the two qubit systems in gray boxes.

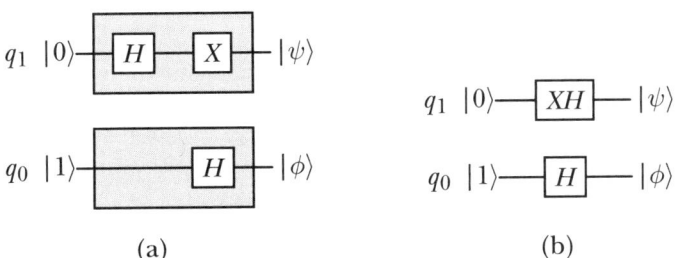

(a)   (b)

Figure 5-15: (a) Combining qugates horizontally. (b) An equivalent circuit.

Because the two qugates applied to q_1 are horizontally aligned, we can combine them with ordinary matrix multiplication. The combined system XH is shown in Equation 5.19.

$$XH = \begin{bmatrix} 0 & 1 \\ 1 & 0 \end{bmatrix} \vee \begin{bmatrix} 1 & 1 \\ 1 & -1 \end{bmatrix} = \vee \begin{bmatrix} 1 & -1 \\ 1 & 1 \end{bmatrix} \tag{5.19}$$

Now we can apply this system XH to the qubit q_0, giving us the output $|\psi\rangle$ in Equation 5.20.

$$|\psi\rangle = (XH)\,|0\rangle = \vee \begin{bmatrix} 1 & -1 \\ 1 & 1 \end{bmatrix} \begin{bmatrix} 1 \\ 0 \end{bmatrix} = \vee \begin{bmatrix} 1 \\ 1 \end{bmatrix} = |+\rangle \qquad (5.20)$$

Since the system for $|\phi\rangle$ is just a single qugate, we don't need to actually compute that system. Just looking at either part of Figure 5-15, we can immediately write $|\phi\rangle = H\,|1\rangle = |-\rangle$.

This is a great set of results. So far, all of our results have agreed that $|\psi\rangle = |+\rangle$ and $|\phi\rangle = |-\rangle$.

Vertical Analysis

Now let's analyze the circuit using purely vertical techniques. Figure 5-16 shows the two vertical groupings of the qubits at the ends and the two vertical groupings of qugates in the computational part of the circuit.

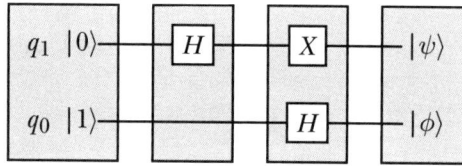

Figure 5-16: A first try at combining qugates vertically. This drawing is flawed.

As this drawing stands, we can't write the math for the output states of the qubits. To see why, start by grouping the input qubits to form $q_1 \otimes q_0$. From their initial values, we get $|0\rangle \otimes |1\rangle$, or $|01\rangle$. We know that this is a four-element column matrix.

But the diagram has this matrix going into a system composed only of the qugate H, represented by a two-by-two matrix. Because the tensored inputs are horizontally aligned with this box, the matrices will be multiplied together. But linear algebra doesn't let us multiply a four-by-one matrix on the left with a two-by-two matrix on the right.

Our vertical approach seems to have created a mathematical dead end. To fix this, I'll include an identity qugate I on qubit q_0. This operator does nothing to its input, so it doesn't affect the computation carried out by the circuit. But now we can form the tensor product of H and I to create a four-by-four matrix, appropriate for modifying the four-by-one column vector from the input.

The new version of the circuit fragment is shown in Figure 5-17, along with a few markers so we can discuss the qubit at different moments along its journey.

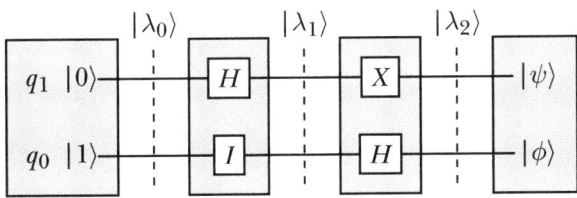

Figure 5-17: Combining qugates vertically. Including an I qugate in the left system fixes the problem in Figure 5-16.

Let's work out the first labeled qubit state, $|\lambda_0\rangle$. That's $q_1 \otimes q_0$, written out in tableau form in Equation 5.21.

$$|0\rangle \otimes |1\rangle = \begin{bmatrix} 1\,|1\rangle \\ 0\,|1\rangle \end{bmatrix} = \begin{bmatrix} 1\begin{bmatrix} 0 \\ 1 \end{bmatrix} \\ 0\begin{bmatrix} 0 \\ 1 \end{bmatrix} \end{bmatrix} = \begin{bmatrix} 0 \\ 1 \\ 0 \\ 0 \end{bmatrix} = |01\rangle \qquad (5.21)$$

This goes into the system $H \otimes I$ (remember that our convention is to always tensor objects reading from top to bottom in the diagram). We find the components of that in Equation 5.22.

$$H \otimes I = \vee \begin{bmatrix} 1I & 1I \\ 1I & -1I \end{bmatrix} = \vee \begin{bmatrix} 1\begin{bmatrix} 1 & 0 \\ 0 & 1 \end{bmatrix} & 1\begin{bmatrix} 1 & 0 \\ 0 & 1 \end{bmatrix} \\ 1\begin{bmatrix} 1 & 0 \\ 0 & 1 \end{bmatrix} & -1\begin{bmatrix} 1 & 0 \\ 0 & 1 \end{bmatrix} \end{bmatrix}$$

$$= \vee \begin{bmatrix} 1 & 0 & 1 & 0 \\ 0 & 1 & 0 & 1 \\ 1 & 0 & -1 & 0 \\ 0 & 1 & 0 & -1 \end{bmatrix} \qquad (5.22)$$

We can now find $|\lambda_1\rangle$, as in Equation 5.23.

$$|\lambda_1\rangle = (H \otimes I)\,|01\rangle = \vee \begin{bmatrix} 1 & 0 & 1 & 0 \\ 0 & 1 & 0 & 1 \\ 1 & 0 & -1 & 0 \\ 0 & 1 & 0 & -1 \end{bmatrix} \begin{bmatrix} 0 \\ 1 \\ 0 \\ 0 \end{bmatrix} = \vee \begin{bmatrix} 0 \\ 1 \\ 0 \\ 1 \end{bmatrix} \qquad (5.23)$$

The state system we just got from Equation 5.23 goes into our next vertical qugate system, $X \otimes H$. We find the matrix form of that system in Equation 5.24.

$$X \otimes H = \begin{bmatrix} 0H & 1H \\ 1H & 0H \end{bmatrix} = \vee \begin{bmatrix} 0\begin{bmatrix} 1 & 1 \\ 1 & -1 \end{bmatrix} & 1\begin{bmatrix} 1 & 1 \\ 1 & -1 \end{bmatrix} \\ 1\begin{bmatrix} 1 & 1 \\ 1 & -1 \end{bmatrix} & 0\begin{bmatrix} 1 & 1 \\ 1 & -1 \end{bmatrix} \end{bmatrix}$$

$$= \vee \begin{bmatrix} 0 & 0 & 1 & 1 \\ 0 & 0 & 1 & -1 \\ 1 & 1 & 0 & 0 \\ 1 & -1 & 0 & 0 \end{bmatrix} \tag{5.24}$$

We can now find $|\lambda_2\rangle$, as in Equation 5.25. Remember to include the $\vee$ from both Equations 5.23 and 5.24.

$$|\lambda_2\rangle = (X \otimes H)\,|\lambda_1\rangle = \vee \begin{bmatrix} 0 & 0 & 1 & 1 \\ 0 & 0 & 1 & -1 \\ 1 & 1 & 0 & 0 \\ 1 & -1 & 0 & 0 \end{bmatrix} \vee \begin{bmatrix} 0 \\ 1 \\ 0 \\ 1 \end{bmatrix} = \vee^2 \begin{bmatrix} 1 \\ -1 \\ 1 \\ -1 \end{bmatrix} \tag{5.25}$$

This output is a single state made up of two qubits. If we want to write the individual qubit values, in this case we can (we'll see later that this is not always possible). This breakdown is shown in Equation 5.26.

$$|\lambda_2\rangle = \vee^2 \begin{bmatrix} 1 \\ -1 \\ 1 \\ -1 \end{bmatrix} = \vee \begin{bmatrix} 1 \\ 1 \end{bmatrix} \otimes \vee \begin{bmatrix} 1 \\ -1 \end{bmatrix} = |+\rangle \otimes |-\rangle \tag{5.26}$$

Equation 5.26 tells us that the output of the circuit can be written as $|+\rangle$ on the upper qubit and $|-\rangle$ on the lower qubit. This exactly matches all of our previous analyses!

To summarize this approach, we can write our original circuit in the form of Figure 5-18.

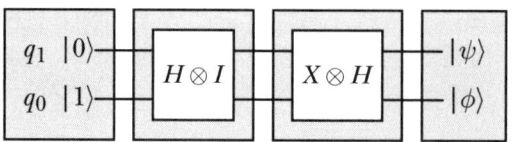

Figure 5-18: Writing Figure 5-14 in vertical form

This is excellent news, as it confirms that creating horizontal systems with matrix multiplication gives the same result as creating vertical systems with the tensor product. (At least it confirms that for this example. If you go through the algebra, you can demonstrate that this example wasn't special, and these two approaches always produce the same result).

Bidirectional Analysis

We can simplify our vertical analysis further by following it up with a horizontal analysis. Because Figure 5-18 contains two operators that are horizontally aligned, we can combine them with ordinary matrix multiplication, using $H \otimes I$ from Equation 5.22 and $X \otimes H$ from Equation 5.24. As always, we write the systems (represented by operators) that appear left to right in the figure in the right-to-left order that we use to compose operators, so we'll compute the matrix product $(X \otimes H)(H \otimes I)$. This gives us the circuit in Figure 5-19, which contains only a single qugate.

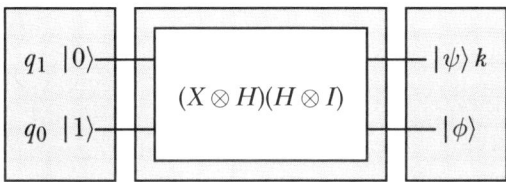

Figure 5-19: Interpreting all of our qugates as a single system formed by the horizontal matrix multiplication of two vertical systems, each formed by a tensor product

The explicit components of this qugate can be found by multiplying out the matrices, as in Equation 5.27.

$$(XH)(HI) = \vee \begin{bmatrix} 0 & 0 & 1 & 1 \\ 0 & 0 & 1 & -1 \\ 1 & 1 & 0 & 0 \\ 1 & -1 & 0 & 0 \end{bmatrix} \vee \begin{bmatrix} 1 & 0 & 1 & 0 \\ 0 & 1 & 0 & 1 \\ 1 & 0 & -1 & 0 \\ 0 & 1 & 0 & -1 \end{bmatrix}$$

$$= \vee^2 \begin{bmatrix} 1 & 1 & -1 & -1 \\ 1 & -1 & -1 & 1 \\ 1 & 1 & 1 & 1 \\ 1 & -1 & 1 & -1 \end{bmatrix} \tag{5.27}$$

Let's check that this one matrix still gives us the same results as before. Equation 5.28 applies $(XH)(HI)$ directly to the input $|01\rangle$.

$$\vee^2 \begin{bmatrix} 1 & 1 & -1 & -1 \\ 1 & -1 & -1 & 1 \\ 1 & 1 & 1 & 1 \\ 1 & -1 & 1 & -1 \end{bmatrix} \begin{bmatrix} 0 \\ 1 \\ 0 \\ 0 \end{bmatrix} = \vee^2 \begin{bmatrix} 1 \\ -1 \\ 1 \\ -1 \end{bmatrix} = |+\rangle \otimes |-\rangle \tag{5.28}$$

Yes! We get the same answer as before. Our tools all work together. We're going to use all of these techniques frequently throughout the book.

We can summarize the horizontal and vertical approaches using a little fragment of four arbitrary qugates, as in Figure 5-20. I've named the qugates this way so that the equation we're about to write reads nicely.

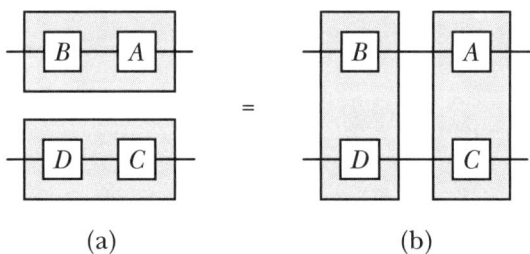

(a)   (b)

Figure 5-20: Four arbitrary qugates. (a) Two horizontal systems that are tensored together vertically. (b) Two vertical systems that are matrix multiplied together horizontally.

We can write the system of qugates in Figure 5-20(a) as a pair of horizontal systems that form a vertical system. Qugates B and A make AB and qugates D and C make CD, giving us the vertical system $(AB) \otimes (CD)$. We can also write the system of qugates as shown in Figure 5-20(b), treating it as a pair of vertical systems that form a horizontal system, or $(A \otimes C)(B \otimes D)$. These give us the same result, as summarized in Equation 5.29.

$$(AB) \otimes (CD) = (A \otimes C)(B \otimes D) \qquad (5.29)$$

Equation 5.29 is a powerful tool that goes by the name of the *mixed-product property*. It tells us that we can think of fragments like the one in Figure 5-20 using either the vertical or the horizontal approach, choosing and mixing these approaches in any way we like. This relationship is the heart of bidirectional analysis, and we'll use it throughout the book.

The No-Cloning Theorem

Now that we know about multi-qubit qugates, we can revisit my earlier claim that we can't branch in a quantum algorithm, which would be equivalent to making a copy of a qubit.

The reason we prohibit branching and copying in our circuits is that nature prohibits it. We fundamentally cannot create a mechanism that can flawlessly copy, or *clone*, an arbitrary quantum state. We're demanding perfection, so we're excluding any approach that could make copies of *some* states or make *approximate* copies of all states.

The statement that we cannot perfectly clone arbitrary quantum states is called the *no-cloning theorem*. Let's look at this important principle.

I'll show you that the theorem is true using a proof technique called *contradiction* [203]. The general idea is that we'll assume that we *can* make a clone of a qubit and see what that implies. We'll find that it leads to two expressions that should be the same but aren't (that is, they contradict each other). Because our only new assumption in this process will be that we can create clones, we'll conclude that this assumption must be wrong.

As with many important insights, there are multiple ways to prove the no-cloning theorem [171]. I'll pick an approach that uses only the tools we already have.

As promised, I'll start by assuming that we *can* clone an arbitrary quantum state. The operation will be embodied in some unitary operator (let's call it K) that can perfectly clone any quantum state from one qubit to another. As shown in Figure 5-21, K takes in two qubits.

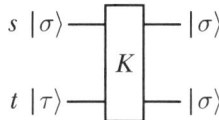

Figure 5-21: The concept of a cloning two-qubit qugate K

Let's call the upper qubit s, for source, and say it starts in state $|\sigma\rangle$. This is the state we want to clone. The other input is t, for target, which starts in state $|\tau\rangle$. The qugate outputs two qubits, both in the state $|\sigma\rangle$. I'll assume that the input states $|\sigma\rangle$ and $|\tau\rangle$ are different, since if they're the same we have no need for K in the first place.

Using our standard shortcut notation to write $|\sigma\rangle \otimes |\tau\rangle = |\sigma\tau\rangle$, we can express the operation we want from K as in Equation 5.30.

$$K|\sigma\tau\rangle = |\sigma\sigma\rangle \tag{5.30}$$

To keep the math manageable, I'm going to assume $|\tau\rangle$ is $|0\rangle$, and we'll see that we can't clone $|\sigma\rangle$. I'll return to this limitation at the end of the section.

Now let's expand Equation 5.30 into component form as shown in Equation 5.31. For this step, it doesn't matter what starting state qubit t is in, as it's immediately replaced by $|\sigma\rangle$.

$$K|\sigma\tau\rangle = |\sigma\sigma\rangle = \begin{bmatrix} \sigma_0 \begin{bmatrix} \sigma_0 \\ \sigma_1 \end{bmatrix} \\ \sigma_1 \begin{bmatrix} \sigma_0 \\ \sigma_1 \end{bmatrix} \end{bmatrix} = \begin{bmatrix} \sigma_0\sigma_0 \\ \sigma_0\sigma_1 \\ \sigma_1\sigma_0 \\ \sigma_1\sigma_1 \end{bmatrix} \tag{5.31}$$

$$= \sigma_0\sigma_0 |00\rangle + \sigma_0\sigma_1 |01\rangle + \sigma_1\sigma_0 |10\rangle + \sigma_1\sigma_1 |11\rangle$$

So far, we've just written down what we want K to do. Let's do it again, only this time instead of expanding the output product state $|\sigma\sigma\rangle$, we'll expand the input product state $|\sigma\tau\rangle$ and then apply K to it. Because K is linear, we should get the same result.

This will take a few steps, shown in Equation 5.32. This should yield the same results we got from Equation 5.31. Recall that I'm setting $|\tau\rangle = |0\rangle$ for convenience for the moment.

$$
\begin{aligned}
K|\sigma 0\rangle &= K\big((\sigma_0|0\rangle + \sigma_1|1\rangle) \otimes |0\rangle\big) && \text{Expand state } |\sigma\rangle \\
&= K(\sigma_0|00\rangle + \sigma_1|10\rangle) && \text{Form product states} \\
&= \sigma_0 K|00\rangle + \sigma_1 K|10\rangle && \text{Move } K \text{ onto each state} && (5.32)\\
& && \text{because it's linear} \\
&= \sigma_0|00\rangle + \sigma_1|11\rangle && \text{Apply } K \text{ to each state}
\end{aligned}
$$

In the last line, I used the definition of K to replace $K|00\rangle$ with $|00\rangle$ and $K|10\rangle$ with $|11\rangle$. That is, I just made the state of the second qubit the same as the first, which is what K does.

We've got a major problem here. Equations 5.31 and 5.32 are completely incompatible. Not only are the coefficients on $|00\rangle$ and $|11\rangle$ different, but also Equation 5.32 is missing the states $|01\rangle$ and $|10\rangle$ that are present in Equation 5.31.

The results of these two approaches can only be equal if they're the same, as shown in Equation 5.33.

$$
\begin{array}{ccc}
\text{Equation 5.31} & & \text{Equation 5.32} \\
\begin{bmatrix} \sigma_0\sigma_0 \\ \sigma_0\sigma_1 \\ \sigma_1\sigma_0 \\ \sigma_1\sigma_1 \end{bmatrix} & \overset{?}{=} & \begin{bmatrix} \sigma_0 \\ 0 \\ 0 \\ \sigma_1 \end{bmatrix}
\end{array} \qquad (5.33)
$$

There are three ways to assign values to these variables that will satisfy this equality. First, we can set $\sigma_0 = 1$ and $\sigma_1 = 0$, corresponding to the input state $|\sigma\rangle = |0\rangle$. Second, we can set $\sigma_0 = 0$ and $\sigma_1 = 1$, corresponding to the input state $|\sigma\rangle = |1\rangle$. Finally, we can set both σ_0 and σ_1 to 0. Not only is this last one a boring vector, but also because it doesn't have a magnitude of 1, it's not even physically possible.

So, our qugate K can "clone" the states $|0\rangle$ and $|1\rangle$, but no others. Thus, K cannot clone general states (that's the contradiction I promised!). The only assumption we made here (besides that $|\tau\rangle = |0\rangle$, which we'll return to in a moment) was that K is a unitary operator that satisfies Equation 5.30. Since that led to a contradiction, something in our assumptions about K must be wrong. The part that says K is unitary can't be wrong because Postulate 2 tells us all operators we use in quantum computing must be unitary, so the assumption that $K|\tau\sigma\rangle = |\sigma\sigma\rangle$ must be wrong. That is, no K can exist that can clone any state.

We've proven that the no-cloning theorem is true!

Or have we? Quantum computing is notoriously counterintuitive. Maybe, for some subtle reason, the choice of how we initialize $|\tau\rangle$ makes a difference.

Maybe the steps in Equation 5.32 would have worked if we hadn't assumed that $|\tau\rangle$ was $|0\rangle$. Using an arbitrary $|\tau\rangle$ creates a lot of bookkeeping that ends up in the same contradiction, confirming that the choice of $|\tau\rangle$ doesn't make a difference. This closes the only loophole, and we're left with the strange restriction that when we write a quantum algorithm, we are unable to make exact copies of arbitrary quantum states.

When we get a surprising result like this, we ought to check our math against the real world. Maybe our math has diverged from how reality works, and this conclusion is a bunch of symbols that don't accurately describe what happens in nature. But like the rest of the math we use in this book, this result has stood up in the face of countless experiments and the attention of countless clever people who have tried to find a flaw. Until we have surprising new observations, the no-cloning theorem is here to stay.

While perfect cloning is forbidden, there are some sneaky ways to do *imperfect* cloning. For example, we can write algorithms that can clone arbitrary states *sometimes*, though we can't predict when they'll do it properly [53] [52]. And under some conditions, we can *approximately* clone quantum states [180]. In the next section, we'll see a qugate that can apparently clone *some* specific states.

Because we can't generally clone any state, however, the lines carrying qubits in circuit diagrams *cannot be split*.

You might be wondering if you could get around the no-cloning theorem by measuring a qubit and then somehow preparing many new qubits in that state. Unfortunately, that loophole is closed to us as well because of how the measurement process itself works, as we'll see in Chapter 6.

Not being able to copy a qubit makes quantum algorithms harder to create, but that challenge is part of what makes this new field so fascinating.

Before we move on, I'll make good on my promise to return to the decision where we set $|\tau\rangle$ to $|0\rangle$. I did that to simplify the math, but it's not a choice that's required to prove the no-cloning theorem. If you're motivated, it's a fun exercise in algebra to prove that the theorem is true even when $|\tau\rangle$ starts out in any arbitrary state. It does take some careful bookkeeping, though, as I mentioned earlier. The general idea is to start with $K|\sigma\tau\rangle$ and write the ket $|\tau\rangle$ as $\tau_0 |0\rangle + \tau_1 |1\rangle$. Then, follow the outline of Equation 5.32 by expanding everything out, applying K to each of the terms, and simplifying. You'll get a more complicated set of conditions on $|00\rangle$ and $|11\rangle$, but there will again be no set of values that satisfies them both simultaneously.

The CX Qugate

We've seen that when we combine qugates vertically, the resulting qugate is a larger unitary square matrix. We can say generally that every matrix that describes a qugate takes as input a state vector that represents one or more qubits, represented as a column vector with the same number of rows as the qugate's matrix, and produces a new state vector of the same number of elements. The output state will be of magnitude 1, or unit norm.

Probably the most important multi-qubit qugate is called the *CX* qugate, where the *X* refers to the *X* qugate. Because the *X* qugate is also called the *NOT* qugate, the *CX* qugate is also called the *CNOT* qugate. The *C* stands for "controlled" or "conditional," so this qugate is also referred to as *controlled-X*, *controlled-NOT*, *conditional-X*, or *conditional-NOT*. I'll stick with the short *CX* in this book.

The *CX* qugate is versatile! We'll use it in four important ways in quantum algorithms: It's a *switch*, it's a narrowly limited *copier*, it's part of creating *entanglement*, and we use it to perform *phase kickback*. We'll look at the first two uses here, and we'll see the third in "Entanglement" on page 146. The fourth application, phase kickback, will come up when we look at Deutsch's algorithm in Chapter 8.

In all of its uses, *CX* takes as input two qubits, and produces two qubits as output.

Figure 5-22 shows three equivalent ways to draw the *CX* qugate.

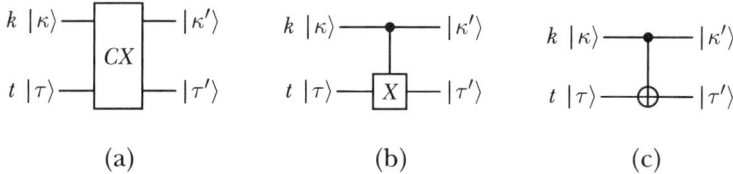

$$\begin{array}{ccc} \text{(a)} & \text{(b)} & \text{(c)} \end{array}$$

Figure 5-22: Three equivalent ways to draw the CX qugate. The labels k and t refer to control and target.

In Figure 5-22(a), I've drawn *CX* as a generic two-qubit qugate. Two qubits come in on the left, and two qubits emerge on the right. Because *CX* is so frequently used, it has two special shortcut versions, shown in parts (b) and (c). In these diagrams, the upper qubit (marked k, with state $|\kappa\rangle$) is called the *control* qubit, and the lower qubit (marked t, with state $|\tau\rangle$) is called the *target* qubit.

The labels "control" and "target," and even the name *CX*, are all somewhat troublesome. The problem is that they really only make sense when we use *CX* as a switch.

By analogy, suppose you meet someone at a party and they tell you a monkey joke. You could then refer to that person for the rest of your life as "the monkey joke teller." That name isn't incorrect, but it's a limiting way to refer to someone who has many other qualities.

With that in mind, our first use of *CX* will be to treat it as a switch. This is the usage that gave us the terms "controlled-X," "control," and "target," so we're on solid ground using them for now.

CX *as a Switch*

In this section, we'll see how to use *CX* as a switch. We'll broaden its interpretation and use as the book continues.

Just for a moment, let's think of CX as a classical gate that uses two conventional bits as input and output. The truth table for this classical version of CX is shown in Figure 5-23.

Inputs		Outputs	
Control	Target	Control	Target
0	0	0	0
0	1	0	1
1	0	1	1
1	1	1	0

$CX\,|00\rangle = |00\rangle$
$CX\,|01\rangle = |01\rangle$
$CX\,|10\rangle = |11\rangle$
$CX\,|11\rangle = |10\rangle$

Figure 5-23: Left: The truth table for CX interpreted as a classical gate with input and output bits. Right: The operation of the quantum CX on two-qubit systems.

Figure 5-23 tells us that the control bit is passed, unchanged, to its output. When the control input is 0, the target bit is also passed to the output without change. But, and here's the key contribution of CX, when the control input is 1, the target output is the *opposite* of its input. That is, when the control is 1, we apply an X (or NOT) operation to the target input to produce the target output.

This technique of first looking at some quantum operations as though they were classical operations working on bits is a nice way to meet some quantum gates. Be careful in such moments, though, because the transition from the classical version to the quantum version sometimes involves some change of behavior. When this happens, I'll point it out.

The transition from bits to the quantum world usually means replacing the classical bit 0 with the quantum state $|0\rangle$, and similarly replacing 1 with $|1\rangle$. You can see this on the right side of Figure 5-23, where I've shown the behavior of the CX qugate by making these replacements.

Now we can see where the name "controlled-X" comes from, because the control qubit determines whether or not the X qugate is applied to the target qubit. In Figures 5-22(b) and (c), the control qubit is marked with a small black circle. It's connected by a vertical line to a qugate on another quantum line. In this case, we either write that qugate as a typical X qugate (that is, a box with X inside) or use the special symbol of a circle with what looks like a plus sign inside (this symbol by itself is also occasionally used for a typical, or uncontrolled, X). Although the target qubit always appears to be acted on by the X qugate, the vertical line coming into that qugate tells us that it's controlled and only gets applied if the control is $|1\rangle$. Otherwise, it's as if the X isn't even there, and the target output is the same as the target input.

When used as a switch in this way, CX is the closest that quantum computing gets to the if-then construction in classical programming languages. That is, Figure 5-23 tells us that we can interpret CX as saying, "*If* the control is $|1\rangle$, *then* apply X to the target."

To see how it does this, let's start with the matrix for CX, given in Equation 5.34. As usual, this matrix uses the convention that we read our diagrams from top to bottom, so the control line is directly above the target line.

$$CX = \begin{bmatrix} 1 & 0 & 0 & 0 \\ 0 & 1 & 0 & 0 \\ 0 & 0 & 0 & 1 \\ 0 & 0 & 1 & 0 \end{bmatrix} = \left[\begin{array}{c|c} I & 0 \\ \hline 0 & X \end{array} \right] \tag{5.34}$$

If you multiply CX by itself, you'll get the four-by-four identity matrix. That is, CX is its own inverse, or $CX = CX^{-1}$. Note that the name CX could be confusing if you think it's describing the product of two qugates in a row, C and X. There's currently no standard qugate named C, so any time you see a C used at the start of a qugate name like this, you can assume it's referencing a controlled version of the qugate named after the C. Thus, CX refers to a single qugate. It's a kind of clumsy piece of notation, but it's ubiquitous now.

On the right side of Equation 5.34, I've written CX as a *block matrix*, where I replaced each two-by-two square with a single symbol. The upper-left block is the two-by-two identity matrix I, the lower-right block is the two-by-two matrix for X, and the other two blocks are all 0s.

The block matrix form is the key to understanding how CX is able to conditionally control whether another qugate is applied. To see why, let's apply CX to an input system when $k = |0\rangle$. I've written the input state in Equation 5.35, ending with a block matrix that summarizes the two vertical blocks, each of shape two by one.

$$|0\tau\rangle = |0\rangle \otimes |\tau\rangle = \begin{bmatrix} 1 \\ 0 \end{bmatrix} \otimes \begin{bmatrix} \tau_0 \\ \tau_1 \end{bmatrix} = \begin{bmatrix} \tau_0 \\ \tau_1 \\ 0 \\ 0 \end{bmatrix} = \left[\begin{array}{c} |\tau\rangle \\ \hline 0 \end{array} \right] \tag{5.35}$$

Notice that in the final block form, the upper two elements are the ket $|\tau\rangle$, while the lower two elements are both 0. I wrote them as a 0 without the ket because this is a column vector of two 0 elements, unlike $|0\rangle$, which has a 1 for the upper element. I hope it's clear from context that the shape of this 0 is a column vector of two elements, both of which are 0, while the 0 in the CX block matrix of Equation 5.34 is a two-by-two matrix where every entry is 0.

Now we can apply the block form of CX from Equation 5.34 to the block form of $|0\tau\rangle$ from Equation 5.35, as shown in Equation 5.36.

$$CX|0\tau\rangle = \left[\begin{array}{c|c} I & 0 \\ \hline 0 & X \end{array} \right] \left[\begin{array}{c} |\tau\rangle \\ \hline 0 \end{array} \right] = \left[\begin{array}{c} I|\tau\rangle \\ \hline 0 \end{array} \right] = \begin{bmatrix} \tau_0 \\ \tau_1 \\ 0 \\ 0 \end{bmatrix} = |0\rangle \otimes |\tau\rangle \tag{5.36}$$

The lower entry in the third term is 0 rather than, say, *X*0, because 0 times anything is 0. Equation 5.36 confirms that when the control is $|0\rangle$, the output system is the same as the input system.

Now let's set the control to $|1\rangle$. The new input system to *CX* is shown in Equation 5.37.

$$|1\tau\rangle = |1\rangle \otimes |\tau\rangle = \begin{bmatrix} 0 \\ 1 \end{bmatrix} \otimes \begin{bmatrix} \tau_0 \\ \tau_1 \end{bmatrix} = \begin{bmatrix} 0 \\ 0 \\ \tau_0 \\ \tau_1 \end{bmatrix} = \begin{bmatrix} 0 \\ \hline |\tau\rangle \end{bmatrix} \qquad (5.37)$$

And now the payoff, applying the system $|1\tau\rangle$ to *CX*, as in Equation 5.38.

$$CX\,|1\tau\rangle = \begin{bmatrix} I & 0 \\ \hline 0 & X \end{bmatrix} \begin{bmatrix} 0 \\ \hline |\tau\rangle \end{bmatrix} = \begin{bmatrix} 0 \\ \hline X\,|\tau\rangle \end{bmatrix} = \begin{bmatrix} 0 \\ 0 \\ \tau_1 \\ \tau_0 \end{bmatrix} = |1\rangle \otimes X\,|\tau\rangle \quad (5.38)$$

Hooray, now *X* has been applied to the second qubit!

Another way to look at what's going on here is to think of the input product state $|\kappa\tau\rangle$ as a kind of binary mask applied to the matrix *CX*. When $|\kappa\rangle = |0\rangle$, the lower elements of the system $|\kappa\tau\rangle$ are both 0, which "turns off" the lower part of the matrix *CX* by multiplying everything by 0. The result is that $|\tau\rangle$ is affected only by the *I* in the upper part of the *CX* block matrix, giving us an output of $I\,|\tau\rangle = |\tau\rangle$. But when $|\kappa\rangle = |1\rangle$, it's the upper part of $|\kappa\tau\rangle$ that becomes 0 and thus "turns off" the upper half of the *CX* matrix by multiplying it by 0. The result is that $|\tau\rangle$ is affected only by the *X* in the lower part of the *CX* block matrix, giving us an output of $X\,|\tau\rangle$.

The *CX* qugate choreographs this dance between the 0 and 1 elements, the tensor products of the input qubits, and the structure of the *CX* matrix itself to give us a valuable tool for selectively enabling or disabling qugates.

With some effort, we can turn this into an if-then-else operation. Figure 5-24 shows two approaches. We can control any single-qubit qugate by placing its matrix in the bottom-right corner of the four-by-four matrix in Equation 5.34. That is, to make, say, a controlled-*H* matrix, we place an *H* matrix where that block matrix uses an *X* matrix. Here, I've used two arbitrary qugates that I've named *A* and *B*.

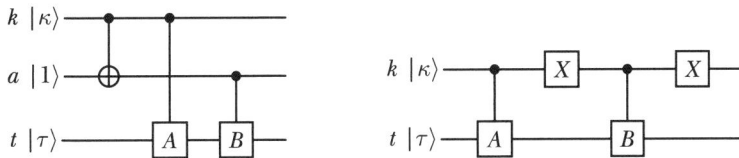

Figure 5-24: Two ways to implement the construction "if $|\kappa\rangle = |1\rangle$, then apply qugate A to qubit t, else apply qugate B to t"

Our goal here is to perform the statement, "If k is 1, then apply A, else apply B."

On the left side of Figure 5-24, the topmost qubit k arrives in state $|\kappa\rangle$. This is our control, which I'll assume is either $|0\rangle$ or $|1\rangle$. The middle qubit, named a for auxiliary, starts out in state $|1\rangle$. Finally, our target t starts out as $|\tau\rangle$.

Suppose that $k = |1\rangle$. Then the CX flips a from $|1\rangle$ to $|0\rangle$. Since $k = |1\rangle$, qugate A is applied to t, and because $a = |0\rangle$, qugate B is not applied. We've achieved the statement "If $k = |1\rangle$, then apply A but not B." Alternatively, if $k = |0\rangle$, then a remains as $|1\rangle$, so A is not applied and B is, giving us "If $k = |0\rangle$, then apply B but not A."

On the right side, I skipped the auxiliary qubit a but included another X qugate. If $k = |1\rangle$, then we apply A. Then k goes through an X, setting it to $|0\rangle$, so B is not applied, and then we use another X to put k back into its original state. If $k = |0\rangle$, then we don't apply A, the X flips k to $|1\rangle$, we apply B, and then again we return k to its starting state.

Both of these little circuits give us the behavior we were after, described by "If $k = |1\rangle$, then apply A, else apply B."

CX as a Copier

In the CX truth table in Figure 5-23, when both inputs are 0, the target output is the same as the control input (that is, they're both 0). When the input control is 1 and the input target is 0, the target gets flipped and is also output as a 1, the same as the input target.

We might be tempted to see CX as an exception to the no-cloning theorem. After all, the target always comes out with the same value as the control!

But CX doesn't clone as we've used the term, because it can't make a perfect copy of *any* input state. It does put the target into the same state as the control, but only when the control is a basis state and the target starts out as $|0\rangle$.

For these reasons, I think of using CX in this way as an extremely special-purpose *copier*. Even that might be too general a term, but "thing that copies a single basis state into a second qubit when that second qubit starts in state $|0\rangle$" is a mouthful!

Entanglement

The third way we use the CX qugate is to help us create a relationship among qubits called *entanglement*. As we saw in Chapter 1, along with superposition, interference, and measurement, entanglement is one of the four qualities that distinguishes quantum computing from classical computing, and we'll see that it plays a vital role in many quantum algorithms.

To study entanglement, I'll begin innocently enough and set the control qubit of a CX qugate to a superposition. To make a superposition, I'll run a qubit initialized to $|0\rangle$ through an H qugate to put it in the state $|+\rangle$. That will be our control, as shown in Figure 5-25 (note that this drawing contains a flaw, which we'll fix once we understand the problem).

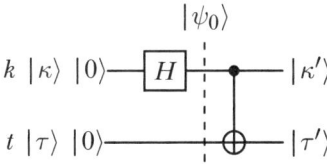

Figure 5-25: A (flawed) drawing that applies CX to a control of $|+\rangle$ and target of $|0\rangle$

We find the components of the input state $|+0\rangle$ at $|\psi_0\rangle$ in Equation 5.39.

$$|\psi_0\rangle = |+0\rangle = |+\rangle \otimes |0\rangle = \vee \begin{bmatrix} 1 \\ 1 \end{bmatrix} \otimes \begin{bmatrix} 1 \\ 0 \end{bmatrix} = \vee \begin{bmatrix} 1 \\ 0 \\ 1 \\ 0 \end{bmatrix} \qquad (5.39)$$

Now we'll apply CX to this, as shown in Equation 5.40. I'll name the resulting state $|\gamma\rangle$.

$$CX\,|+0\rangle = \begin{bmatrix} 1 & 0 & 0 & 0 \\ 0 & 1 & 0 & 0 \\ 0 & 0 & 0 & 1 \\ 0 & 0 & 1 & 0 \end{bmatrix} \left(\vee \begin{bmatrix} 1 \\ 0 \\ 1 \\ 0 \end{bmatrix} \right) = \vee \begin{bmatrix} 1 \\ 0 \\ 0 \\ 1 \end{bmatrix} = |\gamma\rangle \qquad (5.40)$$

If we can pull $|\gamma\rangle$ apart into the tensor product of two states, then we can use those states to label the outputs in Figure 5-25. Let's try to do that.

Entangled Pairs

Spoiler: We won't be able to write $|\gamma\rangle$ as a product state. Given any four-element complex column vector with a norm of 1 (that is, any two-qubit state), there is an algorithm for checking whether it's a product state [109]. It uses some ideas that we haven't covered, so instead I'll show that the specific state $|\gamma\rangle$ can't be written as the tensor product of two smaller states.

Like the no-cloning theorem, this proof hinges on a contradiction. I'll assume that $|\gamma\rangle$ *is* a product state. Then I'll write out the terms of that tensor product, and we'll find that we get boxed into a corner where no numbers satisfy the equations. I'll start with the generic product state in Equation 5.41.

$$|\psi\rangle \otimes |\phi\rangle = (\psi_0\,|0\rangle + \psi_1\,|1\rangle) \otimes (\phi_0\,|0\rangle + \phi_1\,|1\rangle)$$
$$= \psi_0\phi_0\,|00\rangle + \psi_0\phi_1\,|01\rangle + \psi_1\phi_0\,|10\rangle + \psi_1\phi_1\,|11\rangle \qquad (5.41)$$

The state $|\gamma\rangle$ can be written as a scaled sum of the four basis states, as shown in Equation 5.42.

$$|\gamma\rangle = \vee \begin{bmatrix} 1 \\ 0 \\ 0 \\ 1 \end{bmatrix} = \vee \left(1 \begin{bmatrix} 1 \\ 0 \\ 0 \\ 0 \end{bmatrix} + 0 \begin{bmatrix} 0 \\ 1 \\ 0 \\ 0 \end{bmatrix} + 0 \begin{bmatrix} 0 \\ 0 \\ 1 \\ 0 \end{bmatrix} + 1 \begin{bmatrix} 0 \\ 0 \\ 0 \\ 1 \end{bmatrix} \right) \tag{5.42}$$

$$= \vee |00\rangle + 0 |01\rangle + 0 |10\rangle + \vee |11\rangle$$

If $|\psi\rangle$ is a product state, we can write it in the form of Equation 5.41. Matching up Equations 5.41 and 5.42 gives us the four expressions in Equation 5.43.

$$\begin{aligned} \psi_0 \phi_0 &= \vee \\ \psi_0 \phi_1 &= 0 \\ \psi_1 \phi_0 &= 0 \\ \psi_1 \phi_1 &= \vee \end{aligned} \tag{5.43}$$

To write $|\gamma\rangle$ as a product state, we need to find the four numbers that satisfy the four equations in Equation 5.43. And that's the problem.

To see why, let's start with the second relation in Equation 5.43. It tells us that $\psi_0 \phi_1 = 0$, so either (or both) of these numbers must be 0. Let's arbitrarily pick ψ_0 and say that it's 0.

Since $\psi_0 = 0$, then $\psi_0 \phi_0$ will be 0 as well. But Equation 5.43 tells us that $\psi_0 \phi_0$ must be $\vee = 1/\sqrt{2}$, so ψ_0 can't be 0.

The only other choice is to set $\phi_1 = 0$. But then $\psi_1 \phi_1$ will be 0, and again Equation 5.43 tells us that this needs to be $\vee$, so we have the same problem.

We're stuck, because we require incompatible products of these four numbers. We are equally stymied if we start with the third relation, $\psi_1 \phi_0 = 0$. There's no way to pick the four elements of $|\psi\rangle$ and $|\phi\rangle$ so that all four relationships in Equation 5.43 are satisfied to make $|\gamma\rangle$.

Something has gone terribly wrong. But we made only one assumption: that $|\gamma\rangle$ is a product state. So that assumption must be wrong, and $|\gamma\rangle$ cannot be written as the tensor product of two states of two elements each.

We call such a state an *entangled state*.

But wait! If the problem is that some expressions in Equation 5.43 seemingly need to be both 0 and not 0, maybe superposition can save us? Unfortunately, it can't. While superposition is a powerful tool, the problem here isn't one that it can fix. The source of the contradiction isn't that some terms need to have different values, but rather that *no* values match all four conditions. That is, there is no consistent set of numbers, even in superposition, that can satisfy all four requirements simultaneously.

But wait again! Suppose we ignore all of this discussion, build the circuit shown in Figure 5-25, and actually measure one or both qubits. This is sure to return *something*. Doesn't that give us the values of the qubits and resolve our problem?

Surprisingly, no. While we can certainly measure either or both qubits, and we do indeed get back results, those measurements won't let us write $|\gamma\rangle$ as a tensor product. We'll see the reason for this strange situation when we discuss measurement in Chapter 6.

An entangled state is truly a different beast than a product state.

Note that it's not the tensor product alone that's creating entanglement. We can tensor together two states, such as $|0\rangle \otimes |1\rangle$, and that system isn't entangled. Entangled systems of two qubits are those that *cannot* be constructed by tensoring together the states of two (or more) qubits.

In any random collection of arbitrary two-qubit systems, there will be far fewer product states than entangled states. In fact, product states are extremely rare. If you make a random four-element complex vector with a norm of 1, it's a valid quantum state, but it will only rarely be a product state. Almost all states are entangled.

I said earlier that something was wrong with our drawing in Figure 5-25. Let's return to that and fix things up. The problem is that when we draw an entangled state, we can't put any meaningful labels on the individual output wires.

If we write names for the states on the output lines, this implies that their tensor product is the output of the circuit. But as we saw, the output state $|\gamma\rangle$ cannot be written as a product state. The flaw in Figure 5-25 was that the two wires were labeled with the named output states $|\kappa'\rangle$ and $|\tau'\rangle$. That implies that the output can be written as the product state $|\kappa'\rangle \otimes |\tau'\rangle$, but we've just seen that this cannot be done.

Although we can't identify these individual states, we know that the system's output is $|\gamma\rangle$, so let's write that as the state of the *pair* of values coming out of this *CX*. The convention for naming entangled qubits in a circuit like this is to join the qubit lines with a curly bracket, and then name the *system* rather than the individual qubits. Figure 5-26 shows the correct version of our circuit. The output of this circuit is an entangled pair, shown with a curly bracket. The wires don't have individual values.

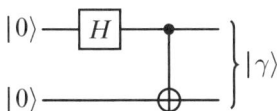

Figure 5-26: A corrected version of Figure 5-25

The curly bracket is also used for showing groupings, as I did in Figure 4-5. Usually its meaning is clear from context, and I'll mention it explicitly when I think there's any ambiguity.

To see entanglement in a slightly different way, let's look at the algebra that creates this new type of state. As in Figure 5-26, Equation 5.44 shows how to find $|\gamma\rangle$ without writing down the elements of a single matrix.

$$CX\,|{+}0\rangle = CX(|{+}\rangle \otimes |0\rangle) \qquad \text{Expand the product state}$$

$$= CX\Big(\vee\,(|0\rangle + |1\rangle) \otimes |0\rangle\Big) \qquad \text{Expand } |{+}\rangle$$

$$= \vee CX\big((|0\rangle \otimes |0\rangle) + (|1\rangle \otimes |0\rangle)\big) \qquad \text{Pull out } \vee \text{ and distribute } \otimes \quad (5.44)$$

$$= \vee CX(|00\rangle + |10\rangle) \qquad \text{Compact the tensor products}$$

$$= \vee(CX\,|00\rangle + CX\,|10\rangle) \qquad \text{Distribute } CX$$

$$= \vee(|00\rangle + |11\rangle) \qquad \text{Use Fig. 5-23}$$

To directly compare this to the result of Equation 5.40, let's write down the component forms of the results, as in Equation 5.45.

$$\vee(|00\rangle + |11\rangle) = \vee \left(\begin{bmatrix} 1 \\ 0 \\ 0 \\ 0 \end{bmatrix} + \begin{bmatrix} 0 \\ 0 \\ 0 \\ 1 \end{bmatrix} \right) = \vee \begin{bmatrix} 1 \\ 0 \\ 0 \\ 1 \end{bmatrix} \qquad (5.45)$$

Yes! Equations 5.42 and 5.45 agree. I find that two-qubit systems (and their four-by-four qugate matrices) are usually tolerable to explicitly write down and multiply, but for three qubits or more, the algebraic approach is much faster and offers less opportunities for minor (but disastrous!) errors when explicitly computing matrix elements.

We've just seen a concrete demonstration of why CX cannot be used as a cloner. Given $|{+}0\rangle$, a cloner should produce $|{++}\rangle$, but the output of $CX\,|{+}0\rangle$ is $\vee(|00\rangle + |11\rangle)$.

Entanglement is a key tool when building quantum algorithms. Entangled states don't require any kind of special handling, and they can be processed by qugates like any other systems of qubits. Conceptually, one feature of entanglement is that it lets us distribute information among multiple qubits. They can then be processed in different ways (or even by different people, in different places at different times!), and the operations we apply to either qubit have an influence on the other. We'll use entanglement in Chapter 7, when we teleport a quantum state from one place to another, as well as in other algorithms in Part II.

We saw the usual way to draw an entangled pair in Figure 5-26, redrawn in Figure 5-27(a). This graphic works well when the entangled qubits are directly above and below one another and are not used in computation later.

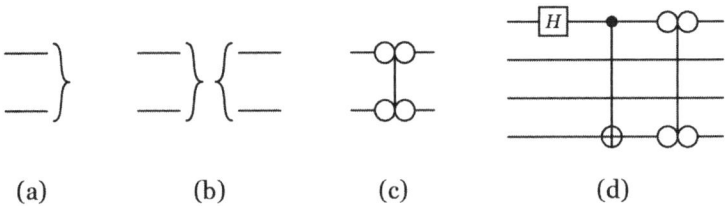

(a)   (b)   (c)   (d)

Figure 5-27: Four ways to indicate entanglement in a quantum circuit

The other three graphics in Figure 5-27 show alternative ways to notate entanglement. Part (b) tells the reader that the pair of wires are entangled and then go on to be used in more computation. Part (c) is my own invention for showing entangled lines. I designed it to solve the problem of showing entanglement when the lines are not immediately one above the other. In that case, the curly bracket can be misleading (it could seem to suggest that the intermediate wires were entangled as well). Connecting the two pairs of adjacent circles with a line lets us assert that two qubits are entangled even if they're not directly adjacent. For example, part (d) shows a circuit with four qubits, where the top and bottom lines are entangled.

Other Controlled Qugates

We can take the block matrix we used to analyze the CX qugate and replace the X in the lower-right corner with any other qugate represented by a two-by-two matrix.

Let's call the matrix for this qugate U (for unitary). Then, a *controlled-U* (CU) qugate could be drawn as in Figure 5-28. Note that I left the outputs bare. If the outputs for a particular set of inputs form a product state, we could give them names, but if they're entangled, we'd want to mark them with one of the symbols in Figure 5-27.

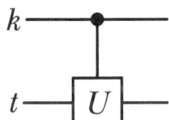

Figure 5-28: A
controlled-U qugate

We can summarize the operation of this qugate in the block matrix of Equation 5.46.

$$CU = \left[\begin{array}{c|c} I & 0 \\ \hline 0 & U \end{array} \right]$$
(5.46)

This works by the same logic as CX. When the control is $|0\rangle$, the U part of the matrix has no effect, because all of its elements are multiplied by 0. When the control is $|1\rangle$, the I part of the matrix is multiplied by 0 and the U block operates on the two elements of the target state.

Like the CX qugate, if we place the control k in an equal superposition $|+\rangle$, the CU qugate will return a new equal superposition of $|0\rangle |\tau\rangle$ and $|1\rangle (U |\tau\rangle)$.

Sometimes we want to control a qugate using the opposite of the normal convention, so it's applied when the control is $|0\rangle$ and bypassed when the control is $|1\rangle$. The usual way to draw this is shown in Figure 5-29.

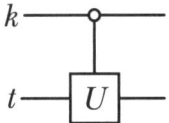

Figure 5-29: Controlling
a qugate so it's applied
only when k is $|0\rangle$

In this diagram, the qugate is being controlled by the qubit k, using an open dot for the control. The open dot means that the U qugate is applied only if the control k is $|0\rangle$.

Other Multi-Qubit Qugates

There are a few other multi-qubit qugates that are good to know about. Some of these I'll use in Part II, but others I won't refer to again. I'll show them to you here because they sometimes appear in algorithms that you'll see in publications and other discussions. I'd like you to have some knowledge of them now, so when you see their names (and graphics), you won't be surprised.

Let's look at a few of these useful multi-qubit qugates.

The *swap* qugate, often written *SWAP*, exchanges two qubits. There are two different but commonly used symbols for the SWAP qugate, shown in Figure 5-30(a) and (b).

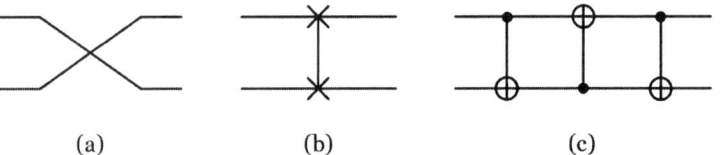

(a)   (b)   (c)

Figure 5-30: (a), (b) Two ways to draw the SWAP qugate. (c) Creating
SWAP from CX qugates.

The matrix form of the SWAP qugate is shown in Equation 5.47.

$$\text{SWAP} = \begin{bmatrix} 1 & 0 & 0 & 0 \\ 0 & 0 & 1 & 0 \\ 0 & 1 & 0 & 0 \\ 0 & 0 & 0 & 1 \end{bmatrix} \tag{5.47}$$

We don't need to copy or clone a qubit to perform the swap. To see this, consider the diagram of three CX qugates shown in Figure 5-30(c). This sequence of qugates implements SWAP. The center CX qugate is inverted from the way we usually draw it. This form has the matrix shown in Equation 5.48, where I've temporarily named it CX' [131].

$$CX' = \begin{bmatrix} 1 & 0 & 0 & 0 \\ 0 & 0 & 0 & 1 \\ 0 & 0 & 1 & 0 \\ 0 & 1 & 0 & 0 \end{bmatrix} \qquad (5.48)$$

We can check that this describes a CX with the control on the lower line by running through all four possible inputs. I've shown the results in Figure 5-31.

Inputs		Outputs	
Control	Target	Control	Target
0	0	0	0
0	1	1	1
1	0	1	0
1	1	0	1

$CX'\,|00\rangle = |00\rangle$
$CX'\,|01\rangle = |11\rangle$
$CX'\,|10\rangle = |10\rangle$
$CX'\,|11\rangle = |01\rangle$

Figure 5-31: Left: The truth table for CX' interpreted as a classical gate with input and output bits. Right: The operation of the quantum CX' on two-qubit systems.

Figure 5-31 shows the inputs and outputs of the middle qugate in Figure 5-30(c), as defined in Equation 5.48. Here, when the second qubit is $|1\rangle$, it flips the first qubit, matching the drawing that puts the control on the lower line and the target on the upper line.

The three CX qugates in Figure 5-30(c) are horizontally aligned, so we know that they can be formed into one system using matrix multiplication. If you multiply the matrices $(CX)(CX')(CX)$, you'll get the same matrix as SWAP in Equation 5.47.

Just as we can control one qugate with a qubit, we can put a qugate under the control of *two* other qubits, so that the controlled qugate is applied only if *both* controls are $|1\rangle$. When we do this for an X (or *NOT*) qugate, we call this a *controlled-controlled-NOT* or *conditional-conditional-NOT* qugate and write it as CCX or $CCNOT$. It's also called a *Toffoli* qugate, after its inventor, Tommaso Toffoli [279]. Its icon is shown in Figure 5-32(a).

We can apply this approach to control the application of any qugate. In particular, if we use it to control a SWAP qugate, we call the result a controlled-SWAP or *Fredkin* qugate, named for its inventor, Edward Fredkin. It's usually drawn as shown in Figure 5-32(b) [263].

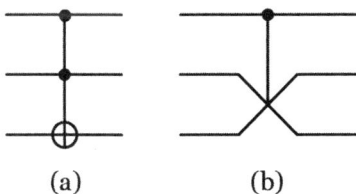

(a) (b)

Figure 5-32: (a) A CCX, or Toffoli, qugate. (b) A Fredkin, or controlled-SWAP, qugate.

I won't write out the matrices for the Toffoli and Fredkin qugates, as they're both eight-by-eight grids of 0 and 1 entries that don't reveal much just by looking at them.

Single-Qubit Qugates

While we're discussing qugates, let's wrap up with a few single-qubit qugates (and one two-qubit qugate) that are commonly used in quantum circuits in addition to the I, X, and H qugates that we already know [98].

For completeness, the matrices for I and H are repeated here in Equation 5.49 (recall that $\vee = 1/\sqrt{2}$).

$$I = \begin{bmatrix} 1 & 0 \\ 0 & 1 \end{bmatrix}, \qquad H = \vee \begin{bmatrix} 1 & 1 \\ 1 & -1 \end{bmatrix} \tag{5.49}$$

The X qugate is an old friend by now. This is actually one of a trio of qugates named X, Y, and Z. These are also called the *Pauli* qugates, after the physicist Wolfgang Pauli [284].

We've seen the X qugate in action many times. I've summarized it in Equation 5.50.

$$X\,|\psi\rangle = \begin{bmatrix} 0 & 1 \\ 1 & 0 \end{bmatrix} \begin{bmatrix} \alpha \\ \beta \end{bmatrix} = \begin{bmatrix} \beta \\ \alpha \end{bmatrix} \tag{5.50}$$

We haven't used the Y qugate yet, and generally speaking, you'll see it more rarely than X. Its matrix and action on a generic state are shown in Equation 5.51.

$$Y\,|\psi\rangle = \begin{bmatrix} 0 & -i \\ i & 0 \end{bmatrix} \begin{bmatrix} \alpha \\ \beta \end{bmatrix} = \begin{bmatrix} -i\beta \\ i\alpha \end{bmatrix} = i\begin{bmatrix} -\beta \\ \alpha \end{bmatrix} \tag{5.51}$$

Finally, the Z qugate negates the amplitude of the $|1\rangle$ basis state. Its matrix and operation are summarized in Equation 5.52.

$$Z\,|\psi\rangle = \begin{bmatrix} 1 & 0 \\ 0 & -1 \end{bmatrix} \begin{bmatrix} \alpha \\ \beta \end{bmatrix} = \begin{bmatrix} \alpha \\ -\beta \end{bmatrix} \tag{5.52}$$

The controlled version of the Z qugate, called *controlled-Z*, *conditional-Z*, or, most often, *CZ*, is particularly interesting. For the two-qubit inputs $|00\rangle$, $|01\rangle$, and $|10\rangle$, the output of CZ is the same as the input. But the input $|11\rangle$ is negated to become $-|11\rangle$, as shown in Equation 5.53.

$$CZ\,|11\rangle = \begin{bmatrix} 1 & 0 & 0 & 0 \\ 0 & 1 & 0 & 0 \\ 0 & 0 & 1 & 0 \\ 0 & 0 & 0 & -1 \end{bmatrix} \begin{bmatrix} 0 \\ 0 \\ 0 \\ 1 \end{bmatrix} = \begin{bmatrix} 0 \\ 0 \\ 0 \\ -1 \end{bmatrix} \tag{5.53}$$

Because CZ has this effect only when both inputs are $|1\rangle$, it doesn't really matter which we call the "control" and which we call the "target." The

CZ qugate has been given a special, symmetrical shortcut because of this property. The generic form of the *CZ* qugate is shown in Figure 5-33(a), and the special symbol is shown in part (b).

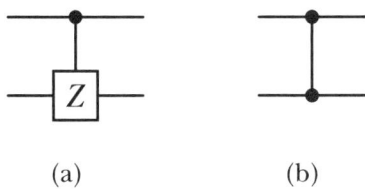

(a)   (b)

Figure 5-33: Two ways to draw the controlled-Z, or *CZ*, qugate

Next up is the *P* qugate, also called the *phase* qugate. When we modify a state $|\psi\rangle = \alpha\,|0\rangle + \beta\,|1\rangle$ with *P*, we multiply β by a complex number with magnitude 1. In other words, we multiply it by $e^{i\theta}$ for some value of an angle represented by the real number θ. The *S* and *T* qugates are versions of the *P* qugate for $\theta = \pi/2$ and $\theta = \pi/4$, respectively. These qugates are shown in Equation 5.54.

$$P = \begin{bmatrix} 1 & 0 \\ 0 & e^{i\theta} \end{bmatrix}, \qquad S = \begin{bmatrix} 1 & 0 \\ 0 & e^{i\pi/2} \end{bmatrix}, \qquad T = \begin{bmatrix} 1 & 0 \\ 0 & e^{i\pi/4} \end{bmatrix} \tag{5.54}$$

Setting $\theta = \pi$ in the phase qugate gives us *Z*. The angles in the *Z*, *S*, and *T* qugates are closely related, so we can write $S = T^2$ and $Z = S^2$.

Finally, we can write down a completely generic two-by-two unitary matrix, often labeled *U*. You'll rarely see this in practice, and I won't use this qugate in this book, but every now and then it does pop up in the literature. Every unitary two-by-two matrix, including all those we've just seen, can be made from this *U* with an appropriate choice of parameters. Different authors write the elements of the *U* matrix differently. One form of the matrix for *U* is given in Equation 5.55 [146].

$$U(\alpha, \beta, \gamma, \delta) = \begin{bmatrix} e^{i(\alpha-\beta-\delta)}\cos\gamma & -e^{i(\alpha-\beta+\delta)}\sin\gamma \\ e^{i(\alpha+\beta-\delta)}\sin\gamma & e^{i(\alpha+\beta+\delta)}\cos\gamma \end{bmatrix} \tag{5.55}$$

You might be wondering if we need all of these different qugates. In fact, only a few qugates are absolutely necessary. A set of qugates that allow any quantum computation at all is called a *universal set* or *universal family* of qugates. Usually these sets are of only theoretical interest, because we often need impractically huge numbers of these qugates to simulate what the other qugates can do. There are multiple universal families, such as the set of *CX* and the generic single-qubit qugate *U* in Equation 5.55; the set of *CX*, *H*, and the phase qugates; and the set of the Toffoli and *H* qugates [225]. The smallest universal set is a single qugate. For example, the Toffoli qugate is universal, so you can build any quantum circuit out of enough Toffoli qugates. In practice, having lots of different qugates at our disposal makes designing and implementing quantum algorithms a lot easier.

Summary

This chapter has taken us from single, isolated qubits and qugates to collections, or systems, of qubits and qugates working together.

We started with the tensor product, a mathematical tool that combines two matrices by creating every combination of their elements in a specific order. When we tensored together two kets describing the state of two quantum objects, we called the resulting ket a product state. Every state, including product states, can be written as a unique sum of the basis vectors.

We looked at quantum algorithms in diagram, or circuit, form. These diagrams typically have multiple horizontal lines, one per qubit, and each qubit flows from the left to the right, transformed by qugates along the way. Qubits may be given names so we can refer to them (usually shown at the far left of each line), and we can insert some vertical dashed lines with labels to discuss the states of the qubits at different locations along the diagram.

We saw that one way to work out what's happening to the qubits in the circuit is to follow each qubit's changes, or evolution, as it flows from the left to the right. We call this a horizontal analysis.

Alternatively, we can collect vertically aligned qubits together into a single system. We can also collect qugates into systems and then follow the evolution of the entire qubit system as it passes from one qugate system to the next. We call this a vertical analysis.

Both techniques give us the same final answer, but in any given diagram one may be more convenient or make more sense to us than the other. We can also use both methods, if that makes the calculation more sensible or convenient.

We saw that we cannot reliably make perfect copies of arbitrary qubits, which is proven by the no-cloning theorem. This means our quantum diagrams can't branch, as that would require making a copy of the state just before the branch.

We saw that the *CX* qugate takes two inputs, a "control" and a "target." When the control is in state $|1\rangle$, the target has the *X* qugate applied to it; otherwise, it is unaffected. This gave us a rudimentary if-then test, which we could extend with a few more qugates (or auxiliary qubits) into an if-then-else test.

Using *CX* and *H* qugates, we saw how to make entangled pairs, which can only be written as a system and not as a tensor product of two qubits.

Finally, we reviewed our single- and multi-qubit quantum gates and briefly noted a few others.

All that remains to build working quantum algorithms is to see how to use measurements to extract a final answer. This is the topic of our next chapter.

6

MEASUREMENT

We read the world wrong and say that it deceives us.
—Rabindranath Tagore, *Stray Birds*, 1916 [216]

There are two worlds: the world we can measure with line and rule, and the world that we feel
with our hearts and imagination.
—Leigh Hunt, *The Farmer's Wife*, 1933 [93]

Up to this point, we've seen that quantum computations begin with qubits in an initial state and are then operated upon repeatedly with quantum gates, changing their states. This is a pretty good model of computation.

To make this model practical, however, we have to include an output step. That is, we need to be able to determine the final states of the qubits at the end of the computation. Obtaining useful output is, after all, the whole purpose of running the computer in the first place!

In a classical computer, this measurement is straightforward. We have devices that can objectively measure the status of a bit and report that it either does or does not have some property. For example, we might measure how much electrical charge is on an object representing a bit. Any charge below some threshold is considered a 0, and otherwise it's a 1.

Notice an important aspect of this process: To measure something, we must interact with it. We typically try to make this interaction so small that it has no noticeable effect on our measurement. In our everyday world, this is often not a problem. For example, suppose someone holds out their hand and offers us a grape. We can determine that they are actually holding a grape by looking at it, which requires photons from some light source to interact with the grape and eventually reach our eyes. That interaction doesn't have much effect on it (the grape might warm up a little, if there are enough photons).

In quantum computing, measurements have a more dramatic impact on the carriers of quantum bits. In fact, the very act of measuring a quantum bit changes it fundamentally.

And as if that weren't unusual enough, remember that a superposition is a new state of being that seems to be best described as a list of possible states and probabilities. The act of measurement causes that list to collapse into a single state. Nobody knows for sure how that happens, or why. We don't even know exactly what constitutes a measurement.

And yet, if quantum computing is to have any practical value to us, we need to somehow find a way to determine if each qubit emerging from the computer is a 0 or a 1.

Figuring out how to make such determinations required the work of many scientists and engineers. And while there are still open questions (like exactly what qualifies as a "measurement"), we now have practical solutions that let us obtain outputs from quantum computers.

Those solutions are the subject of this chapter.

The Main Ideas of Measurement

Let's start with the big ideas that will guide the rest of this chapter. Different kinds of quantum computing hardware use different technologies to represent qubits. In this book, we're not concerned with these technologies, but only with their shared ability to represent a qubit. For the sake of simplicity, I'll refer to the carrier of a quantum bit as a quantum particle, but don't take the "particle" part too literally. There are lots of different ways to represent qubits, and not all of them look anything like what we'd call a particle. So this is not a literal term, but just a catch-all for any technology representing a qubit.

Suppose we have some physical device that can detect the state of a qubit. Each type of technology that is used to represent qubits will require its own specific type of measuring device, constructed specifically to measure the properties of whatever that technology uses to represent a qubit.

In quantum computing, we abstract all of this away and say that every quantum computer comes equipped with one or more measuring devices, each of which we generically call a *meter*.

When a qubit enters a meter, we say that the meter *measures* the qubit or that we're performing a *measurement* on, or an *observation* of, the qubit. We sometimes say that we *obtain* the result of the measurement or observation, though we also refer to *finding*, *getting*, or *seeing* the result. We also say that the meter *reports* or *shows* its result. I'll treat all of these variations as synonyms.

While the word *measure* is probably familiar to you, quantum measurements are decidedly unlike those we're used to on an everyday, human scale.

A quantum meter provides a binary output. That is, it has only two output states. Following the usual language of information theory, we refer to a meter's output as a "bit" and label its states 0 and 1 [193]. We use these same two labels regardless of what physical process they're actually referring to (for example, whether an electric charge has or has not been detected, or whether light is or is not present in some location).

When a quantum bit is measured, no matter what goes into the meter, one of two things happens: either the meter reports a 0 and the qubit leaves the meter in state $|0\rangle$, or the meter reports a 1 and the qubit leaves the meter in state $|1\rangle$. That's it.

Well, it can get more complicated, but not if we're using the computational basis, which is the most common basis and the only one I use in this book (in other bases, we still get 0 or 1 from the meter, but the possible states of the qubit after measurement might not be the states $|0\rangle$ and $|1\rangle$).

If the qubit going into the meter is in the basis state $|0\rangle$, then the meter will *always* report a 0 and the qubit will leave the meter unchanged, in the basis state $|0\rangle$. Similarly, if the qubit going into the meter is in the basis state $|1\rangle$, then the meter will *always* report a 1 and the qubit will leave the meter unchanged, in the state $|1\rangle$.

Suppose instead that the qubit is in a superposition $\alpha|0\rangle + \beta|1\rangle$, where neither α nor β is 0. This is the general case and so will be our main focus in this chapter. When a state in superposition arrives at a meter, *we cannot predict with certainty what the meter will report, or what state the qubit will be in after measurement*. Either the meter will report 0 and the qubit will emerge in state $|0\rangle$, or the meter will report 1 and the qubit will emerge in state $|1\rangle$, but we can't say for sure which of these will happen.

I mean this literally. Nobody can accurately state what the output of the meter will be when the input is in superposition. Not now, not ever. Not future humans, not sentient gas clouds [176], not even hyperintelligent pandimensional creatures [3]. It's not a question of lacking some information or not being able to compute a simulation accurately enough. In this situation, *the universe is probabilistic*, which means we can't state with certainty just what will happen next. It is inherently, fundamentally, inescapably unpredictable.

You may find this hard to believe. Albert Einstein famously refused to accept it, arguing that "He does not play dice with the universe" (referring not to any specific religious being but rather a "personification of nature" [175]).

Luckily, just as with physical dice, we can make statistical predictions about what we'll see. When we roll a normal (but ideal) die, every one of its six faces has an equal probability of coming up on top. In a quantum measurement, we can compute the exact *probability* of the meter reporting either 0 or 1, but until we perform the measurement, we can't be sure which output will emerge. Remarkably, even when a qubit enters the meter in a superposition, it emerges as only either $|0\rangle$ or $|1\rangle$, corresponding to the meter's output.

These facts make the design and use of quantum algorithms a challenging task. The rest of this chapter focuses on how to determine these probabilities and how to think about making measurements of some or all of the qubits in a multi-qubit circuit.

Measuring Qubits

Let's look more closely at how classical and quantum measurements differ.

Classical algorithms are *repeatable*. As long as the hardware is working properly, running the same program with the same starting conditions over and over will always produce the same outputs. We also describe this by saying that classical algorithms are *deterministic*, because the inputs and the program itself completely determine everything that is computed, including the outputs.

Determinism in this context is a good thing. If you're tracking someone's banking records, working out the servicing schedule on a major bridge, or planning medical treatment, you want the same inputs to produce the same outputs every single time. Determinism is a cornerstone of classical computing.

Quantum computing is different. While the computation is deterministic, the output is not. That's because of the nature of measurement.

In this chapter, we'll see that when we measure the output of a quantum circuit, in all but the simplest circumstances the values we get back are not deterministic. They are *probabilistic*, or *nondeterministic*.

I use the word *probabilistic* rather than *unpredictable* because the latter suggests that we don't know anything about what we'll measure. We do know the probabilities of the different outcomes, so they're not totally unpredictable. We just don't know specifically which outcome we'll see after any given run of the algorithm. Even though words like *predictable* and *unpredictable* are imprecise in this context, many people use them when discussing measurement, and I'll do the same in this book.

The probabilistic nature of the outcomes we measure gives a *random* quality to our measurements. This isn't a simulated, *pseudorandom* process like that provided by libraries on classical computers. This is truly random, or a result that nobody, anywhere, with any technology, can determine beforehand.

All of these words are sometimes used to describe quantum measurement. *Unpredictable* and *nondeterministic* are probably the best fits, but they're bulky. Many people prefer to use the more compact word *random*, and I'll use that as well, but keep in mind that we might know the possible outcomes and their probabilities.

The nondeterministic nature of measurement is a fundamental and inescapable trait of quantum measurement. It cannot be avoided. Too much unpredictability would make our outputs useless, so we're fortunate that its effects here are narrow and specific. Even so, we'll see that this unpredictability has profound implications for how we design quantum algorithms and how we use quantum computers.

To stay focused, I'm going to ignore some advanced topics in measurement, such as accounting for the influence of the environment on our hardware, or representing qubits that are in one specific state (that is, not a superposition) that is guaranteed to be an element in a list of states but where we don't know which one it is. Skipping these topics lets us avoid a bunch of new math and stay focused on the core ideas.

Happily, our simpler approach is ultimately just as powerful as the more complicated methods, so we're not losing anything by taking this gentler path [146, p. 87].

Postulate 4

We'll begin with our fourth and final postulate of quantum mechanics, stated in terms that will be useful for us [146].

> **Postulate 4** A measurement of a state $|\psi\rangle \in \mathbb{C}^d$ is created from a set of projection operators Π_k for $k \in [d]$.

This postulate has three components:

1. The probability of measuring outcome k is $\langle\psi|\,\Pi_k\,|\psi\rangle$.
2. The projection operators are complete: $\displaystyle\sum_{k\in[d]} \Pi_k = I$.
3. If outcome k is observed, the system immediately has the state
$$\frac{\Pi_k\,|\psi\rangle}{\sqrt{\langle\psi|\,\Pi_k\,|\psi\rangle}}.$$

Don't be spooked by this complicated postulate! It's not as ferocious as it appears. In this chapter, we'll unpack everything.

This postulate will be a practical tool for us. It tells us how to predict the different outcomes of a measurement, which helps us design our algorithms so that the outcomes we want are more likely.

We'll also see that it leads to a powerful principle whereby we can measure just some of the qubits from a calculation. This act narrows down the possible values of the qubits we didn't measure. Knowing this, we can apply additional operators to those unmeasured qubits so that they'll give us even more useful information when we ultimately do measure them.

Meters

Meters in quantum circuits are often represented with an icon like one of the shapes in Figure 6-1. The binary output of the meter is shown with two parallel lines leaving the right side of the icon, representing a classical bit. There's no standard notation for the qubit that emerges from the meter.

(a) (b)

Figure 6-1: (a) My icon for a meter. (b) An alternative icon.

Although the drawing in Figure 6-1(b) is popular, I find it misleading, because it looks like the kind of meter you'd find on a piece of audio equipment, swinging back and forth with the volume of its input. This action nicely represents real numbers, but not the binary bits we get from quantum measurements. I prefer the triangular icon on the left because it suggests that the general quantum state $\alpha \, |0\rangle + \beta \, |1\rangle$ is narrowed down into either $|0\rangle$ or $|1\rangle$, with no other options.

Experiments

I'm here to talk about music with lots of saxophone, a little saxophone, and . . . nope, I'm looking here, and those are the only two types of music.

—Kirk Hamilton, *Strong Songs* podcast, 2023 [88]

A common description of the scientific method starts with a *question*: "What is going on with this thing that's caught my attention?" To answer that question, we perform *experiments*, make *observations*, and *collect data*. Then we look at the data and think, and try one idea after another, in an attempt to work out some story, or *hypothesis*, that makes sense of the data. Usually in physics that hypothesis is phrased in mathematical form. So we concoct a description (which we might call a *model* or *theory*), often composed of a bunch of equations, that enables us to make a *prediction*. That is, we predict the result of performing a brand-new experiment with a given set of starting conditions. Then we perform the experiment and see if our prediction was correct.

Usually, when we're just getting started, we'll find that our prediction is wrong, by either a little or a lot. So we run the loop again, make more observations, gather more data, mess around with our math, make a new prediction, run the new experiment, and see what happens. We do that again and again, and eventually, with luck, persistence, and skill, we may end up with a body of math that makes accurate predictions.

For our study of quantum computing, we are fortunate because we have working devices available today that make it possible to run our own experiments, collect data, and develop theories. So, rather than me just telling you how measurement works, we'll discover it.

In this chapter, I'll present actual data. IBM provides limited but free online access to a few small quantum computers where you can run your own algorithms [95]. I used one of their five-qubit machines to run most of the algorithms in this chapter and produce the data in the figures.

The value of running quantum programs on real quantum hardware is that there is no simulation and no guesswork. It doesn't matter what our math might or might not say. We get to see what actually happens in the physical world.

Measuring Hello, World!

Let's start small, with the *Hello, World!* algorithm from Chapter 4. I've repeated the circuit on the left side of Figure 6-2, and the results I got from the actual hardware are shown on the right. To generate this figure, I ran the algorithm 1,024 times. Each run is often called a *shot*.

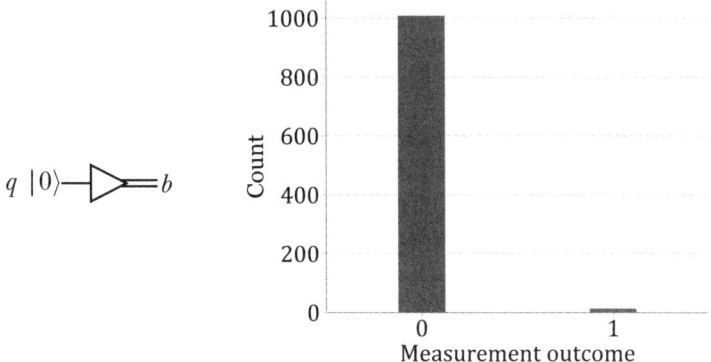

Figure 6-2: The Hello, World! *quantum circuit, and the results of running this circuit 1,024 times*

The circuit starts with a quantum bit that I named q. It's initialized to the state $|0\rangle$ and then goes directly into a meter. I said earlier that every time the meter receives a qubit in state $|0\rangle$, it reports the bit 0 and the qubit leaves the meter in state $|0\rangle$. But if we look at the figure, we see that we don't *always* get 0. What's happening here?

The problem is that I described an idealized, perfect quantum computer, and the hardware we have now is not perfect (and may never be perfect). Every now and then, something goes wrong. The problems are due to a wide variety of internal and external influences that we often lump together and call *noise*. Sources of noise include heat, vibrations, flaws in the computer's components, a star exploding in another galaxy, someone sneezing near the hardware, a spike in the power supply, and a seemingly endless collection of other causes.

Classical computers also make errors. For example, the designers of the computers aboard the Apollo spacecraft that took humans to the moon and back in the 1960s and '70s knew that cosmic rays and other physical phenomena could disrupt the computer's memory and processing units, which could lead to giving the astronauts incorrect, and possibly fatal, output. To reduce this risk, the spacecraft carried three identical copies of the onboard computer. Every calculation was run simultaneously on all three computers, and the results were fed into yet another computer. If there was ever disagreement between the three inputs, this final computer chose the most popular answer and passed that on to the rest of the system and the astronauts. This approach was called *triple modular redundancy* [219] [280]. Today, errors in classical computers still happen regularly, due to many different causes, but these errors are detected and corrected by a wide variety of hardware and software techniques.

Finding and correcting errors in quantum computers is the focus of a field called *quantum error correction (QEC)*. We won't get into those techniques here. The important thing is that although the situation is improving, it's hard to get away from errors when using today's hardware. People some-times refer to our current era of quantum computing as the *NISQ* era, an acronym that stands for *noisy intermediate-scale quantum* [273]. This means that most of our quantum computers are of an "intermediate" size (usually, up to a few hundred qubits) and make uncorrected errors due to noise.

One way to compensate for noise is to run a circuit many times. The hope is that the noise will be *uncorrelated*, or unrelated, from one run to the next. Generally speaking, we hope that while some errors will probably creep in, no one specific type of error will happen significantly more fre-quently than any other. Thus, the theoretically correct result will be seen most frequently, amid a sea of less frequent erroneous results.

That's just what we're seeing in the graph in Figure 6-2, where after 1,024 shots, the theoretical answer 0 dominates the erroneous 1. We've suc-cessfully run *Hello, World!* on an actual quantum computer! From now on, while every result will have some noise, I'll focus on the dominant results.

Measuring $X \ket{0}$

Suppose we expand our *Hello, World!* circuit from Figure 6-2 by inserting an X qugate, as shown on the left side of Figure 6-3. The input to the meter should be $\ket{1}$ every time, and thus the meter should report 1 every time. Though there's some noise, the graph on the right side of Figure 6-3 con-firms this expectation.

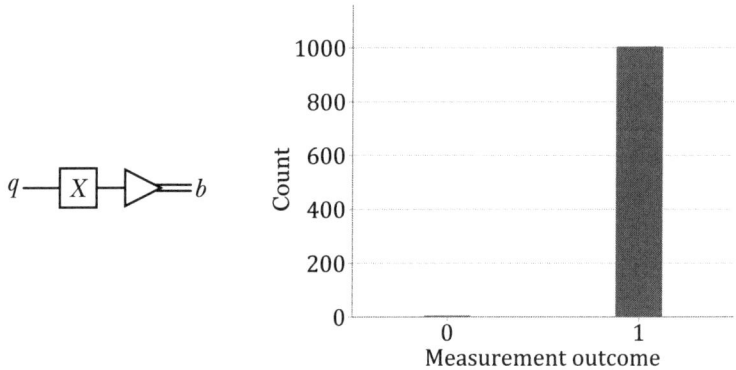

Figure 6-3: Sending $\ket{1}$ to the meter, and the results of running this circuit 1,024 times

Measuring $H \ket{0}$

Now let's get a little more ambitious and replace the X qugate with an H. This means the meter will receive the superposition $H \ket{0} = \ket{+}$. The circuit and its output are shown in Figure 6-4.

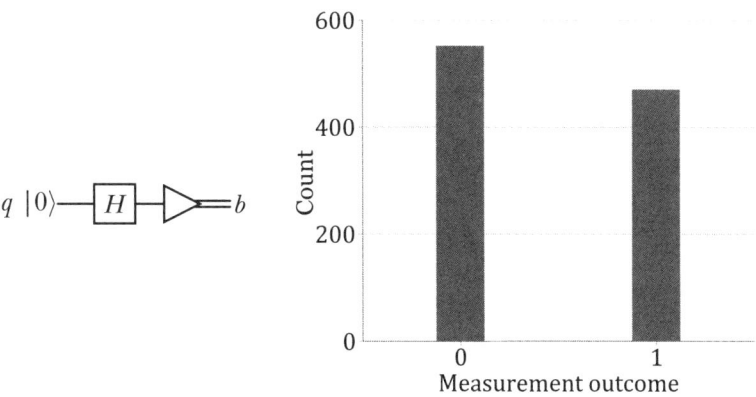

Figure 6-4: Inserting an H qugate into the circuit of Figure 6-2 before measurement, and the results of running this circuit 1,024 times

Does this make sense? The output of $H\left|0\right\rangle$ is shown in Equation 6.1.

$$H\left|0\right\rangle = \left|+\right\rangle = \vee\left|0\right\rangle + \vee\left|1\right\rangle \tag{6.1}$$

This looks symmetrical to me, so I'd expect symmetrical measurements as well. Why do we see so many more $\left|0\right\rangle$ results than $\left|1\right\rangle$?

Remember that we're trying to discover how measurement works. When we get surprising data, we need to look more closely at it to either confirm the surprise (and then account for it) or declare it to be an anomaly.

To decide how much we should trust these results, let's run the circuit many more times and collect all that new data. I ran it for 20,000 shots, getting the results in Figure 6-5(a).

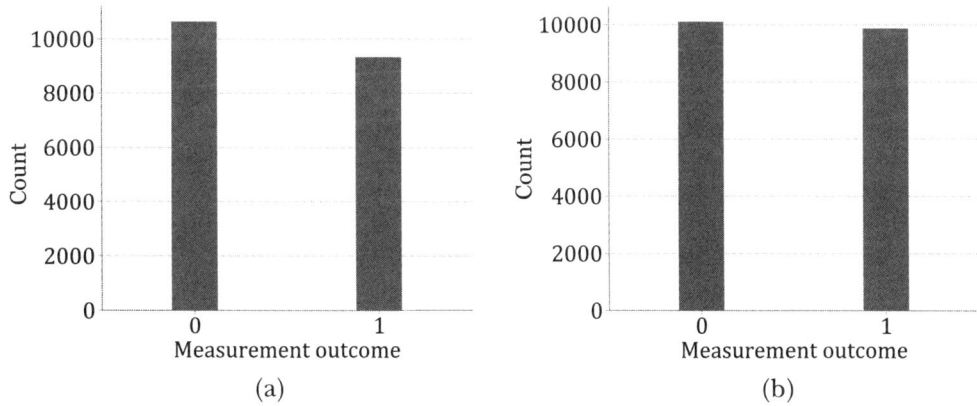

Figure 6-5: (a) Running the circuit of Figure 6-4 for 20,000 shots on a five-qubit computer. (b) Running it on a seven-qubit computer.

That's not much more symmetrical. Perhaps part of the problem is due to known sources of error that we can address. Today's quantum computers are highly sensitive instruments that can require regular *calibration* to keep their errors in check. So, I ran the same circuit on a larger, seven-qubit computer that had been more recently calibrated. This produced the results in Figure 6-5(b).

As you can see, measuring $|+\rangle$ on a larger, more recently calibrated quantum computer for 20,000 shots returns about equal numbers of 0 and 1 measurements. I'll take that as our most accurate observation for this experiment.

But what is this telling us? Why is measuring the superposition returning a roughly equal number of 0 and 1 outputs? Let's dig deeper.

Measuring $HH\,|0\rangle$

We don't observe superpositions themselves. Whether or not they are real depends on whether you think the math is itself reality or just a tool we use to describe it.

—Jon Cartwright, *An Inventory of the Quantum Realm*, 2023 [31]

We can use our actual hardware to look more closely at the nature of a superposition. That is, when we say that a physical quantum object has the state $|+\rangle = \vee(|0\rangle + |1\rangle)$, what does that really mean? And what happens when we measure it? As a first hypothesis, let's suppose that this is a mathematical way of writing our own uncertainty.

As an analogy, suppose that we have just flipped a coin and it has landed on a table in front of us. Before we look at it, we know that it's showing either heads or tails. That's the reality of the coin, independent of us or our observations. When we do finally look at the coin to see if it's showing heads or tails, we only discover the coin's preexisting condition. Our knowledge increased, but our observation had no effect on the coin's status. Objectively, in the real world, the coin was either showing heads or tails. Our choice of when (or even if) to look is irrelevant to the coin's real, existing status.

In the same way, the approach we're tentatively considering says that our qubit is not actually in a superposition, but rather it's in one of the basis states $|0\rangle$ or $|1\rangle$. Then, writing its state as a superposition is just a mathematical way to say that we haven't looked yet, so we don't know which of these states it's in.

Let's test this hypothesis! I'll make a circuit with two H qugates in a row, as in Figure 6-6. The right-hand side of this figure shows the results of running the circuit 1,024 times on a five-qubit computer.

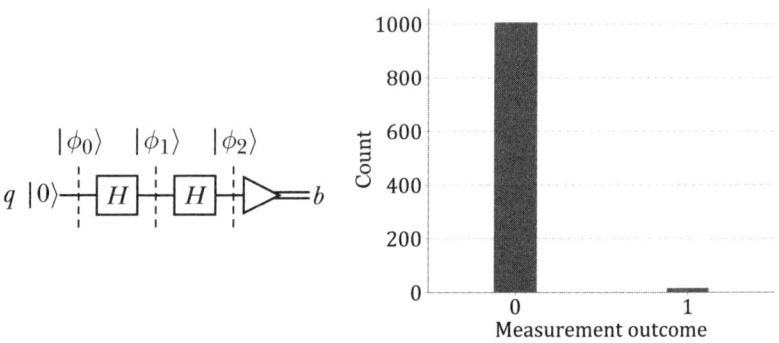

Figure 6-6: The circuit of Figure 6-4 with a second H qugate, and the measurements from 1,024 shots

The theory we're considering says that $|\phi_1\rangle$ is definitely either $|0\rangle$ or $|1\rangle$, just as the coin was either showing heads or tails, but we don't know which. Let's suppose it's $|0\rangle$. Then, the value $|\phi_2\rangle$ going into the meter is given by Equation 6.2.

$$|\phi_2\rangle = H |\phi_1\rangle = H |0\rangle = |+\rangle \tag{6.2}$$

So this theory predicts that the meter is receiving $|+\rangle$. We just ran that experiment and got the results in Figures 6-4 and 6-5. Compare those to the results of our new circuit, shown on the right side of Figure 6-6. Even in the presence of noise, they aren't close. This looks just like Figure 6-2, which told us that (ignoring noise) we always measure $|0\rangle$.

These observations on real-world devices tell us that our tentative interpretation of $\vee(|0\rangle + |1\rangle)$ must be wrong. This cannot describe something like the coin, which is either $|0\rangle$ or $|1\rangle$, and we just don't know which. This superposition state is . . . something else.

This is all good news! Using actual hardware, we've learned something important. A qubit in superposition is nothing like a coin on a table. It is not either $|0\rangle$ or $|1\rangle$, with only our ignorance preventing us from knowing which.

Instead, a superposition is some kind of subtle, novel kind of existence involving those two states. And measuring it isn't as simple as looking at a coin on a table.

Let's see if we can figure out what happens when we measure a superposition by running more experiments, observing their output, and then thinking about that data to form a theory.

Measuring an Unequal Superposition

In the state $|+\rangle$, both $|0\rangle$ and $|1\rangle$ have the same amplitude, $\vee$. And we saw that in the long run, with the seven-qubit computer, we measured 0 and 1 about the same number of times. Could the measurement results be related to the amplitudes?

Let's try measuring a qubit in an unequal superposition (that is, where the amplitudes aren't the same) and see what happens. I'll arbitrarily pick the superposition in Equation 6.3 (we'll soon see why those square roots are in there).

$$|\psi\rangle = \sqrt{0.4} |0\rangle + \sqrt{0.6} |1\rangle \tag{6.3}$$

I wrote a small circuit that initializes a qubit in this state and then immediately feeds that into a meter. Let's run this program many times and plot the running ratio of the results. For this experiment, I used IBM's high-quality quantum simulator [95].

Figure 6-7 shows the results for two different runs of 100 and 1,000 shots. The dotted line shows the running percentage of measurements that were 0, and the solid line shows the same thing for 1.

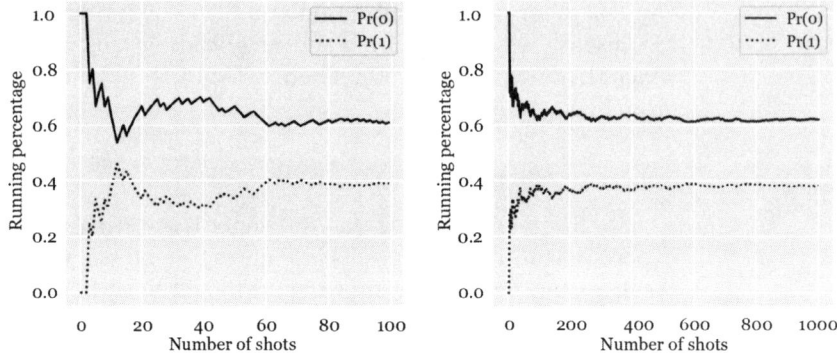

Figure 6-7: Measuring $|\psi\rangle = \sqrt{0.4}\,|0\rangle + \sqrt{0.6}\,|1\rangle$ on a quantum simulator; (left) 100 shots, (right) 1,000 shots

Remember, right now we're trying to find a hypothesis. Our task is to look at the data and devise an explanation, which we'll then test.

When I look at the results in Figure 6-7, it seems that the probability of measuring a 0 is settling in at about 0.4 and the probability of measuring a 1 is about 0.6.

Let's do a quick probability check. Every measurement produces one of two possible results, so the sum of these probabilities must be 1. Sure enough, 0.4 + 0.6 = 1, so these results make sense so far.

I also notice the seeming coincidence that the amplitude of $|0\rangle$ is $\sqrt{0.4}$ and the probability of measuring a 0 is about 0.4. Furthermore, the amplitude of $|1\rangle$ is $\sqrt{0.6}$, and the probability of measuring a 1 is about 0.6.

So, I'll suggest the hypothesis (that is, I'll guess) that the probability of measuring either state is given by the square of its amplitude.

Let's test this hypothesis. I'll try the new state given in Equation 6.4.

$$|\psi\rangle = \sqrt{0.75}\,|0\rangle + \sqrt{0.25}\,|1\rangle \qquad (6.4)$$

I predict we'll find that the probability of measuring state $|0\rangle$ will be $(\sqrt{0.75})^2 = 0.75$, and the probability of measuring state $|1\rangle$ will be given by $(\sqrt{0.25})^2 = 0.25$. The output of the simulator is shown in Figure 6-8.

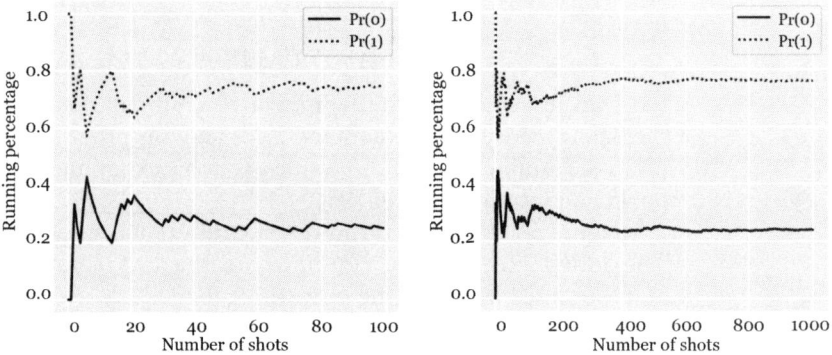

Figure 6-8: Measuring the superposition $|\psi\rangle = \sqrt{0.75}\,|0\rangle + \sqrt{0.25}\,|1\rangle$ on a quantum simulator; (left) 100 shots, (right) 1,000 shots

Yes! I'll tentatively accept this hypothesis as true. In the real world, we'd run many more experiments, but you and I both know the answer we're working toward, and I'm not leading you astray.

Does our theory work if the amplitudes are complex numbers? Suppose that we replace the amplitude of $|0\rangle$ with $i\sqrt{0.75}$, as in Equation 6.5.

$$|\psi\rangle = i\sqrt{0.75}\,|0\rangle + \sqrt{0.25}\,|1\rangle \tag{6.5}$$

Because the amplitude of $|1\rangle$ hasn't changed, our hypothesis says that the probability of getting a 1 is still 0.25, and so the probability of getting a 0 must still be 0.75. That is, the coefficient $i\sqrt{0.75}$ has to give us the same probability that we got from $\sqrt{0.75}$, somehow making the factor of i irrelevant.

The same argument lets us also put an i in front of the amplitude of $|1\rangle$. Since the probability of getting a $|0\rangle$ is still 0.75, the probability of getting a 1 from the amplitude $i\sqrt{0.25}$ must also still be 0.25.

Do we need to refine our hypothesis so that $\sqrt{0.25}$ and $i\sqrt{0.25}$ both give us a probability of 0.25? Yes, because $(i\sqrt{0.25})^2 = -0.25$. Not only does a negative probability make no sense, it also differs from the positive 0.25 that we measured in Figure 6-8. We need to get rid of that minus sign.

There are many different mathematical operations we could try to remove the influence of that i, but rather than go down lots of dead ends, I'll jump to the approach that works best in the long run. We'll find the probability of measuring 0 or 1 by *squaring the magnitude* of the amplitudes of $|0\rangle$ and $|1\rangle$, respectively, rather than the amplitudes themselves. For a complex amplitude α, that squared magnitude is $|\alpha|^2$.

Recall from Equation 2.11 on page 39 that $|\alpha|^2 = \overline{\alpha}\alpha$. We can also compute the magnitude explicitly from the components of α in either rectangular or polar form.

To stress-test this new wrinkle to our hypothesis, I'll make up a state using arbitrary complex numbers for α and β. For no reason, I chose $\alpha = 0.3 + 0.7i$ and 0.4 for the real part of β. We know that the probabilities have to add up to 1 (that is, $|\alpha|^2 + |\beta|^2 = 1$), so we can find the imaginary part of β by multiplying everything out. Then $\sqrt{1 - (0.3^2 + 0.7^2 + 0.4^2)} \approx 0.5$, so I'll write β as $0.4 + 0.5i$. Our test state $|\psi\rangle$ is given in Equation 6.6.

$$|\psi\rangle = (0.3 + 0.7i)\,|0\rangle + (0.4 + 0.5i)\,|1\rangle \tag{6.6}$$

Let's predict what we'll see. The amplitude of $|0\rangle$ is $\alpha = 0.3 + 0.7i$. Computing $\overline{\alpha}\alpha$ gives us $|\alpha|^2 = 0.58$. The amplitude of $|1\rangle$ is $\beta = 0.4 + 0.5i$, and $\overline{\beta}\beta$ gives us $|\beta|^2 = 0.42$. As we'd expect, the probabilities add up to 1.

This tells us that if we prepare many qubits in the state $|\psi\rangle$ and measure them, about 58 percent of the measurements will be 0 and about 42 percent will be 1. Figure 6-9 shows the results.

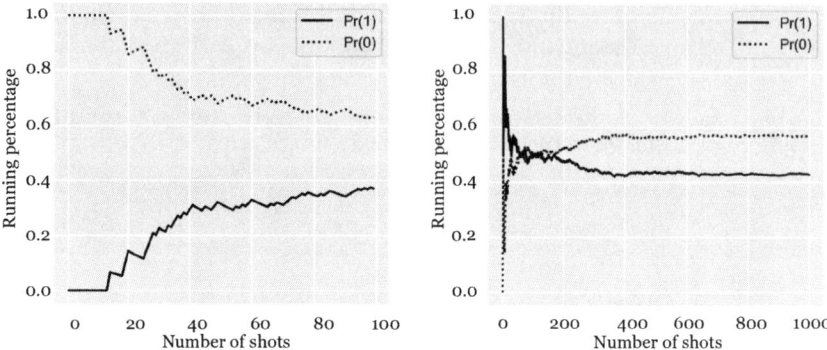

Figure 6-9: Measuring the state $(0.3 + 0.7i) \, |0\rangle + (0.4 + 0.5i) \, |1\rangle$; (left) 100 shots, (right) 1,000 shots

Our prediction matches the experiment!

These few examples are not remotely numerous, precise, or rigorous enough to let us claim that our hypothesis is correct. But in the century or so since this math was first proposed, it has been tested countless times, and, except for noise, it has always correctly predicted actual measurements.

Let's summarize this discussion. Given a state $|\psi\rangle = \alpha \, |0\rangle + \beta \, |1\rangle$, I'll write the probability of measuring a 0 as $\mathrm{Pr}_{|\psi\rangle}(0)$ and the probability of measuring a 1 as $\mathrm{Pr}_{|\psi\rangle}(1)$. Then, the probabilities of measuring these outputs for a state $|\psi\rangle = \alpha \, |0\rangle + \beta \, |1\rangle$ can be written as in Equation 6.7.

$$\mathrm{Pr}_{|\psi\rangle}(0) = |\alpha|^2$$
$$\mathrm{Pr}_{|\psi\rangle}(1) = |\beta|^2 \tag{6.7}$$

Nobody can say exactly why Equation 6.7 describes actual observations. Maybe this math is modeling some mechanism in nature that we're not aware of. Maybe something else is going on. As of today, it's a beautiful mystery. But whatever the reason, this math consistently predicts how nature behaves. Chalk one up for mathematics!

The probabilities expressed in Equation 6.7 are the consequence of something called the *Born rule* [254]. These probabilities must be *normalized*, or add up to 1. After all, *something* has to happen as the result of a measurement. We can write this normalization requirement in Equation 6.8.

$$|\alpha|^2 + |\beta|^2 = 1 \tag{6.8}$$

Nature doesn't have to obey our logic, but every test of quantum theory has upheld this relationship. We sometimes say that nature only *allows* quantum states that satisfy this normalization rule and *does not allow* any others.

Amplitudes from Projection

In the previous section, I referred to the amplitudes of a state $|\psi\rangle$ as α and β. That's fine, but it won't scale well when we move to systems with lots of qubits.

Measuring multiple qubits is going to require a few new ideas, so this section will be a little denser than usual. I'll explain what's going on as we go, so take your time.

Let's find a systematic way to write the coefficients (or amplitudes) in qubit values without having to give them unique names. Because α and β are the coefficients of $|\psi\rangle$ with respect to the basis states $|0\rangle$ and $|1\rangle$, we can find them by *projecting* $|\psi\rangle$ onto each of these states.

Recall from Equation 2.56 that we can write the projection of state *from* onto state *onto* with the inner product $\langle onto, from\rangle$. Equation 6.9 uses the braket form of this operation to project $|\psi\rangle$ onto $|0\rangle$.

$$
\begin{aligned}
\langle 0|\psi\rangle &= \langle 0|(\alpha\,|0\rangle + \beta\,|1\rangle)) && \text{Expand } |\psi\rangle \\
&= \langle 0|\,\alpha\,|0\rangle + \langle 0|\,\beta\,|1\rangle && \text{Distribute } \langle 0| \\
&= \alpha\underbrace{\langle 0|0\rangle}_{1} + \beta\underbrace{\langle 0|1\rangle}_{0} && \text{From orthonormality, } \langle j|k\rangle = \delta_{j,k} \quad (6.9)\\
&= \alpha
\end{aligned}
$$

In the third line, I used the fact that the computational basis is orthonormal, so the inner product of any basis state with itself is 1 and the inner product with any other basis state is 0. Equation 6.9 tells us that $\langle 0|\psi\rangle$ is another name for α. Computing $\langle 1|\psi\rangle$ following the same process shows us that it's another name for β.

Because $\langle 0|\psi\rangle$ gives us α, and the probability of obtaining a measurement of 0 from $|\psi\rangle$ is $|\alpha|^2$, we can put these two statements together. The probability of measuring state k (where k is either 0 or 1) is given by Equation 6.10.

$$
\Pr\nolimits_{|\psi\rangle}(k) = |\langle k|\psi\rangle|^2 \qquad (6.10)
$$

This is a valuable result, because not only does it tell us the probabilities of measuring $k = 0$ or $k = 1$ for a single qubit, but it also gives us the probability of measuring a larger binary number k from a system of qubits. We'll get to that soon.

For now, let's take a closer look at Equation 6.10. I'll expand the right side and regroup, as shown in Equation 6.11.

$$
\begin{aligned}
|\langle k|\psi\rangle|^2 &= \overline{\langle k|\psi\rangle}\,\langle k|\psi\rangle && \text{Use Eq. 2.11} \\
&= \langle \psi|k\rangle\,\langle k|\psi\rangle && \text{Use Eq. 2.75} \\
&= \langle \psi|\left(\,|k\rangle\langle k|\,\right)|\psi\rangle && \text{Regroup}
\end{aligned} \qquad (6.11)
$$

What's that term in the middle, between the parentheses?
Let's take a closer look.

The Outer Product

Since both bras and kets are matrices, an expression like $|k\rangle \langle k|$ is a matrix multiplication. To understand its result, consider the general case of multiplying two matrices. If the first matrix is of shape m by n (that is, it has m rows and n columns), then the second matrix must be of shape n by p (that is, n rows and p columns). We might write this multiplication in terms of shapes as $(m, n) \times (n, p)$. Matrix multiplication requires that the "inner" dimensions (here, n) must be the same. The resulting matrix has a size given by the "outer" dimensions, or m by p. Let's apply this idea to the products of bras and kets.

When we formed an inner product of a bra and a ket, because the bra was a 1-by-d row matrix and the ket was a d-by-1 column matrix, we got back a matrix with only a single element (which we usually interpret as a number, rather than a matrix of size 1 by 1).

When the order is reversed, we're multiplying a ket of shape d by 1 with a bra of shape 1 by d, giving us a matrix with the shape d by d.

We can visualize this construction with a picture. Equation 6.12 shows the basic idea for a ket $|\psi\rangle$ with components a and b and a bra $\langle\phi|$ with components c and d.

$$|\psi\rangle\langle\phi| \to \begin{bmatrix} a \\ b \end{bmatrix}\begin{bmatrix} c & d \end{bmatrix}\begin{bmatrix} \cdot & \cdot \\ \cdot & \cdot \end{bmatrix} \to \begin{bmatrix} a \\ b \end{bmatrix}\begin{bmatrix} c & d \end{bmatrix}\begin{bmatrix} ac & ad \\ bc & bd \end{bmatrix} \to \begin{bmatrix} ac & ad \\ bc & bd \end{bmatrix} \quad (6.12)$$

In words, we make an empty 2-by-2 matrix, place $|\psi\rangle$ to its left, and $\langle\phi|$ above. Now each term in the matrix is formed by the product of the element in the row matrix to its left and the element in the column matrix above it.

We call an expression of the form $|\psi\rangle\langle\phi|$ the *outer product*, or *ketbra*, of $|\psi\rangle$ and $|\phi\rangle$. As we've seen, *the outer product is a matrix*.

Let's return to Equation 6.11 and write the matrix for the outer product $|k\rangle\langle k|$, as shown in Equation 6.13.

$$\mathrm{Pr}_{|\psi\rangle}(k) = \langle\psi| \left(|k\rangle\langle k| \right) |\psi\rangle \qquad \text{Eq. 6.11}$$

$$\approx \langle\psi| \left(\begin{bmatrix} k_0 \\ k_1 \end{bmatrix}\begin{bmatrix} \overline{k_0} & \overline{k_1} \end{bmatrix}\begin{bmatrix} \cdot & \cdot \\ \cdot & \cdot \end{bmatrix} \right) |\psi\rangle \qquad \begin{array}{l} \text{Structure of the} \\ \text{outer product} \end{array} \quad (6.13)$$

$$= \langle\psi| \begin{bmatrix} k_0\overline{k_0} & k_0\overline{k_1} \\ k_1\overline{k_0} & k_1\overline{k_1} \end{bmatrix} |\psi\rangle \qquad \text{Compute outer product}$$

This shows an interpretation of Equation 6.11 as the product of three matrices (not four): the bra $\langle\psi|$, the outer product $|k\rangle\langle k|$, and the ket $|\psi\rangle$.

Now that we know how to form the outer product, let's take a deeper look at it. I'll start with a few of its useful properties, and then I'll show how the outer product lets us define a new operator that will play a central role in writing quantum measurements.

One useful property arises when each state in an outer product has dimensionality d. In this case, the outer product $|\psi\rangle\langle\phi|$ is a square matrix of dimensions d by d. If that matrix is unitary, we could use it as an operator.

If the two states we're using are both basis states, their outer product matrix takes on a special, simple form. Let's find the outer product of $|r\rangle\langle c|$, where c and r are integers, so $|c\rangle$ and $|r\rangle$ are both basis states. The result, shown in Figure 6-10, is a matrix that is 0 everywhere except for a single 1 at row r and column c.

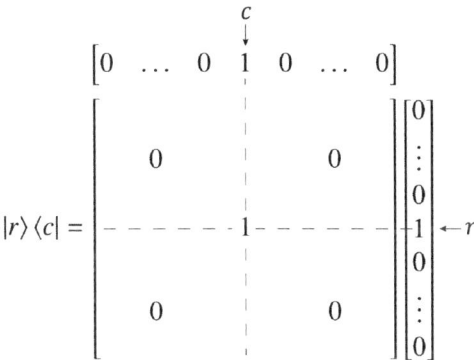

Figure 6-10: The outer product of the basis states $|r\rangle$ and $|c\rangle$

We can express this relationship in a few other useful ways.

If we "sandwich" a matrix M between $\langle r|$ and $|c\rangle$, we extract the element $\mu_{r,c}$. This is like running the picture in Figure 6-10 in reverse. We can write this in Equation 6.14. Recall that I write the elements of matrix M as μ to emphasize that they're complex numbers.

$$\langle r|\, M\, |c\rangle = \mu_{r,c} \tag{6.14}$$

I sometimes call an expression like $\langle r|\, M\, |c\rangle$ a *bramket*, because it's a braket with a matrix (generically M) in the middle (hey, the term "braket" itself is a pun!).

If $M = |r\rangle\langle c|$, then we can confirm that element $\mu_{c,r}$ is 1, as shown in Equation 6.15.

$$\langle r|\,M\,|c\rangle = \langle r|\,\underbrace{\big(\,|r\rangle\langle c|\,\big)}_{M}\,|c\rangle \quad\text{Because we said } M = |r\rangle\langle c|$$

$$= \underbrace{\langle r|r\rangle}_{1}\,\underbrace{\langle c|c\rangle}_{1} \qquad\qquad\text{Regroup} \qquad\qquad (6.15)$$

$$= 1 \qquad\qquad\qquad\qquad\text{Basis states have magnitude 1}$$

In the third line, I used the fact that in this book we only use computational basis states, which are orthonormal, so the inner product of any basis state with itself is 1. We can show that every other element is 0 in the same way, as the inner product of any basis state with any other basis state is 0. Let's do this. I'll replace $|r\rangle$ with $|k\rangle$ (where $r \neq k$) while keeping the same M, giving us Equation 6.16.

$$\langle k|\,M\,|c\rangle = \langle k|\,\underbrace{\big(\,|r\rangle\langle c|\,\big)}_{M}\,|c\rangle = \underbrace{\langle k|r\rangle}_{0}\,\underbrace{\langle c|c\rangle}_{1} = 0 \qquad\qquad (6.16)$$

We can summarize this result for any c and r in the single statement of Equation 6.17.

$$\langle r|\,\big(\,|r\rangle\langle c|\,\big)\,|c\rangle = \langle r|\,M\,|c\rangle = \delta_{r,c} \qquad\qquad (6.17)$$

Finally, we come to the special case when c and r have the same value. Let's call that k. Then we're finding $|k\rangle\langle k|$, which is what got us started on this topic in Equation 6.11. This outer product is a matrix that is 0 everywhere, except for a single 1 at $\mu_{k,k}$.

Particularly useful instances of this special case are the matrices when $k = 0$ and $k = 1$, because those correspond to our basis states $|0\rangle$ and $|1\rangle$. I'll give these matrices the special names Π_0 and Π_1. (The capital Greek letter pi, Π, is conventional in the field for these matrices. As we'll see later, the choice of the Greek equivalent of "P" is motivated because these are called *projection* matrices.) Their definitions are given in Equation 6.18.

$$\Pi_0 \overset{\Delta}{=} |0\rangle\langle 0| = \begin{bmatrix} 1 & 0 \\ 0 & 0 \end{bmatrix}, \qquad \Pi_1 \overset{\Delta}{=} |1\rangle\langle 1| = \begin{bmatrix} 0 & 0 \\ 0 & 1 \end{bmatrix} \qquad (6.18)$$

More generally, the outer product of any basis state $|k\rangle$ with itself is named Π_k and is defined as in Equation 6.19.

$$\Pi_k \overset{\Delta}{=} |k\rangle\langle k| \qquad\qquad (6.19)$$

Projection operators (and their matrix representations) have a special property: Applying them two or more times is the same as applying them once. That is, once the operator has done its job, repeated applications make no additional changes. It's like snuffing out a candle flame by pinching the wick with your fingers. Once you've extinguished the flame, pinching the wick again doesn't change anything.

If this reminds you of the projection operation we looked at in Chapter 2, you're on the right track! We'll see that these projection matrices do a job much like our projection of an arrow onto another arrow.

There's a formal term for operators that, once applied, make no further changes on repeated applications. We call them *idempotent*. We can capture the idea symbolically for any projection matrix Π as in Equation 6.20, which says that applying Π_k twice (or three times, or four, or more) is the same as applying it once.

$$\Pi_k{}^2 = \Pi_k \qquad (6.20)$$

We can see why this is so in Equation 6.21.

$$\Pi_k{}^2 = \Pi_k \Pi_k \qquad \text{Expand } \Pi_k{}^2$$

$$= \Big(|k\rangle\langle k| \Big) \Big(|k\rangle\langle k| \Big) \qquad \text{Write each } \Pi_k \text{ as in Eq. 6.19}$$

$$= |k\rangle \underbrace{\langle k|k\rangle}_{1} \langle k| \qquad \text{Regroup center terms} \qquad (6.21)$$

$$= |k\rangle\langle k| \qquad \text{Because } \langle k|k\rangle = 1$$

$$= \Pi_k \qquad \text{From Eq. 6.19}$$

There's another observation that will prove useful: A projection matrix is the same as its adjoint. The reason why is shown in Equation 6.22.

$$\Pi_k{}^\dagger = \overline{\Pi_k}{}^T \qquad \text{Definition of the adjoint}$$

$$= \Pi_k{}^T \qquad \text{Because every element of } \Pi_k \text{ is real} \qquad (6.22)$$

$$= \Pi_k \qquad \text{Because } \Pi_k \text{ is 0 off of the main diagonal}$$

Remember that we can use a square matrix as an operator in a quantum circuit only if it's unitary. As you may have noticed, *projection matrices are not unitary*. We can see that because they cannot be undone, or reversed. If we apply a projection matrix such as Π_0 or Π_1 to a two-element quantum state, either α or β goes to 0. There's no way to undo that operation and recover the original value of that coefficient. Thus, we cannot use projection matrices as a computational part of a quantum algorithm.

So why have we just discussed them? It's because these matrices will be just right for representing the meters that can come at the very end of a quantum circuit. As the meters aren't part of the computational process, it's okay to model them with operators that can't be used for computation.

Because we'll use projection matrices to model measurements, let's get a little more experience with them. I'll see if we can use them to find the vector representing the projection of some state $|\psi\rangle$ onto some basis state $|k\rangle$.

As we've seen, we can write any state $|\psi\rangle$ as a sum of weighted basis states, as Equation 6.23.

$$|\psi\rangle = \sum_{k\in[d]} \psi_k |k\rangle = \sum_{k\in[d]} |k\rangle \psi_k \qquad (6.23)$$

Each coefficient ψ_k is the amount of $|\psi\rangle$ that is due to basis vector $|k\rangle$. In other words, it's the projection of $|\psi\rangle$ onto $|k\rangle$, or $\langle k|\psi\rangle$. Multiplying that by $|k\rangle$ gives us $|k\rangle\langle k|\psi\rangle$, the scaled version of that basis vector. I've written that expression out in tableau form in Equation 6.24. I compute the inner product first by multiplying the two rightmost matrices, and then I use that to scale $|k\rangle$. In each of these matrices, the 1 element is at position k.

$$|k\rangle\langle k|\psi\rangle = \begin{bmatrix} 0 \\ \vdots \\ 1 \\ \vdots \\ 0 \end{bmatrix} \begin{bmatrix} 0 & \cdots & 1 & \cdots & 0 \end{bmatrix} \begin{bmatrix} \psi_0 \\ \vdots \\ \psi_k \\ \vdots \\ \psi_{d-1} \end{bmatrix} \qquad \text{Write out the matrices}$$

(6.24)

$$= \begin{bmatrix} 0 \\ \vdots \\ 1 \\ \vdots \\ 0 \end{bmatrix} \psi_k = \begin{bmatrix} 0 \\ \vdots \\ \psi_k \\ \vdots \\ 0 \end{bmatrix} = \psi_k\,|k\rangle \qquad \text{Multiply right matrices}$$

And that's our projection. Now let's try it again, but this time using outer products. Instead of computing the inner product $\langle k|\psi\rangle$ first by multiplying the matrices on the right, in Equation 6.25 I'll compute the outer product $|k\rangle\langle k|$ (or Π_k) first by multiplying the matrices on the left. The 1 element in the outer product matrix is at position k, k.

$$|k\rangle\langle k|\psi\rangle = \begin{bmatrix} 0 & \cdots & & 0 \\ & & & \\ \vdots & & 1 & \vdots \\ & & & \\ 0 & \cdots & & 0 \end{bmatrix} \begin{bmatrix} \psi_0 \\ \vdots \\ \psi_k \\ \vdots \\ \psi_{d-1} \end{bmatrix} \qquad \text{Form outer product } |k\rangle\langle k| = \Pi_k$$

(6.25)

$$= \begin{bmatrix} 0 \\ \vdots \\ \psi_k \\ \vdots \\ 0 \end{bmatrix} = \psi_k\,|k\rangle \qquad \text{Multiply matrices}$$

Both approaches give us the same result, the basis vector $|k\rangle$ scaled by ψ_k.

As I mentioned, we're going to use the outer product to create the operators that let us make measurements. Let's get right into that.

Back to Measurement

Recall that Equation 6.10 showed us that when we measure a qubit, the probability of the meter reporting a 0 or a 1 depended on the state's amplitudes on the basis states $|0\rangle$ and $|1\rangle$.

We expanded Equation 6.10 into Equation 6.11, which introduced the outer product $|k\rangle\langle k|$. Now we can use Equation 6.19 to replace that $|k\rangle\langle k|$ with Π_k, giving us Equation 6.26. This tells us the probability of measuring output k if we measure $|\psi\rangle$.

$$\Pr{}_{|\psi\rangle}(k) = \langle\psi|\,\Pi_k\,|\psi\rangle \qquad (6.26)$$

If Equation 6.26 looks familiar, that's because it's the first statement in Postulate 4!

We went through a lot of work to write the probability of measuring k in this form. All that work will pay off when we see how nicely this generalizes to characterizing the probabilities of measuring systems of multiple qubits.

Now we can knock out the other two statements of Postulate 4 as well.

Statement 2 tells us that if a measurement has m possible outcomes, the sum of all the Π_k, one for each outcome, is the identity matrix I.

This is a generalization of the normalization rule, telling us that the sum of the probabilities for all measurements must be 1. But now our probabilities are those given by Equation 6.26. For a single qubit, there are only two projection matrices, Π_0 and Π_1. Using their definitions in Equation 6.18, we can add them together directly, as in Equation 6.27.

$$\Pi_0 + \Pi_1 = \begin{bmatrix} 1 & 0 \\ 0 & 0 \end{bmatrix} + \begin{bmatrix} 0 & 0 \\ 0 & 1 \end{bmatrix} = \begin{bmatrix} 1 & 0 \\ 0 & 1 \end{bmatrix} = I \qquad (6.27)$$

Later, we'll have systems of qubits where there are d possible outcomes. Remember that each matrix Π_k has a 1 at index (k, k) and is 0 everywhere else. So each Π_k contributes a single 1 along the main diagonal from the upper left to the lower right. If we sum up the d matrices Π_0 through Π_{d-1}, we'll get a d-by-d matrix that is 0 everywhere except for a 1 at each entry along the main diagonal. That's a description of the identity matrix!

The third statement of the postulate is another fact of nature, not something we can derive or work out by reasoning. That is, it's a statement of what the world shows us.

When we measure a system of qubits, each qubit enters its own meter. The output of that meter depends only on that qubit's probabilities. The result is that each meter produces one of two results that we call 0 and 1.

In addition to presenting a bit at the meter's output, another effect of measurement is that the qubit emerges in a state consistent with the meter's output. This means that every time we measure a 0, the qubit exits the meter in the state $|0\rangle$, and every time we measure a 1, the qubit exits the meter in the state $|1\rangle$. No matter what state a qubit is in when it enters the meter, it leaves as either $|0\rangle$ or $|1\rangle$, matching the 0 or 1 shown by the meter. That's what nature does. We'll see soon how the math models this.

We say that the superposition *collapses* to one of its basis states.

Be wary of interpreting this to mean that there's some kind of mechanism inside the meter that first collapses the qubit to $|0\rangle$ or $|1\rangle$ and then outputs the corresponding bit 0 or 1. Maybe this happens, or something like it. But maybe not. Nobody really knows the mechanism that causes superpositions to collapse upon measurement. All we do know is that, somehow, the qubit emerges as $|0\rangle$ or $|1\rangle$, and the meter reports the corresponding bit.

I'd like to write this experimentally observed behavior of the meter symbolically. I'll start by using the projection operators we met recently.

I said earlier that projection operators aren't used in quantum computations, and it's essential that you keep this in mind. *Projection operators are never used within quantum circuits. They appear only at the end.* (There are special exceptions to this rule, but we won't use them in this book.)

They're not qugates. As we discussed, they're not even unitary. They're a special mathematical operation that we use only for the purpose of computing probabilities, so we can characterize what outputs we're likely to see. This is vital to remember, so I'll say it again: *Projection operators are never used in quantum circuits.*

But! They are a convenient way to describe measurements that happen at the very end of these circuits. That's how we'll get to the third statement of Postulate 4.

I'll begin by observing that if we measure $|\psi\rangle$ and get outcome k (for a single qubit, that's either 0 or 1), then that output is *almost* the same as $\Pi_k |\psi\rangle$, as shown in Equation 6.28 for Π_0.

$$\Pi_0 \, |\psi\rangle = \begin{bmatrix} 1 & 0 \\ 0 & 0 \end{bmatrix} \begin{bmatrix} \alpha \\ \beta \end{bmatrix} = \begin{bmatrix} \alpha \\ 0 \end{bmatrix} = \alpha \, |0\rangle \qquad (6.28)$$

This is almost but not quite the same as just $|0\rangle$.

In the same way, $\Pi_1 \, |\psi\rangle = \beta \, |1\rangle$. We can more generally say that the operation $\Pi_k \, |\psi\rangle$ gives us a state of the same dimensionality as $|\psi\rangle$, whose elements are 0 everywhere except at location k, where it has value ψ_k, as illustrated in Figure 6-11.

$$\Pi_k \, |\psi\rangle = \begin{bmatrix} 0 \\ 0 \\ \vdots \\ \psi_k \\ \vdots \\ 0 \\ 0 \end{bmatrix} \longleftarrow \text{index } k$$

Figure 6-11: The operation $\Pi_k \, |\psi\rangle$

Unfortunately, unless $|\psi\rangle$ is itself a basis state, the result $\Pi_k \, |k\rangle$ will not have a magnitude of 1 and thus cannot exist in nature.

But that's easily fixed! To get the state $\psi_k \, |k\rangle$ to have a magnitude of 1, we need only divide it by ψ_k, the coefficient of $|\psi\rangle$ with respect to basis state $|k\rangle$. That will leave us with just $|k\rangle$, which is a basis vector and therefore has a magnitude of 1. If we know ψ_k we can just do the division, but what if we don't know that value? Can we still normalize the result without using an explicit value for ψ_k?

We know that ψ_k is the amplitude for state k, so $|\psi_k|^2$ is the probability that we'll observe outcome k. And we know, at this point in our discussion, that we *did* observe outcome k.

So, to turn any $\Pi_k \, |\psi\rangle$ into a physically realizable state, we can divide by the square root of $|\psi_k|^2$. The probability $|\psi_k|^2$ of measuring state k is given by $\langle \psi | \, \Pi_k \, |\psi\rangle$, as we saw in Equation 6.26. I've put all of this together in Equation 6.29.

$$\frac{\Pi_k \, |\psi\rangle}{|\Pi_k \, |\psi\rangle \, |} = \frac{\Pi_k \, |\psi\rangle}{\sqrt{\mathrm{Pr}_{|\psi\rangle}(k)}} = \frac{\Pi_k \, |\psi\rangle}{\sqrt{\langle \psi | \, \Pi_k \, |\psi\rangle}} \tag{6.29}$$

And that is the mathematical part of the third statement of Postulate 4. It's a lot of symbols to represent the basis state $|k\rangle$, which is the state the qubit has when it leaves a meter that reported k. In words, if we know that we have a meter output of k when measuring $|\psi\rangle$, then we can write this as a projection of $|\psi\rangle$ onto $|k\rangle$, or $\Pi_k \, |\psi\rangle$, divided by the amplitude of that projection, $\sqrt{\langle \psi | \, \Pi_k \, |\psi\rangle}$.

Since the end result of all this work is just the basis vector $|k\rangle$, why go through all the complicated stuff in Equation 6.29? The value of that form is that it generalizes to systems with multiple qubits. So let's look at such systems!

Measuring Multiple Qubits

We've seen how to measure one qubit. But one-qubit quantum computers aren't any more attractive than one-bit classical computers. So let's get some more qubits in there, and measure them all!

Figure 6-12 shows a circuit made up of three qubits. They all start as $|0\rangle$ and get processed by a quantum circuit, and then we measure each qubit.

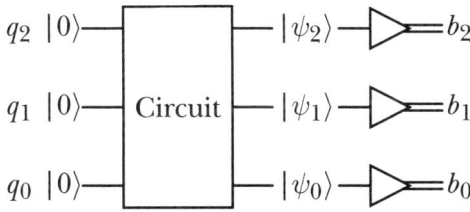

Figure 6-12: A three-qubit computation that starts as $|000\rangle$ and is operated upon; then we measure the output qubits

Assembling the output bits from top to bottom, we can combine them into the three-bit binary number $\mathbf{b} = b_2 b_1 b_0$. For example, if our measurements gave us 1 for b_2 and b_1 and 0 for b_0, that would correspond to the binary number 110, or 6 in our everyday (base 10) decimal notation.

This is why I've been numbering qubits in descending order as we move from the top line in a diagram to the bottom, putting q_0 on the bottom-most line.

You could number them the other way, but then if we measured a 6 at the output it would be represented by the binary string 011. Computer scientists say that putting the most significant bit at the left end, or writing 6 as 110, is the *big-endian* way to write the number, while writing the most significant bit at the right is the *little-endian* way. Most people use the big-endian approach, and that's what I use in this book. In case you're curious, the terms big-endian and little-endian come from the classic satire *Gulliver's Travels* [214]. One day, the son of the king of Lilliput cut his fingers while breaking an egg at the little end. In response, the king issued an edict that all citizens must henceforth break their eggs at the big end. A long and terrible war broke out between supporters of this law and rebels who still preferred to break their eggs at the little end. The terms were introduced to computer science as part of a plea to prevent a similarly meaningless catastrophe over how to write binary numbers [40].

Suppose we want the output of Figure 6-12 to be 5, or 101 in binary. Given the three states $|\psi_2\rangle$, $|\psi_1\rangle$, and $|\psi_0\rangle$, what's the probability that we'll get 101 from a measurement?

Since the measurements in this circuit come after all computation has been done, the qubits are no longer influencing one another. We say that they're now *independent* of one another.

The rules of probability tell us that the probability of getting multiple *independent* results (also called *uncorrelated* or *unrelated* results) is the product of the probability of each result. For example, if we want to know the probabilities that it's a sunny day, *and* that there's a reggae song playing on a radio nearby, *and* that we've recently had a haircut, then we find the probability for each of these individual conditions and multiply them together.

Because at this point our qubits are independent, the probability of measuring 101 from our circuit can be expressed as shown in Equation 6.30, where $\times$ means scalar multiplication.

$$\Pr(b_2 = 1) \times \Pr(b_1 = 0) \times \Pr(b_0 = 1) \tag{6.30}$$

Let's generalize this to a bitstring $\mathbf{b}$ of m bits. Each bit b_k is either 0 or 1. From Postulate 4 (or Equation 6.26), we can find the probability that each state $|\psi_k\rangle$ gives us a measured value b_k from $\langle \psi_k | \Pi_{b_k} | \psi_k \rangle$. Multiplying those probabilities together gives us Equation 6.31. (The large Π at the start is the multiplicative version of the big sigma Σ at the start of a summation. It works like that sigma, telling us to run a loop and multiply together all the elements it generates.)

$$\Pr_{|\psi\rangle}(\mathbf{b}) = \langle\psi_0|\,\Pi_{b_0}\,|\psi_0\rangle\,\langle\psi_1|\,\Pi_{b_1}\,|\psi_1\rangle\cdots\langle\psi_{m-1}|\,\Pi_{b_{m-1}}\,|\psi_{m-1}\rangle$$

$$= \prod_{k\in[m]}\langle\psi_k|\,\Pi_{b_k}\,|\psi_k\rangle \qquad (6.31)$$

Equation 6.31 tells us the probability that our output will, upon measurement, give us the binary number **b**, but it requires us to loop over every qubit. It would be nice to find an expression that doesn't require this explicit looping.

One way to interpret Equation 6.31 is that it's basically Figure 6-12 interpreted using our horizontal grouping strategy from Chapter 5. That's because we're not treating the qubits as a system. Instead, we're finding the probability that each qubit will produce the result we want, and then we're multiplying those probabilities together.

Let's see this visually. I've drawn the output of a circuit of m qubits in Figure 6-13(a). These m qubits have states $|\psi_{m-1}\rangle$ through $|\psi_0\rangle$. To compute the probability that each qubit will, upon measurement, produce the corresponding bit from **b**, I've shown each qubit going "into" a circle labeled with the projection matrix for that bit. Remember, *the projection matrices are not qugates*! They're just a tool we use for computing probabilities. This tells us that Figure 6-13(a) is not a quantum circuit diagram but only a visualization of how we compute a probability. That's also why I've put these operators in a circle rather than the boxes we use for qugates.

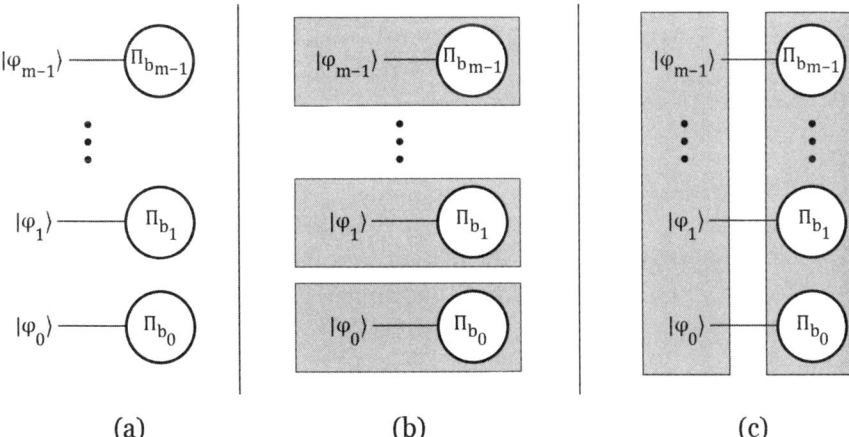

Figure 6-13: Computing the probability of the output of a circuit in three ways

In Figure 6-13(b), I've grouped the operations horizontally. This is a way to visualize Equation 6.31, explicitly running each qubit into its own projection matrix.

Let's compare this grouping with a vertical interpretation, as shown in Figure 6-13(c). We know that we can create a system of qubits by tensoring the states from top to bottom, as in Equation 6.32. I've given the resulting system state the name $|\phi\rangle$.

$$|\phi\rangle = |\psi_{m-1}\rangle \otimes |\psi_{m-2}\rangle \otimes \cdots \otimes |\psi_1\rangle \otimes |\psi_0\rangle \qquad (6.32)$$

Each projector Π can be written as a matrix, so we can tensor them together, too, in the same top-to-bottom order, as in Equation 6.33. I've given the resulting system state the name $\Pi_{\mathbf{b}}$.

$$\Pi_{\mathbf{b}} = \Pi_{b_{m-1}} \otimes \Pi_{b_{m-2}} \otimes \cdots \otimes \Pi_{b_1} \otimes \Pi_{b_0} \qquad (6.33)$$

We can summarize the vertical interpretation as telling us that $|\phi\rangle$ is a column matrix of dimension 2^m (that is, it contains 2^m complex numbers) and $\Pi_{\mathbf{b}}$ is a square matrix of 2^m elements on a side.

To find $\mathrm{Pr}_{|\phi\rangle}(\mathbf{b})$, the probability that the qubit system $|\phi\rangle$ will, when measured, produce the bitstring $\mathbf{b}$, we need only apply Postulate 4. The expression is shown in Equation 6.34.

$$\mathrm{Pr}_{|\phi\rangle}(\mathbf{b}) = \langle\phi|\,\Pi_{\mathbf{b}}\,|\phi\rangle \qquad (6.34)$$

In other words, when we multiply together three matrices (a row, a square, and a column), the result is a number between 0 and 1 indicating the probability that, if we measure a qubit system in the state $|\phi\rangle$, we'll get the bitstring $\mathbf{b}$. One bramket does the whole job.

This is why we went to all of the trouble of getting to Equation 6.29. In addition to being part of Postulate 4, it also paved the way to this compact and efficient way to find the probability of measuring the bitstring $\mathbf{b}$ from multiple qubits.

I promised you that writing Postulate 4 in a general form was going to pay off!

What state are the qubits in, after measurement, if the results match bitstring $\mathbf{b}$? It's the state $|\phi'\rangle$ made by tensoring together $|0\rangle$ and $|1\rangle$ according to the bits in $\mathbf{b}$. If we want to be formal about it, we can use the third statement in Postulate 4 to write that state, as in Equation 6.35.

$$|\phi'\rangle = \frac{\Pi_{\mathbf{b}}\,|\phi\rangle}{\sqrt{\langle\phi|\,\Pi_{\mathbf{b}}\,|\psi\rangle}} \qquad (6.35)$$

The state $|\phi'\rangle$ is the quantum state equivalent of the bitstring $\mathbf{b}$. Because each qubit is now either $|0\rangle$ or $|1\rangle$, and thus not in a superposition, measuring our qubits again will give us the same bitstring $\mathbf{b}$. This new measurement leaves the qubits in the same state, so everything now is stable. We can measure as many times as we like; we'll always get back the same $\mathbf{b}$, and the qubits after measurement will always be in the same states as before measurement.

Measuring Some Qubits

Quantum measurement has a cool trick up its sleeve just waiting for us to discover it. The trick is that if we measure just *some* of the qubits in a system, we can force the other, unmeasured qubits to partially collapse their superpositions! If we perform this action carefully, our measurements can push those unmeasured qubits into superpositions that contain only those states that are more likely to give us a useful output.

Let's see how this trick is done. In this section, our mathematical tools will really pay off.

Figure 6-14(a) shows a visualization of how we'd compute the probability that, if we measure only qubits 4, 1, and 0 of a five-qubit system (leaving the others untouched), we'll find them to have the desired bit values b_4, b_1, and b_0.

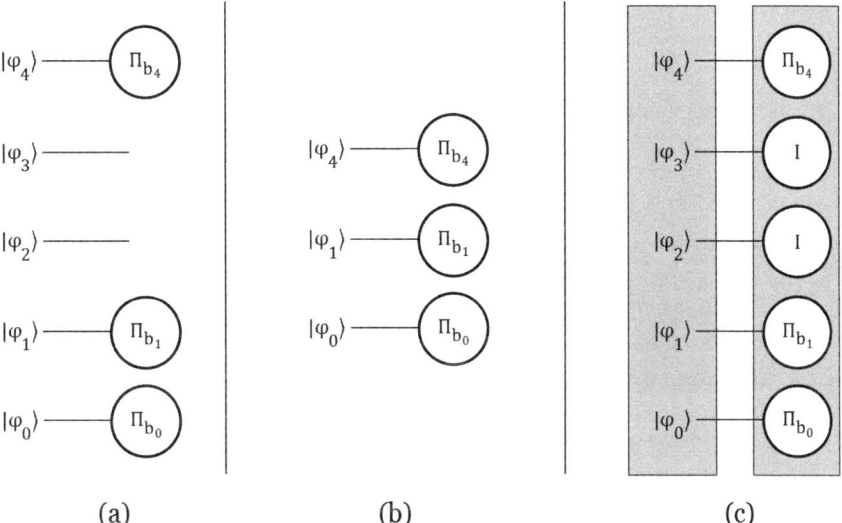

Figure 6-14: Computing the probability of the output of a circuit. (a) The output of a circuit of five qubits, measuring only bits 4, 1, and 0. (b) Interpreting part (a) horizontally. (c) Interpreting part (a) vertically.

If we're thinking horizontally, as in Figure 6-14(b), then we can just ignore states $|\psi_3\rangle$ and $|\psi_2\rangle$. We'll measure the three qubits we're interested in, get back the probabilities that they will have the values we want, multiply those probabilities together, and we'll have our result.

Thinking vertically, we can construct a system like that of Figure 6-14(c) by placing measurement operators on the qubits we want to measure and identity operators elsewhere. We use identity matrices I to indicate we're doing nothing to qubits $|\psi_3\rangle$ and $|\psi_2\rangle$. The thinking is similar to the situation in Figure 5-17, where we needed to insert an identity operator to build a complete vertical system.

Even though identity operators are valid qugates, in this context I've drawn them with circles to indicate that we're not actually applying them as qugates. Rather, in this context they're placeholders where projection operators could have been.

To evaluate Figure 6-14(c), we can first build each vertical system. So, we'd make an input system by tensoring together all the qubits, and then assemble a measurement system by tensoring together the measurement operators and identity operators.

To find the probability that the three qubits we're interested in will produce the bits we desire, we can use Equation 6.35.

Let's take these ideas out for a run. Suppose a three-qubit quantum circuit has produced the output state $|\psi\rangle$ given in Equation 6.36.

$$|\psi\rangle = \frac{i}{2} |010\rangle + \frac{1}{\sqrt{8}} |011\rangle + \frac{1+i}{2\sqrt{2}} |100\rangle + \frac{\sqrt{3}}{2\sqrt{2}} |110\rangle \qquad (6.36)$$

I've deliberately chosen complicated-looking coefficients, but you can see from their squared magnitudes (that is, their probabilities) in Figure 6-15 that their probabilities aren't too messy. Note that only these four states in $|\psi\rangle$ have nonzero probabilities.

k	0	1	2	3	4	5	6	7
ψ_k	0	0	$\frac{i}{2}$	$\frac{1}{\sqrt{8}}$	$\frac{1+i}{2\sqrt{2}}$	0	$\frac{\sqrt{3}}{2\sqrt{2}}$	0
$|\psi_k|^2$	0	0	1/4	1/8	1/4	0	3/8	0

Figure 6-15: Amplitudes and squared magnitudes of the four states in a superposition $|\psi\rangle$

The bottom row of Figure 6-15 adds up to 1, so this is a valid quantum state.

What's the probability that, when we measure this three-qubit system, we'll get back a 1 for the middle qubit?

According to our previous discussion, our measurement system has an I on the first and third qubits and a Π_1 for the second qubit, as we want the probability that the qubit will return a 1. The measurement system is thus $I \otimes \Pi_1 \otimes I$.

These are each two-by-two matrices, so their tensor product is an eight-by-eight matrix. Writing out huge matrices isn't that informative, so I'll just summarize the results using $|\psi\rangle$ from Equation 6.36 and the measurement system $I \otimes \Pi_1 \otimes I$ that we just made. That result is in Equation 6.37, where I used the squared magnitudes from Figure 6-15.

$$
\begin{aligned}
&\langle\psi| \left(I \otimes \Pi_1 \otimes I\right) |\psi\rangle \\
&= 0 |\psi_0|^2 + 0 |\psi_1|^2 + 1 |\psi_2|^2 + 1 |\psi_3|^2 + \\
&\quad 0 |\psi_4|^2 + 0 |\psi_5|^2 + 1 |\psi_6|^2 + 1 |\psi_7|^2 \\
&= 0 + 0 + \frac{1}{4} + \frac{1}{8} + 0 + 0 + \frac{3}{8} + 0 \\
&= \frac{3}{4}
\end{aligned}
\qquad (6.37)
$$

Does this make sense? Writing the states in binary, Equation 6.37 sums the probabilities for finding the states $|\psi_2\rangle = |010\rangle$, $|\psi_3\rangle = 011$, $|\psi_6\rangle = 110$, and $|\psi_7\rangle = 111$. Those are all the states with a 1 in the middle. Three of those states have nonzero probabilities. So, this says that the probability of measuring a 1 for the middle qubit is the sum of the probabilities of all the three-qubit states that have a 1 in the middle. That's three states out of four, matching our result of 3/4. It worked!

Note that I'm adding probabilities here, but in Equation 6.31 I multiplied them. Why the switch? Probability is a big subject, but let's make a quick digression to see why we add probabilities in some situations and multiply them in others.

Computing Probabilities

Let's look at probabilities using a decidedly non-quantum example, so we can focus on the ideas.

Suppose you and a friend are visiting an ice cream shop. The shop sells three flavors, vanilla, chocolate, and strawberry, and they offer one mix-in with each serving: either cookie dough, nuts, or toffee. Your friend is spontaneous and likes all the flavors and mix-ins equally, so you can never be sure what they'll order.

Let's first ask for the probability that your friend will choose chocolate *and* cookie dough. We can draw this situation graphically as in Figure 6-16, where each combination of flavor and mix-in has its own cell, and I've marked the one for both chocolate and cookie dough.

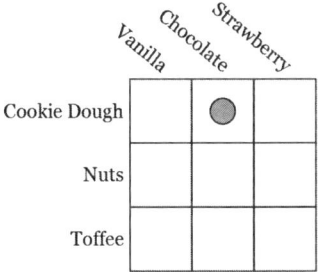

Figure 6-16: Of the nine choices, your friend has selected chocolate ice cream and a cookie dough mix-in.

There are nine possible combinations of flavor and mix-in, so the probability of your friend picking any particular combination is 1 in 9, or 1/9.

Let's consider this a different way. Looking at Figure 6-17, we see that three combinations involve chocolate, so there's a 3/9 = 1/3 chance that your friend will order chocolate. And three of the combinations involve a cookie dough mix-in, so there's a 1/3 chance of your friend choosing cookie dough.

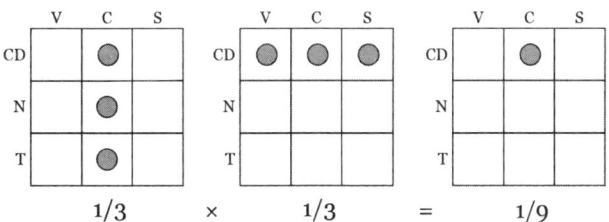

Figure 6-17: The probability that your friend will pick chocolate is 1/3, and the probability of cookie dough is 1/3. The probability of both is 1/3 × 1/3 = 1/9.

To find the probability of your friend ordering both chocolate *and* cookie dough, we multiply the two probabilities we just found: $1/3 \times 1/3 = 1/9$, as we found before.

The key elements of this situation are that the flavor and mix-in choices are *uncorrelated*, or independent of one another. Knowing either one doesn't tell you anything about the other. We call this a *joint probability*, and we compute it by *multiplying* the individual probabilities. We can recognize this kind of situation by the use of the word *and* in the setup.

Let's ask another question. What's the chance that your friend's choice of flavor will be either vanilla *or* chocolate, regardless of their choice of mix-in? I've drawn this situation in Figure 6-18.

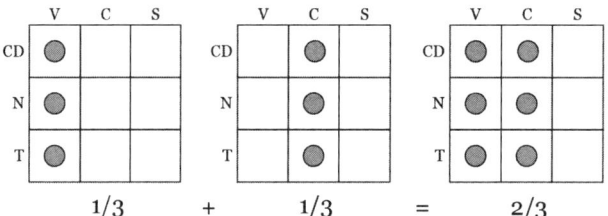

Figure 6-18: The probability of your friend ordering vanilla is $1/3$, and the probability of them ordering chocolate is also $1/3$. The probability of either is $1/3 + 1/3 = 2/3$.

Because your friend can only choose one or the other, we say that these options are *mutually exclusive*. The figure shows that the chance of them ordering vanilla is $1/3$ and the chance of them ordering chocolate is $1/3$, so the chance of your friend ordering one or the other comes from *adding* the probabilities, giving us $1/3 + 1/3 = 2/3$. We can recognize this kind of situation by the use of the word *or* in the setup.

Finally, let's ask for the probability that your friend will order either strawberry *or* the toffee mix-in. I've drawn this in Figure 6-19.

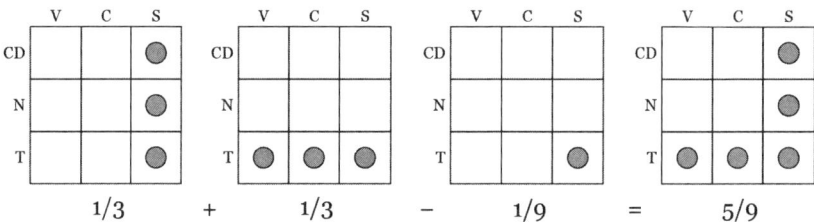

Figure 6-19: The probability of your friend ordering strawberry is $1/3$, and the probability of them ordering toffee is $1/3$. The probability of them ordering either one is $5/9$, because we remove $1/9$ to prevent double counting the combination of the two choices.

We can see that the probability of strawberry is $1/3$, and the probability of toffee is also $1/3$. It might be tempting to add the probabilities as we did in the previous example, but in this case the options are not mutually exclusive.

If we added the probabilities, the combination of strawberry and toffee would be included twice, so we need to subtract one of those instances from the total. Looking at Figure 6-19, we want to count the bottom-right cell only once. That choice has a probability of 1/9, so to avoid double counting that cell we have to remove it once. Thus, the probability of your friend ordering either strawberry *or* toffee is 1/3 + 1/3 − 1/9 = 5/9.

Let's apply this reasoning to our measurement. We wanted the probability that the middle state was a 1. That means we wanted the probability that the state we measured was the first state, *or* the second state, *or* the fourth. Therefore, we add up those three probabilities. Because these options are mutually exclusive, there was nothing to subtract since no conditions were double counted.

Returning to Measurement

Let's run the measurement of Figure 6-14(a) on a simulator and see if the results match our discussion. Equation 6.37 tells us to expect to see a state with a 1 in the middle three-fourths of the time, or with a probability of 0.75. Figure 6-20 shows the results of 10,000 repetitions of preparing a qubit in state $|\psi\rangle$ from Equation 6.36 and then measuring it. The counts, reading left to right, are 2,465 for state $|010\rangle$, 1,206 for state $|011\rangle$, 2,501 for state $|100\rangle$, and 3,828 for state $|110\rangle$.

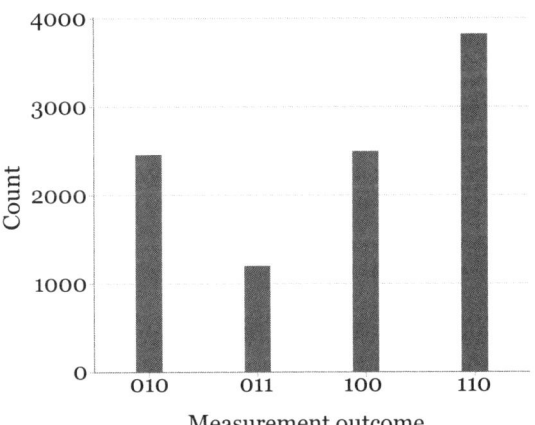

Figure 6-20: Evaluating the state $|\psi\rangle$ of Equation 6.36 for 10,000 shots on a quantum simulator

When we add up the number of times we measured the states 010, 011, and 110 and divide by the total number of shots, we get the expression (2,465 + 1,206 + 3,828) / 10,000 = 0.7499. Pretty close to 0.75!

If we run through the same process but use Π_0 to find the probability that qubit q_1 is measured as a 0, we get Equation 6.38.

$$
\begin{aligned}
\langle\psi| \left(I \otimes \Pi_0 \otimes I\right) |\psi\rangle \\
= 1\,|\psi_0|^2 + 1\,|\psi_1|^2 + 0\,|\psi_2|^2 + 0\,|\psi_3|^2 + \\
1\,|\psi_4|^2 + 1\,|\psi_5|^2 + 0\,|\psi_6|^2 + 0\,|\psi_7|^2 \\
= 0 + 0 + \frac{1}{4} + 0 \\
= \frac{1}{4}
\end{aligned}
\tag{6.38}
$$

Here, we're summing the probabilities of all three-qubit states that have a 0 in the middle. The only state in $|\psi\rangle$ like that is $|100\rangle$, with squared magnitude $|\psi_4|^2 = 1/4$, or a probability of 0.25. The results in Figure 6-20 gave us this state with probability $2{,}501\,/\,10{,}000 = 0.2501$. Again, pretty close to 0.25!

The second statement in Postulate 4, which confirms that all of our probabilities add up to 1 (just as we found $3/4 + 1/4 = 1$), is called the *completeness relation*.

You can perform the same calculations for measuring any number of qubits by placing projectors where you want to measure and identity operators everywhere else. This technique for finding the probability related to measuring just some of the qubits in a system is called the *partial measurement rule*, the *principle of partial measurements*.

Note that each of the probabilities in Equations 6.37 and 6.38 involved $|\psi_k|^2$. As we've discussed, another way to write this amplitude is $|\Pi_k\,|\psi\rangle|^2$. The square of the amplitude written this way matches the form that's given in Postulate 4. The steps are shown in Equation 6.39.

$$
\begin{aligned}
|\Pi_k\,|\psi\rangle|^2 &= \Big\langle \Pi_k\,|\psi\rangle \Big| \Pi_k\,|\psi\rangle \Big\rangle && \text{Magnitude as inner product,} \\
&&& \text{Eq. 2.71} \\[4pt]
&= \left(\Pi_k\,|\psi\rangle\right)^{\dagger} \left(\Pi_k\,|\psi\rangle\right) && \text{Write bra as ket adjoint from} \\
&&& \text{Eq. 3.18} \\[4pt]
&= \left(\,|\psi\rangle^{\dagger}\,\Pi_k{}^{\dagger}\right)\left(\Pi_k\,|\psi\rangle\right) && \text{Apply adjoint from Eq. 3.22} \\[4pt]
&= \langle\psi|\,\Pi_k{}^{\dagger}\,\Pi_k\,|\psi\rangle && \text{Remove parentheses} \\[4pt]
&= \langle\psi|\,\Pi_k\,\Pi_k\,|\psi\rangle && \text{Since } \Pi_k{}^{\dagger} = \Pi_k \text{ from Eq. 6.22} \\[4pt]
&= \langle\psi|\,\Pi_k\,|\psi\rangle && \text{Because } \Pi_k{}^2 = \Pi_k \text{ from Eq. 6.21}
\end{aligned}
\tag{6.39}
$$

This is a pretty fantastic result! It tells us something important: There are two quite different ways to write the probability of getting the number k when we measure the state $|\psi\rangle$. As we develop quantum algorithms, we'll want to manipulate our qubits so that we're most likely to measure the results that are useful to us. These expressions give us two different ways to evaluate the probability that we'll observe any specific k, so we can check that our algorithms are working the way we want, even before running them.

I love how Equation 6.39 brings together so many ideas that we've seen throughout the book.

Because this relationship is so important, I've summarized it in Equation 6.40.

$$|\Pi_k\,|\psi\rangle\,|^2 = \langle\psi|\,\Pi_k\,|\psi\rangle \qquad (6.40)$$

In the pictures we've seen so far, I've been drawing all the measurements as if they're happening at the same time (or nearly the same time), but there's no need for that. The probabilities associated with a measurement don't depend on when we make that measurement. So whether we measure a certain qubit now, or a week from now, or a million years from now, the probabilities of its results are the same. This tells us that we can move all of our measurements to the end of a computation. This is called the *principle of deferred measurement* [146, p. 186]. As we'll see in Part II, this can be useful for making some algorithms simpler to draw or analyze.

Partial Measurement

Some quantum algorithms use a technique where they measure only some of the qubits partway through the circuit. Remarkably, this can tell us something about the qubits we didn't measure!

Suppose we have two groups of qubits, which I'll call Data and Helper. The Data group contains the information we care about for solving our problem, and the Helper group contains helper, or auxiliary, qubits that are used as part of the process to get the Data qubits into the right states, but whose final values we don't care about. I've drawn this setup in Figure 6-21. The curly brackets here do not indicate entanglement, but just group the Data and Helper qubits together.

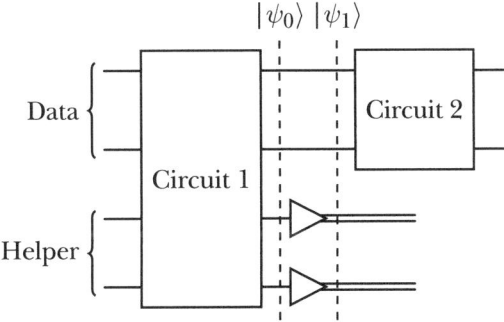

Figure 6-21: Measuring the Helper qubits, then continuing to compute. The curly brackets indicate grouping.

Here, I'm measuring the Helper qubits before the data qubits are further processed.

Let's suppose that the output of the first compute block, marked as $|\psi_0\rangle$, is an equal superposition of the four states $|0011\rangle$, $|1011\rangle$, $|0101\rangle$, and $|1010\rangle$, as in Equation 6.41.

$$|\psi_0\rangle = \frac{1}{2}(|0011\rangle + |1011\rangle + |0101\rangle + |1010\rangle) \qquad (6.41)$$

Now suppose that when we measure the Helper qubits, we find that both meters report 1.

At this point, even though we haven't measured them, we know that the Data qubits are either $|00\rangle$ or $|10\rangle$! That's because $|\psi_0\rangle$ contained only two states that had $|11\rangle$ for the Helper qubits, and they were the ones with $|00\rangle$ or $|10\rangle$ in the Data qubits. We say that the superposition has now *partially collapsed*, so that the state at $|\psi_1\rangle$ contains only those states consistent with the Helper qubits being $|11\rangle$. The superposition $|\psi_1\rangle$ is shown in Equation 6.42, where I'm writing only the Data qubits, as the Helper qubits have been measured.

$$|\psi_1\rangle = \vee(|00\rangle + |10\rangle) \qquad (6.42)$$

In this case, we know even more about the second Data qubit: It must be $|0\rangle$, because the only states in the superposition at $|\psi_0\rangle$ that are consistent with a measurement of 1 for both Helper qubits are $|00\rangle$ and $|10\rangle$. Without doing anything with the second Data qubit, we've learned that it's $|0\rangle$. Wow. We can use that information in further computations.

We can say that the measurement of the Helper qubits *eliminated*, or *filtered out*, the states in the superposition that were not consistent with a measurement of 1 for both Helper qubits. This is what we mean by saying that the superposition of the four qubits has partially collapsed. Alternatively, we can view this as saying that the measurement *selected* only the states that were consistent with measuring 1 for both Helper qubits, suggesting that these states *survived*, or were *retained*, after that measurement.

Partial measurement is powerful! We typically use it to filter the Data qubits so that they have some specific structure. We'll see examples of this technique in action in our survey of quantum algorithms in Part II.

Measurement and Entanglement

When we discussed entanglement, I promised that we'd see more of the mystery and power of that important phenomenon when we got to measurement. Well, here we are, so let's dig in!

Let's start by meeting and naming the four most famous entangled quantum pairs. They will be important to us now, and again in Part II when we discuss quantum algorithms.

Introducing Bell States

In 1964, the physicist John Bell wrote about a set of four entangled two-qubit states. They are now called *Bell states* in his honor [250] [165]. The four states are defined in Equation 6.43.

$$|\Psi^+\rangle = |\beta_s{}^+\rangle = \vee(|00\rangle + |11\rangle)$$
$$|\Psi^-\rangle = |\beta_s{}^-\rangle = \vee(|00\rangle - |11\rangle)$$
$$|\Phi^+\rangle = |\beta_m{}^+\rangle = \vee(|01\rangle + |10\rangle)$$
$$|\Phi^-\rangle = |\beta_m{}^-\rangle = \vee(|01\rangle - |10\rangle)$$
(6.43)

The notations using Φ and Ψ are common, but I find them hard to remember. I created the β (for Bell) versions to be more descriptive. Their subscript is either s or m, referring to whether the two qubits have the *same* or *mixed* values, respectively. In both cases, the first qubit in the first system is always $|0\rangle$. The superscript + or − tells us if the two states are to be added or subtracted.

Another naming scheme writes the states as $\beta(x, y)$ for two single-bit variables x and y. I'll temporarily use a tilde above one of these variables to indicate "not," so if $x = 0$, then $\tilde{x} = 1$, and vice versa. With this, we can write each of the four Bell states from the single formula in Equation 6.44.

$$|\beta(x, y)\rangle = \vee\left(|0y\rangle + (-1)^x |1\tilde{y}\rangle\right)$$
(6.44)

This version of the definition is worth knowing about because it can sometimes be convenient, and it sometimes appears in the literature. I won't use it in this book.

For completeness, Equation 6.45 shows all three naming schemes in the same place.

$$|\beta_s{}^+\rangle = |\Psi^+\rangle = |\beta(0, 0)\rangle = \vee\left(|00\rangle + |11\rangle\right)$$
$$|\beta_m{}^+\rangle = |\Phi^+\rangle = |\beta(0, 1)\rangle = \vee\left(|01\rangle + |10\rangle\right)$$
$$|\beta_s{}^-\rangle = |\Psi^-\rangle = |\beta(1, 0)\rangle = \vee\left(|00\rangle - |11\rangle\right)$$
$$|\beta_m{}^-\rangle = |\Phi^-\rangle = |\beta(1, 1)\rangle = \vee\left(|01\rangle - |10\rangle\right)$$
(6.45)

In a Bell state, it's the superposition that's special, not so much the individual states in that superposition.

That's because the states that make up a Bell state are familiar product states. For example, $|01\rangle$ is the product state $|0\rangle \otimes |1\rangle$. But when we make $|\beta_m{}^+\rangle$ by combining $|01\rangle$ and $|10\rangle$, we get an entangled state, because there's no way to write $|\beta_m{}^+\rangle$ as a tensor product of two smaller states.

A pair of qubits in an entangled state is, in many ways, like any other pair of qubits in superposition. We can operate on them with qugates and measure them, with each measurement producing either a 0 or a 1.

What's special about being entangled is that if we measure either qubit in an entangled pair, *we know the value of the other qubit, whether we measure it or not*. In quantum language, after measuring either qubit of a Bell pair, we know the value of the other qubit *with certainty*.

This is a startling result that can be challenging to get used to, so let's express it again another way. When we measure one qubit in an entangled pair (it doesn't matter which one), that qubit collapses probabilistically to either $|0\rangle$ or $|1\rangle$. At that moment, *we know what we'd get from measuring the other qubit*. We can think of this as if that qubit's state had collapsed as well.

If the state we're measuring is $|\beta_s{}^+\rangle$ or $|\beta_s{}^-\rangle$, the other qubit will *always* be the same as the one we measured. If the state is $|\beta_m{}^+\rangle$ or $|\beta_m{}^-\rangle$, the other qubit will always be the other basis state (that is, if we measured 0, measuring the other qubit will produce 1, and vice versa).

One way to say this is that measuring either qubit tells us the state of the other qubit. This is immediate. It doesn't matter if that other qubit is a nanometer away, on the other side of the Earth, or orbiting a distant star. Experiments have proven time and again that this collapse happens *instantaneously*. The collapse of the second qubit doesn't happen soon after the measurement of the first qubit, or even in the time it would take light to travel from the first qubit to the second (as far as we know, the top speed that *anything*, material or otherwise, can move). The collapse is *immediate*, no matter the distance. Of course, as with all experimental conclusions, we really only know this to be true to the precision and accuracy of our current instruments and methodology. There may be a delay between measuring one qubit and the collapse of the other, but if so, it's faster than anyone has ever been able to observe or measure.

Another way to look at an entangled pair is to say that they *share a single superposition*. That is, the two qubits are inextricably linked together in a single superposition. When we measure one qubit, that shared superposition collapses, which collapses the other qubit.

In an entangled quantum pair, the two objects are not in a preexisting state that we are discovering. They are not like a flipped coin, where the upper side is objectively heads or tails before we've looked. Before any measurement, they are in a shared superposition. *Neither qubit has a definite state.* It is, after all, in a superposition, which we've seen is a novel state of existence described by a set of measurement outcomes, each with a probability. Only upon measurement of one of the qubits does that qubit, *and its entangled partner*, collapse to specific states.

Instead of a superposition describing just one qubit, in an entangled pair that superposition describes two qubits.

Suppose that the superposition is $|\beta_s{}^+\rangle = \vee(|00\rangle + |11\rangle)$. Then we know that the *entire system* of two qubits is either $|00\rangle$ or $|11\rangle$. If we measure either qubit and it's $|0\rangle$, then the other qubit must collapse to $|0\rangle$, because both qubits belong to the same system. Think of this like a partial measurement, where we've measured $|0\rangle$ for the first qubit. The entire system can only be $|00\rangle$ or $|11\rangle$. The only system state consistent with a measurement of $|0\rangle$ for the first qubit is for the second qubit to also be $|0\rangle$. The system describing the two qubits simply does not include a state where one of them is measured as 0 and the other is 1.

Because the collapse of an entangled object happens instantaneously when its companion is measured, no matter how far apart the objects are, entanglement might sound like a great way to send a signal faster than the speed of light. We call this idea *superluminal transport*, and everything we know says that it's impossible [293]. Entanglement doesn't do the impossible, because we cannot use it to send information faster than the speed of light [71].

Here's the general argument. Suppose you and a friend generate an entangled pair in your basement (say, $|\beta_s^-\rangle$) and then your friend goes to Pluto for a vacation and takes one of the qubits with them. You both agree that you won't look at the qubits for a year, but then you can measure them any time afterward. So a year passes, and then on some arbitrary day when you feel the urge, you measure your qubit. That means you know the state of your friend's qubit. In fact, you don't know that your friend hasn't measured their qubit already, causing *your* qubit to collapse, maybe an hour ago, or last week. The only way for the two of you to work out what went on is to share some information in a classical way, such as with a radio signal or by getting back together for lunch and comparing notes.

So, although measuring one entangled qubit instantaneously collapses the other, wherever it may be, there's no way to harness this phenomenon in any way to communicate information unless you also trade some other information classically. And your communications are still limited, at best, by the speed of light. The details of this argument can become subtle, but if you dig into them you can prove that there aren't any loopholes in this process. You just can't send information without also sending some additional information by classical means [293].

The idea that measuring either qubit in an entangled pair also collapses the other qubit is a radical notion. We now have all the tools we need to express entanglement and its measurement precisely, so let's do it.

Measuring Bell States

So far we've been treating each measurement of a qubit as something that affects only that qubit. But with entangled pairs, measuring just one qubit tells us *everything* about the other. We know, *with certainty*, what we'll find for that other qubit if and when we measure it.

Suppose we have a two-qubit system in state $|\beta_s^+\rangle = \vee(|00\rangle + |11\rangle)$, and we measure either qubit. Before we make any measurements, each qubit measured alone has an equal probability of showing a 0 or a 1 on the meter. But this story changes the instant we measure one of the qubits. At that moment, we know with certainty that the other qubit will produce the same result, *whether or not we measure it*. So if we measure one qubit and find that it's 0, then we can use the other qubit in later computations, confident that it, too, has collapsed and is now in state $|0\rangle$.

The key thing to remember is that prior to any measurement, the qubits have no fixed value. Each is in a superposition and could produce either 0 or 1 when measured. But after one qubit has been measured, the state of the other qubit is immediately known, no matter where it is.

Let's run some measurements to illustrate this phenomenon. I'll arbitrarily use $|\beta_s^+\rangle = \vee(|00\rangle + |11\rangle)$.

We'll start by measuring just one qubit at a time. We can measure either of two qubits and look for either of two values, giving us a total of four possible measurements. I've shown them in Table 6-1.

Table 6-1: The measurement probabilities of $|\beta_s^+\rangle$ when we measure only one qubit

Diagram	Formula	Mag.	Prob.
$|\beta_s^+\rangle \left\{ \begin{array}{c} \Pi_0 \\ I \end{array} \right.$	$(\Pi_0 \otimes I)\,|\beta_s^+\rangle = \begin{bmatrix} 1 & 0 & 0 & 0 \\ 0 & 1 & 0 & 0 \\ 0 & 0 & 0 & 0 \\ 0 & 0 & 0 & 0 \end{bmatrix} \begin{bmatrix} \vee \\ 0 \\ 0 \\ \vee \end{bmatrix} = \begin{bmatrix} \vee \\ 0 \\ 0 \\ 0 \end{bmatrix}$	$\vee$	$1/2$
$|\beta_s^+\rangle \left\{ \begin{array}{c} \Pi_1 \\ I \end{array} \right.$	$(\Pi_1 \otimes I)\,|\beta_s^+\rangle = \begin{bmatrix} 0 & 0 & 0 & 0 \\ 0 & 0 & 0 & 0 \\ 0 & 0 & 1 & 0 \\ 0 & 0 & 0 & 1 \end{bmatrix} \begin{bmatrix} \vee \\ 0 \\ 0 \\ \vee \end{bmatrix} = \begin{bmatrix} 0 \\ 0 \\ 0 \\ \vee \end{bmatrix}$	$\vee$	$1/2$
$|\beta_s^+\rangle \left\{ \begin{array}{c} I \\ \Pi_0 \end{array} \right.$	$(I \otimes \Pi_0)\,|\beta_s^+\rangle = \begin{bmatrix} 1 & 0 & 0 & 0 \\ 0 & 0 & 0 & 0 \\ 0 & 0 & 1 & 0 \\ 0 & 0 & 0 & 0 \end{bmatrix} \begin{bmatrix} \vee \\ 0 \\ 0 \\ \vee \end{bmatrix} = \begin{bmatrix} \vee \\ 0 \\ 0 \\ 0 \end{bmatrix}$	$\vee$	$1/2$
$|\beta_s^+\rangle \left\{ \begin{array}{c} I \\ \Pi_1 \end{array} \right.$	$(I \otimes \Pi_1)\,|\beta_s^+\rangle = \begin{bmatrix} 0 & 0 & 0 & 0 \\ 0 & 1 & 0 & 0 \\ 0 & 0 & 0 & 0 \\ 0 & 0 & 0 & 1 \end{bmatrix} \begin{bmatrix} \vee \\ 0 \\ 0 \\ \vee \end{bmatrix} = \begin{bmatrix} 0 \\ 0 \\ 0 \\ \vee \end{bmatrix}$	$\vee$	$1/2$

In each row of Table 6-1, we start out with $|\beta_s^+\rangle$ and we measure one qubit with either Π_0 or Π_1. We apply the identity operator to the other qubit, because right now we care only about the qubit we're measuring. In the first row, $\Pi_0 \otimes I$ is the system we use to compute the probability that the first qubit is 0, regardless of the second. The remaining rows represent the probabilities of measuring the first qubit as 1, the second qubit as 0, and the second qubit as 1, each independently of the other qubit. The Formula column writes out the algebra, matrices, and result for each diagram, interpreting the operations in circles as a vertical system. The Mag. and Prob. columns give the magnitude and probability of that row's measurement, respectively.

Table 6-1 tells us that if we measure either qubit alone from $|\beta_s^+\rangle$, we have an equal probability of getting back a 0 or a 1. This is correct, but incomplete. What this isn't telling us is that because the qubits are entangled, *we can't use the same calculations for the second qubit*. The probabilities for the second qubit are *not* given by applying Table 6-1 again to that second qubit.

The problem with applying Table 6-1 again is that it would ignore the information we have: The qubits are entangled, and the first qubit is now $|0\rangle$ as a result of being measured as 0. We can't just ignore that information. One way to think about this is that each row of Table 6-1 begins with $|\beta_s^+\rangle$, but after we've measured one of the qubits, it's collapsed to either $|0\rangle$ or $|1\rangle$. Therefore, we don't have the state $|\beta_s^+\rangle$ anymore, and Table 6-1 doesn't apply.

Happily, we can learn about both qubits in an entangled pair (like a Bell pair) by measuring them both. That is, we ask for the probability of observing an entire state, rather than its individual qubits. With two qubits, there are four possible system states to consider. The calculations are shown in Table 6-2.

Table 6-2: The measurement probabilities of $|\beta_s^+\rangle$ when we measure both qubits

Diagram	Formula	Mag.	Prob.
$|\beta_s^+\rangle \left\{ \begin{array}{l} \Pi_0 \\ \Pi_0 \end{array} \right.$	$(\Pi_0 \otimes \Pi_0)\,|\beta_s^+\rangle = \begin{bmatrix} 1 & 0 & 0 & 0 \\ 0 & 0 & 0 & 0 \\ 0 & 0 & 0 & 0 \\ 0 & 0 & 0 & 0 \end{bmatrix} \begin{bmatrix} \vee \\ 0 \\ 0 \\ \vee \end{bmatrix} = \begin{bmatrix} \vee \\ 0 \\ 0 \\ 0 \end{bmatrix}$	$\vee$	$1/2$
$|\beta_s^+\rangle \left\{ \begin{array}{l} \Pi_0 \\ \Pi_1 \end{array} \right.$	$(\Pi_0 \otimes \Pi_1)\,|\beta_s^+\rangle = \begin{bmatrix} 0 & 0 & 0 & 0 \\ 0 & 1 & 0 & 0 \\ 0 & 0 & 0 & 0 \\ 0 & 0 & 0 & 0 \end{bmatrix} \begin{bmatrix} \vee \\ 0 \\ 0 \\ \vee \end{bmatrix} = \begin{bmatrix} 0 \\ 0 \\ 0 \\ 0 \end{bmatrix}$	0	0
$|\beta_s^+\rangle \left\{ \begin{array}{l} \Pi_1 \\ \Pi_0 \end{array} \right.$	$(\Pi_1 \otimes \Pi_0)\,|\beta_s^+\rangle = \begin{bmatrix} 0 & 0 & 0 & 0 \\ 0 & 0 & 0 & 0 \\ 0 & 0 & 1 & 0 \\ 0 & 0 & 0 & 0 \end{bmatrix} \begin{bmatrix} \vee \\ 0 \\ 0 \\ \vee \end{bmatrix} = \begin{bmatrix} 0 \\ 0 \\ 0 \\ 0 \end{bmatrix}$	0	0
$|\beta_s^+\rangle \left\{ \begin{array}{l} \Pi_1 \\ \Pi_1 \end{array} \right.$	$(\Pi_1 \otimes \Pi_1)\,|\beta_s^+\rangle = \begin{bmatrix} 0 & 0 & 0 & 0 \\ 0 & 0 & 0 & 0 \\ 0 & 0 & 0 & 0 \\ 0 & 0 & 0 & 1 \end{bmatrix} \begin{bmatrix} \vee \\ 0 \\ 0 \\ \vee \end{bmatrix} = \begin{bmatrix} 0 \\ 0 \\ 0 \\ \vee \end{bmatrix}$	$\vee$	$1/2$

Table 6-2 indicates that if we measure both qubits, half of the time we'll get back the bitstring 00 and the other half of the time we'll get 11. We'll never get 01 or 10, because the probability of each of those outcomes is 0.

This tells us how to interpret the results of measuring a single qubit following the approach of Table 6-1. It correctly shows that half the time we'll get back either a 0 or a 1 for either qubit, but because it doesn't take into account entanglement, it doesn't tell us about the other qubit. Table 6-2 takes both qubits into consideration at once. It tells us that when the qubits are entangled, we find that after measuring either qubit we know what the other qubit's measurement will be, whether or not we actually measure it someday.

What we've confirmed is that, somehow, measuring one qubit of a Bell state causes not just a collapse of the qubit we're measuring, but also a completely predictable measurement of the other qubit.

You could apply the principle of deferred measurement to put off the other measurement until some later time, as shown in Figure 6-22.

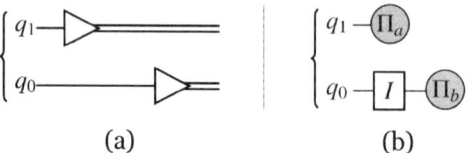

Figure 6-22: (a) Measuring q_1 first, then q_0. (b) A way to model the measurement.

As Figure 6-22(b) indicates, we can model that measurement with the vertical system ($\Pi_a \otimes I$) followed by a projector Π_b on the lower qubit. So, delaying the measurement in time doesn't affect the result that knowing the first qubit's value immediately tells us the other qubit's value, even if we measure it much later.

The properties of entanglement may not sit well with you, and if so, you're in good company. In 1935, Albert Einstein and colleagues Boris Podolsky and Nathan Rosen published a now famous and influential paper (known as the "EPR paper" for their initials) [55] in which they argued that because quantum mechanics allowed one to create entangled states, something was wrong. Either quantum mechanics was "incomplete" because it was missing some additional piece that would make sense of entangled states, or there was some other fundamental omission or error in the theory. Near the end of their paper, they recap the problem of entangled states and conclude, "No reasonable definition of reality could be expected to permit this."

Wow.

In these quotes, they're talking about *objective reality*, the notion that objects have definite properties regardless of our attention. In other words, measurement only reveals those properties that existed before we looked [202] [293]. As children grow, at some point they usually conclude that the world doesn't actually cease to exist when they close their eyes, and then suddenly pop back into existence when they open them again. Yet this is just what entanglement, like superposition, is telling us: Before measurement, the qubits do not have well-defined values. Until we turn our attention to it, an entangled qubit has no single preexisting value. It may be hard to believe, but it is the *very act of observing* that causes one of the qubit's potential values to become a reality *and* causes us to definitely know what we'd find from measuring the other qubit.

The idea that observing the world brings it into being is a challenging one. The physicist Abraham Pais tells a lovely anecdote about the reality of the world persisting when we close our eyes: "I recall that during one walk Einstein suddenly stopped, turned to me and asked whether I really believed that the moon exists only when I look at it" [156].

You're now well equipped to read the EPR paper (it's short and well written) and evaluate their arguments for yourself.

Few people today argue that entangled states don't exist, or that they don't behave as we've discussed. In fact, entangled states are used in a wide variety of applications [71]. In Part II, we'll see that the algorithms we create for quantum computers can use entangled states, and when we run those algorithms on actual hardware, they work as described. Developing a better understanding of entanglement is so difficult, and so important, that three physicists involved in that work won the Nobel Prize in Physics in 2022 [179] [293, §6.A].

What entanglement reveals to us about reality, as discussed in the EPR paper and countless follow-ups, is still a matter of philosophical debate [215].

I've been pretty casual in this discussion, and you might see loopholes or sneaky ways to send information at superluminal speeds with entanglement. Before going off too far in that direction, I encourage you to dig deeper into the details and mechanics, and the experiments that people have performed to understand entanglement [81]. After a century of searching, nobody has found a way, in either theory or practice, to use this instantaneous collapse at any distance to send information.

Phase

There's one more topic that is important both when we're running quantum programs and when we measure their outcomes.

Recall from Chapter 2 that in a complex number $e^{i\theta}$, the word *phase* refers to the value of θ. The term $e^{i\theta}$ is often called a *phase factor*, though sometimes people casually refer to it as just the *phase*. Generally, it's clear from context whether the word "phase" applies just to the angle θ or to the complex exponential $e^{i\theta}$.

We usually distinguish two types of phase, called global and relative. Let's look at them both.

Global Phase

We've seen that experimental measurements and reasoning about probabilities have led us to say that every quantum state must have a magnitude of 1. Unitary operators change a state but maintain this magnitude.

Another way to change a state without affecting its magnitude is to multiply it by 1 or −1. More generally, we can multiply a state by any complex number $e^{i\theta}$ and it won't change in magnitude, because $|e^{i\theta}| = 1$ (in this section, θ will always refer to a real, not complex, number). The real numbers 1 and −1 that I just mentioned are special cases of $e^{i\theta}$ for the values $\theta = 0$ ($|e^{i0}| = |e^0| = 1$) and $\theta = \pi$ ($|e^{i\pi}| = |-1| = 1$).

When we multiply a state $|\psi\rangle$ by $e^{i\theta}$ to create $e^{i\theta}|\psi\rangle$, we call θ (or $e^{i\theta}$) a *global phase* because it's applied "globally" to every amplitude in the state, even if it's a superposition.

To see what effect this phase factor has on measurements, let's look at a single qubit. I'll define two states that differ only by a global phase. The states, which I've named $|\psi\rangle$ and $|\phi\rangle$, are shown in Equation 6.46, where, as always, $|\alpha|^2 + |\beta|^2 = 1$.

$$|\psi\rangle = \alpha|0\rangle + \beta|1\rangle$$
$$|\phi\rangle = e^{i\theta}|\psi\rangle = e^{i\theta}(\alpha|0\rangle + \beta|1\rangle) \qquad (6.46)$$

We know that the probabilities of measuring $|\psi\rangle$ as 0 or 1 are $|\alpha|^2$ and $|\beta|^2$, respectively. In Equation 6.47, we find the probability of measuring 0 for $|\phi\rangle$, which has the global phase.

$$\Pr\nolimits_{|\phi\rangle}(0) = \langle\phi|\,\Pi_0\,|\phi\rangle \qquad \text{From Eq. 6.26}$$

$$= \begin{bmatrix} e^{-i\theta}\overline{\alpha} & e^{-i\theta}\overline{\beta} \end{bmatrix} \begin{bmatrix} 1 & 0 \\ 0 & 0 \end{bmatrix} \begin{bmatrix} e^{i\theta}\alpha \\ e^{i\theta}\beta \end{bmatrix} \qquad \text{Expand the terms}$$

$$= \begin{bmatrix} e^{-i\theta}\overline{\alpha} & e^{-i\theta}\overline{\beta} \end{bmatrix} \begin{bmatrix} e^{i\theta}\alpha \\ 0 \end{bmatrix} \qquad \begin{array}{l}\text{Multiply the rightmost} \\ \text{two matrices}\end{array} \qquad (6.47)$$

$$= \left(e^{-i\theta}\overline{\alpha} \right)\left(e^{i\theta}\alpha \right) \qquad \text{Multiply the matrices}$$

$$= e^{i(\theta-\theta)}\overline{\alpha}\alpha \qquad \text{Gather the exponents}$$

$$= \overline{\alpha}\alpha \qquad \text{Because } e^{i(\theta-\theta)} = e^{i0} = e^0 = 1$$

$$= |\alpha|^2 \qquad \text{As in Eq. 2.11}$$

So, multiplying a state by a global phase has no effect on the probability of measuring a 0 for that state.

Writing out the same steps with Π_1 shows that the probability of measuring $|\phi\rangle$ as 1 is $|\beta|^2$. Because the probabilities of measuring 0 and 1 are the same for both $|\phi\rangle$ and $|\psi\rangle$, we've found that multiplying a state $|\psi\rangle$ by a global phase factor $e^{i\theta}$ has *no effect* on the probabilities of the measurements of $|\psi\rangle$.

This is a big deal.

It means that two expressions that differ only by a global phase describe exactly the same state. A global phase has no effect on the observable (or measurable) aspects of a qubit.

There's another way to write quantum states called the *density matrix*. In this notation, multiplying a state by a global phase has no effect on the matrix describing the state. While the density matrix is more general than Dirac notation, it's also more complicated. I haven't used it here because we don't need that extra power and complexity in this book.

Let's illustrate the difference between these notations with a metaphor. Suppose that you're writing a script for an actor to read for a radio commercial, so you're only concerned with how things sound. If the actor isn't experienced in English, they might be confused by the words "knight" and "night." These homonyms look distinct, but they sound alike. The actor might interpret the k in "knight" as being acoustically meaningful, and try to pronounce it in some way. In the more specialized notation of the International Phonetic Alphabet, both words would be written /naɪt/, and the illusion of their difference (and the extraneous silent k) would disappear.

In this case, the "simple" representation (in everyday English) included some information that was not just superfluous, but possibly misleading, and obscured the fact that these two words sound alike. Using a more complex representation made this illusory difference disappear.

Just as I've written this book in English, and not the International Phonetic Alphabet, so too I've written the math using Dirac notation. Despite obscuring the fact that two states that differ only by global phase are the same state, it's still the clearest notation for everything we'll do here.

Relative Phase and Interference

As we've seen, a global phase is a factor that we apply to both amplitudes in a state. But what if we apply a factor $e^{i\theta}$ to just one of the amplitudes, rather than both? We call this a *relative phase* or *local phase*.

Like global phase, relative phase cannot be detected by a direct measurement. But unlike global phase, two states that differ only by relative phase *are* different states, and can lead to measurable differences.

The mechanism for this influence is based on the interaction of the relative phases in the different states.

We've already seen this effect, which we call *interference*. We know that $H|0\rangle = |+\rangle$ and $H|1\rangle = |-\rangle$. Both states $|+\rangle$ and $|-\rangle$ have equal probabilities of returning 0 or 1 when measured. The only difference between $|+\rangle$ and $|-\rangle$ is that $|-\rangle$ has a relative phase of $e^{i\pi} = -1$ applied to the $|1\rangle$ state. Even though no immediate measurement (that is, one applied right away) can distinguish $|+\rangle$ from $|-\rangle$, the relative phase -1 on $|-\rangle$ can be inferred by applying the H qugate again and then measuring, as detailed in Equation 6.48 for both $|+\rangle$ and $|-\rangle$.

$$H|+\rangle = \vee \begin{bmatrix} 1 & 1 \\ 1 & -1 \end{bmatrix} \vee \begin{bmatrix} 1 \\ 1 \end{bmatrix} = \vee^2 \begin{bmatrix} 2 \\ 0 \end{bmatrix} = \begin{bmatrix} 1 \\ 0 \end{bmatrix} = |0\rangle$$

(6.48)

$$H|-\rangle = \vee \begin{bmatrix} 1 & 1 \\ 1 & -1 \end{bmatrix} \vee \begin{bmatrix} 1 \\ -1 \end{bmatrix} = \vee^2 \begin{bmatrix} 0 \\ 2 \end{bmatrix} = \begin{bmatrix} 0 \\ 1 \end{bmatrix} = |1\rangle$$

Unlike states that differ only by relative phase, two states that differ only by a local phase are *different states*. They behave differently in circuits and can result in different measurements. If we were to compute the density matrices for $|+\rangle$ and $|-\rangle$ we'd find that they are different, confirming that they are indeed different states, despite the fact that we cannot distinguish them by an immediate measurement.

Let's look at a few other aspects of relative phase.

When writing a single qubit, the convention is to write the coefficient on $|0\rangle$ with no explicit phase, instead placing any relative phase information into the coefficient of $|1\rangle$. This makes it easier to visually compare states, since we only need to look at one phase term rather than two (you can always write phase terms on both basis states if you want).

Let's see why this convention works, using the two superposition states shown in Equation 6.49. The state $|\psi\rangle$ has a phase of $e^{i\theta}$ on $|0\rangle$, while $|\phi\rangle$ has the complementary phase, $e^{-i\theta}$, applied to $|1\rangle$.

$$|\psi\rangle = e^{i\theta}\alpha|0\rangle + \beta|1\rangle$$
$$|\phi\rangle = \alpha|0\rangle + e^{-i\theta}\beta|1\rangle$$

(6.49)

In Equation 6.50, we work through the probability of measuring a 0 for $|\psi\rangle$.

$$\Pr{}_{|\psi\rangle}(0) = \langle \psi | \Pi_0 | \psi \rangle \qquad \text{Probability of measuring } 0$$

$$= \begin{bmatrix} e^{-i\theta}\overline{\alpha} & \overline{\beta} \end{bmatrix} \begin{bmatrix} 1 & 0 \\ 0 & 0 \end{bmatrix} \begin{bmatrix} e^{i\theta}\alpha \\ \beta \end{bmatrix} \qquad \text{Expand the matrices}$$

$$= \begin{bmatrix} e^{-i\theta}\overline{\alpha} & \overline{\beta} \end{bmatrix} \begin{bmatrix} e^{i\theta}\alpha \\ 0 \end{bmatrix} \qquad \text{Multiply right matrices} \tag{6.50}$$

$$= e^{i(\theta-\theta)}\overline{\alpha}\alpha \qquad \text{Multiply and combine exponents}$$

$$= \overline{\alpha}\alpha \qquad \text{Probability of measuring } 0$$

Now let's apply the same process to $|\phi\rangle$, as shown in Equation 6.51.

$$\Pr{}_{|\phi\rangle}(0) = \langle \phi | \Pi_0 | \phi \rangle \qquad \text{Probability of measuring } 0$$

$$= \begin{bmatrix} \overline{\alpha} & e^{i\theta}\overline{\beta} \end{bmatrix} \begin{bmatrix} 1 & 0 \\ 0 & 0 \end{bmatrix} \begin{bmatrix} \alpha \\ e^{-i\theta}\beta \end{bmatrix} \qquad \text{Expand the matrices}$$

$$= \begin{bmatrix} \overline{\alpha} & e^{i\theta}\beta \end{bmatrix} \begin{bmatrix} \alpha \\ 0 \end{bmatrix} \qquad \text{Multiply right matrices} \tag{6.51}$$

$$= \overline{\alpha}\alpha \qquad \text{Probability of measuring } 0$$

Though this isn't a proof, it is an example of the general case. If we have a phase on $|0\rangle$, we can apply the opposite phase to $|1\rangle$ instead and not change the probability of measuring a 0 or a 1. So, we can always collect any relative phase terms onto the coefficient of $|1\rangle$ and leave $|0\rangle$ without a phase term. This is the usual way to write a state with relative phase.

When we look at quantum algorithms in Part II, we'll see that relative phases often make important contributions to creating the desired final measurement.

As we saw earlier, the two states $|+\rangle$ and $|-\rangle$ differ by a relative phase, and are different states, but we can't distinguish them by an immediate measurement (that is, placing the meter immediately after the creation of the state with no additional processing). This is an odd state of affairs: the qubit contains a piece of information that we can't directly access.

We can roughly capture this idea with a metaphor. If you've written programs for conventional computers, you're probably familiar with the idea of a *class*. Speaking broadly, a class is a blueprint for objects made of variables and functions [256]. A programmer can manufacture objects from the blueprint, modify their variables, and call their functions. Most classes offer a special variety of variable called a *private variable* that can be used only by objects created from that class's blueprint. No other part of the program can modify a private variable. In fact, no other part of the program can even detect that such variables exist. They are, truly, private to the object. (Since this is a metaphor, it's imperfect. Sometimes there are sneaky ways to get at private variables from outside the class.)

A relative phase is like a private variable. A qubit's relative phase is a purely internal piece of information that it can use, and which can affect calculations that the state is involved in. Yet, we cannot make any immediate measurement that will reveal that information. In terms of immediate measurements, relative phase is a secret, known only to the qubit.

As to how nature manages to pull this off, nobody knows.

Summary

In this chapter, we looked at how we model the measurement of a quantum system and what happens to that system as a result of that measurement.

We saw that experiments consistently show that when we measure a state given by $\alpha \left| 0 \right\rangle + \beta \left| 1 \right\rangle$, the probability of measuring 0 is $|\alpha|^2$ and the probability of measuring 1 is $|\beta|^2$. These are idealized results, though, as quantum computers (like all computers) occasionally make errors. We use the term "noise" to refer to these errors collectively. Because noise is mostly uncorrelated, we often run algorithms many times (we say we run many shots), and we expect that the correct answer will dominate.

We also looked at measurement of a superposition. This confirmed that superpositions are not states that have preexisting single identities but rather a new and abstract kind of existence, described by a list of possible states and associated probabilities.

To form a more general way to talk about measurements, we defined projection operators. Though these are matrices, they are not unitary and may not be used in a quantum algorithm. They're strictly a mathematical convenience for calculating and discussing measurements.

Along the way, we saw that the outer product, or ketbra, is a matrix described by a ket followed by a bra.

We saw that we could measure just some of the qubits in a system and find the probabilities for different partial measurements.

We looked at entangled states, including Bell states, and saw that they cannot be described as a tensor product state. The entangled particles appear to share a single, common superposition. When that superposition collapses, we know the state of the other qubit. While this collapse is instantaneous, we cannot use it to send information faster than the speed of light.

Finally, we looked at global and relative phase. Relative phases cannot be detected by a direct measurement, but they can affect computations.

Whew. We have covered a *lot* of stuff since the beginning of the book! Now you know about qubits and qugates, how we modify qubits, how we make systems of multiple qubits and multiple qugates, and how to measure qubits.

You've seen the four fundamental postulates of quantum mechanics, expressed in forms useful for quantum computing.

That's it for the preparations! Now we have everything we need to move on to Part II, where we'll discuss some of the most famous quantum algorithms and see how they work. You'll be able to use these circuits for your own work, or invent your own original quantum algorithms!

PART II

QUANTUM ALGORITHMS

Don't you understand that we need to be childish in order to understand? Only a child sees things with perfect clarity, because it hasn't developed all those filters which prevent us from seeing things that we don't expect to see.
—Douglas Adams, *Dirk Gently's Holistic Detective Agency*, 1987 [2]

It is not because things are difficult that we do not dare, but because we do not dare, things are difficult.
—Seneca the Younger, *Moral and Political Essays*, 40 [191]

Quantum programming is exciting, because writing quantum programs is nothing like anything we're used to, including programming.

Early conventional programs were written in *assembly language*. These programs invoked specific actions executed by the hardware and offered little in the way of larger conceptual operations. As a result, they were difficult to read and write. In 1957, IBM released the first version of the Formula Translating System, also called the FORTRAN programming language [89] [262]. This changed programming forever.

Most of us now think of ideas like variables, conditional tests, subroutines, and so on as basic conceptual tools of programming, and even of procedural thinking in general. These ideas were part of FORTRAN and have been adopted and extended by many other programming languages ever since. As a result of the increasing expressive power of programming languages, and their integration with the computer's underlying operating systems, most of us never need to think about complexities like memory management, process scheduling, and reliable parallel execution. This freedom has helped lead to the stunning evolution of computers from "number crunchers" to devices that help us create images, write music, stay healthy, manage businesses and governments, and enjoy a real-time global communications network.

Quantum programming is an entirely different way to think about algorithms. Few of the old rules apply, and almost none of the features of modern programming languages are available to us (at least, not yet). We're currently at the equivalent of the assembly language phase, where our programs are usually sequences of instructions that correspond to specific actions executed by the hardware.

But quantum programs have abilities that conventional computers can't touch. Perhaps the most prominent is *quantum parallelism*. A quantum computer can apparently perform computations on arbitrary numbers of different inputs, simultaneously. It lets us search for a needle in a haystack by looking at, and evaluating, every object in the haystack at the same time. No matter how large the haystack is, we can examine every piece of it in the same amount of time it takes to examine a single piece.

Our goal is to embrace these new capabilities and use them to their fullest extent.

As quantum programmers, we face at least two key challenges. First, we need to find ways to use the available building blocks to create an algorithm that solves the problem at hand. Second, we have to extract useful information from the results of that algorithm.

The building blocks are quantum gates, like the I, X, H, and CX qugates we've already met. Because any unitary square matrix can be used as a quantum gate, there are an infinite number of possibilities to draw from. In practice, however, we usually restrict ourselves to one or two dozen qugates that have well-understood and useful properties.

The extraction process involves somehow narrowing the sometimes astronomical number of possible results created by an algorithm down to just a small number of useful ones, and making those results likely to be measured at the computer's output.

We can view the development of quantum programming so far as a series of breakthroughs. These arose from different people coming to grips with different aspects of these problems and using those insights to produce functioning algorithms of increasing complexity, subtlety, and utility.

I think the best way to learn quantum programming is to follow in their footsteps. In this part of the book, I'll present several important quantum algorithms, in roughly their chronological order of development and publication. By studying each algorithm, we can become familiar with the new ideas introduced by each and internalize those ideas into our understanding of what it means to create a quantum algorithm.

By the time we're done, we'll be familiar with not just a new set of programming tools, but a new set of fundamental *ideas* about how we can represent, process, and extract information from machines that exploit the particular properties of quantum objects.

This is exciting, heady stuff. I find that quantum computing offers the kind of invigorating stimulation that kicks my neurons into a higher gear, breaks patterns that had calcified without my notice, and sparks sudden and unexpected ideas.

Reaping these benefits takes some effort. I encourage you to take your time with these algorithms and let their structures and mechanisms sink in until they become comfortable for you. The reward is that you'll then be able to use them in creative and original ways for your own projects.

Chapter 7 starts us off with quantum teleportation, which enables us to transfer the state of one quantum object to another. Then, in Chapters 8 through 11, we'll look at a series of algorithms that each take as an input a small quantum circuit that we're not allowed to look at, but that we can feed inputs to and get back outputs from. Each of these algorithms is designed to reveal something about an increasingly interesting set of mystery circuits.

Chapter 12 presents a general-purpose routine for finding which elements in a database satisfy some criterion we're searching for. After that, Chapter 13 looks at an algorithm that can break the most popular security used on the internet today. Chapter 14 wraps things up with a discussion and pointers to useful quantum computing resources.

If you'd like to run the algorithms as we discuss them, Chapter 14 also describes how you can run small programs on real quantum computers on the web for free [95] [6]. Alternatively, you can download a free quantum computer simulator for almost any operating system or language and run it on your home computer [166] [44]. These simulators can only process a handful or so of qubits, but that's enough to run real quantum programs and build up your programming skills.

7

TELEPORTATION

Hey, hey, hey.
You know, don't be mean.
We don't have to be mean.
'Cuz remember, no matter where you go . . .
there you are.
—Earl Mac Rauch, *The Adventures of Buckaroo Banzai Across the 8th Dimension*,
1984 [170]

In this chapter, we'll meet our first complete quantum algorithm! This algorithm doesn't perform a calculation or give us the answer to a specific problem. Instead, it performs a unique, fascinating task: moving the quantum state of one qubit to another qubit, located anywhere in the universe. And it does this *instantaneously*.

Wait, the no-cloning theorem from Chapter 5 tells us we can't do this, right? The theorem does say that we can't make a copy of a quantum state, but it doesn't prohibit us from *moving* a state from one qubit to another, leaving the original qubit in a different state from how it started.

This process has an exciting name: *quantum teleportation*, or just *teleportation*. It's not quite teleportation the way the term is used in science fiction like *Star Trek*, though. We really do communicate the states of a quantum bit from *here* to *there*, but there are four big differences.

The first is that we're not transferring any kind of matter. We're only communicating the state of a qubit. Even if we transferred the state of enormous numbers of qubits, we still don't have any means for assembling the physical objects that have been put into those states into a grumpy but humane doctor, an exploding warp drive, or even a rock.

The second difference is that we can't send the description of the qubit anywhere we like. We can only transfer the state of a qubit to another qubit it's already been entangled with, and which is already present at the receiving site.

The third difference is that to reliably transfer the state of a quantum bit from one place to another, we must also exchange two classical bits over normal, classical channels, such as radio. That means we can't rely on this method to share information unless we also share some classical bits over conventional channels.

Finally, the fourth difference is that when we move a quantum state from one qubit to another, the state of the original qubit is changed, and we can't recover its original value.

Given all of these qualifiers, it might be better to call this quantum state *transfer* rather than teleportation. It's not *Star Trek* by a long shot, but transferring the state of one quantum particle to another quantum particle is still pretty cool.

Three features that real quantum teleportation has over the fictional version are that *distance doesn't matter, nothing can interfere with the process*, and *the original must be destroyed to be transported*. The first property means that the source and target can be literally anywhere in the universe. The second two properties protect us from ever accidentally creating an "evil Spock" [176].

In this chapter, we'll work a lot with explicit qubit states and the matrices of the operators that modify them. This is unusual. Most of the time, when we analyze a quantum algorithm, we work entirely (or nearly so) with algebra and rarely get down to the level of coefficients. Most of the rest of this book follows that approach. But sometimes working with the actual coefficients can be illuminating, bringing us a little closer to the mechanics of quantum computing. It also allows us to view an algebraic result in a different way, if there aren't too many qubits involved. For these reasons, in this chapter we'll spend most of our time with components, and we'll see explicitly how the operator matrices manipulate the elements of the ket matrices. You won't need to memorize any of these eight-by-eight matrices, as they're all built up from the smaller two-by-two matrices that we're already familiar with.

Okay, enough prep. Let's get teleporting!

The Teleportation Thought Experiment

A great way to think about teleportation is in terms of a story, or what physicists call a *thought experiment*. This story involves two characters. In physics thought experiments with two characters, they are almost always named Alice and Bob, so I'll carry on that tradition here.

In this story, we imagine that Alice and Bob are separated by a great distance: Maybe Alice is on Earth, and Bob is on Mars. Alice has run some algorithm that produces a quantum state, which I'll call the *signal*, described by a ket s with the state $|\sigma\rangle = \alpha |0\rangle + \beta |1\rangle$, where as always $|\alpha|^2 + |\beta|^2 = 1$. Producing this state is only the first part of a two-part computation. Bob is ready to take over from here and finish the computation, so he needs to have a qubit in the state $|\sigma\rangle$.

Alice could send her physical qubit to Bob, so he can work with it. But let's say that Bob is so remote, and sending things is so slow and expensive, that there's no practical way for Alice to physically send her qubit to Bob.

To get around this limitation, suppose that Bob has taken a qubit named b to Mars. Taken together, Alice's qubit s and Bob's qubit b form a two-qubit system $s \otimes b$. When we think of s and b as a system, it doesn't matter that the qubits are far apart from one another.

Because s is in the state $|\sigma\rangle$, it would be great if there were some sequence of operations that Alice, or Bob, or both of them could follow that would give them the qubit system state $|\sigma\rangle \otimes |\sigma\rangle$. Then Bob's qubit b would also be in the state $|\sigma\rangle$, and they'd have teleported the signal! Unfortunately, this means making a copy of $|\sigma\rangle$, and we know that the no-cloning theorem prohibits that.

Maybe we can avoid cloning if Alice's qubit is changed during teleportation. For example, it might go from $|\sigma\rangle$ to some other state, $|\omega\rangle$. Now if we can put $|\sigma\rangle$ onto Bob's qubit, the new system will be $|\omega\rangle \otimes |\sigma\rangle$. There will be no cloning, and they'll have teleported the signal!

That would be great, but nobody has found a way to do it.

A way that does work requires giving Alice one more qubit. This extra qubit (let's call it a, for auxiliary) will help us perform teleportation.

But how does this help Alice, in her lab on Earth, modify Bob's qubit on Mars? The answer is to link all three qubits together, so that when Alice manipulates her qubits s and a, those operations have an effect on Bob's qubit b.

We know how to do that: Use entanglement! If Alice and Bob created an entangled pair before Bob left, and each kept one qubit of the pair with them, operations on either qubit could affect the other.

We still need some way for Alice to move $|\sigma\rangle$ onto Bob's qubit. The key idea is to create a system state of three qubits that I call the *teleportation state*, which I'll write as $|\tau\rangle$. This is the heart of the whole algorithm. Figure 7-1 shows the teleportation algorithm in two steps.

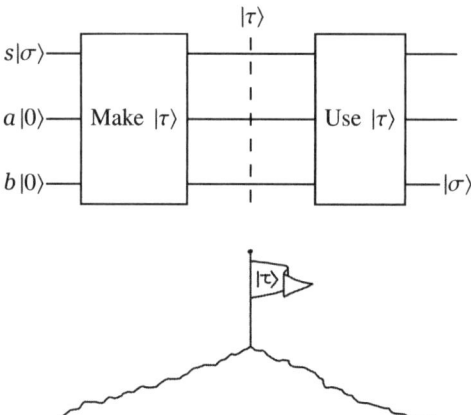

Figure 7-1: The teleportation algorithm can be viewed as two steps.

We can view the first step as climbing a hill to put the qubits into the teleportation state $|\tau\rangle$. We plant our flag (the state $|\tau\rangle$) at the top of the hill and then head down the other side, using $|\tau\rangle$ to put Bob's qubit b into the state $|\sigma\rangle$ that Alice's qubit s was initially in. We don't care about the final states of qubits s and a, so I've left them blank in the figure.

Because the teleportation state $|\tau\rangle$ is at the center of the whole process, let's take a closer look at it.

The Teleportation State $|\tau\rangle$

Let's write the qubits s, a, and b in the teleportation state. A 3-qubit state has $2^3 = 8$ elements. Creating the teleportation state takes only a few quantum gates, and I'll show you that circuit later in this chapter. For now, I'll ask you to take it on faith that Alice and Bob can create $|\tau\rangle$.

The teleportation state involves the qubits s, a, and b. Let's write these in order from top to bottom, as in Figure 7-2. We'll see that the three qubits are entangled together, so their output is the single entangled state $|\tau\rangle$.

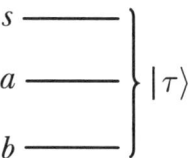

Figure 7-2: The three qubits s, a, and b arranged from top to bottom, making up the entangled state $|\tau\rangle$

The teleportation state $|\tau\rangle$ is an equal superposition of four states. Each is the original $|\sigma\rangle$, perhaps transformed by one or two specific operators.

The state $|\tau\rangle$ is shown in Equation 7.1.

$$|\tau\rangle = \frac{1}{2} \Big(|00\rangle\, I\, |\sigma\rangle + |01\rangle\, X\, |\sigma\rangle + |10\rangle\, Z\, |\sigma\rangle + |11\rangle\, XZ\, |\sigma\rangle \Big) \qquad (7.1)$$

At first glance, it looks like we've cloned $|\sigma\rangle$ not just once but three times. But a closer look reveals that there's been no cloning. What we've done is create additional states involving α and β in a single superposition.

We've been doing this kind of thing for several chapters now. For example, suppose we apply an H qugate to $|\sigma\rangle$. The resulting state is shown on the right side of Equation 7.2 (recall our convention that $\vee = 1/\sqrt{2}$).

$$H\,|\sigma\rangle = \vee \begin{bmatrix} 1 & 1 \\ 1 & -1 \end{bmatrix} \begin{bmatrix} \alpha \\ \beta \end{bmatrix} = \vee \begin{bmatrix} \alpha + \beta \\ \alpha - \beta \end{bmatrix} \qquad (7.2)$$

The final state is an equal superposition of $(\alpha + \beta)\,|0\rangle$ and $(\alpha - \beta)\,|1\rangle$. Both coefficients α and β appear twice, but we haven't cloned anything. Eventually, we'll make a measurement, causing this superposition to collapse, and only one of the states will be associated with the qubit. The same is true of $|\tau\rangle$.

Let's return to $|\tau\rangle$ in Equation 7.1. It's composed of four states in an equal superposition.

In the first state, $|00\rangle\, I\, |\sigma\rangle$, the $|00\rangle$ term refers to the two-qubit state $|0\rangle \otimes |0\rangle$. We then tensor this with $I\,|\sigma\rangle$. This is really just $|\sigma\rangle$, but I included the I for consistency with the other states.

The second state, $|01\rangle\, X\, |\sigma\rangle$, tells us to first form $|0\rangle \otimes |1\rangle$ and then tensor that with $X\,|\sigma\rangle$, or the result of applying the X qugate to the original $|\sigma\rangle$.

The third state, $|10\rangle\, Z\, |\sigma\rangle$, is like the previous one. We first form the state $|1\rangle \otimes |0\rangle$ and then tensor that with $Z\,|\sigma\rangle$.

Finally, $|11\rangle\, XZ\, |\sigma\rangle$ tensors together $|1\rangle \otimes |1\rangle$ with the state made by applying Z and then X (in that order) to $|\sigma\rangle$ (remember that we read algebraic operators from right to left).

It's the structure of $|\tau\rangle$ that enables teleportation. In Figure 7-2, the first two qubits of $|\tau\rangle$ correspond to s and a and the third to b.

If Alice measures qubits s and a, then as usual she'll get back a single bit for each. Let's say she finds $s = 1$ and $a = 0$. Then the law of partial measurement tells us that the superposition describing the *entire system* must collapse to contain only those states that are consistent with Alice's measurement. There is only one such state in $|\tau\rangle$, $|10\rangle\, Z\, |\sigma\rangle$, and therefore, with certainty, Bob's qubit b is now in the state $Z\,|\sigma\rangle$. Bob knows that Z is its own inverse, or $ZZ = I$, so he can apply Z to this state to get $ZZ\,|\sigma\rangle = |\sigma\rangle$.

Voilá, Bob's qubit b has the state $|\sigma\rangle$. Teleportation achieved!

As promised, there's been no cloning. In order for Alice's signal $|\sigma\rangle$ to make it to Bob's qubit, Alice had to measure both s and a. The process of measuring s collapsed it to either $|0\rangle$ or $|1\rangle$, destroying Alice's copy of $|\sigma\rangle$, thereby enabling us to move $|\sigma\rangle$ to Bob's qubit without cloning.

The Teleportation Process

We can think of teleportation as a four-step process. This will enable us to view the big picture in four smaller chunks that we can then assemble at the end.

The four steps are how Alice and Bob make the teleportation state $|\tau\rangle$, how Alice measures the state of the qubits, how Alice tells Bob which of the four states $|\tau\rangle$ collapsed to, and how Bob applies the correct qugates to get the original $|\sigma\rangle$. Let's take these in order.

Building $|\tau\rangle$

Alice and Bob build the teleportation state $|\tau\rangle$ in three steps. The first step entangles the qubits a and b, the second entangles s with the other two qubits, and the third performs one final step of processing.

Alice and Bob start everything off while they're still together on Earth. After lunch one day, they head to Alice's lab to make two qubits, named a and b, both in state $|0\rangle$. They entangle a and b in the same way that we saw in Figure 5-26. The traditional way to write this in a teleportation circuit is for Bob to apply H to b and then apply a CX using b as a control and a as a target, as shown in Figure 7-3(a).

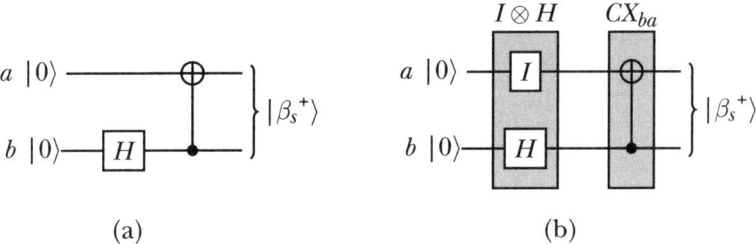

Figure 7-3: (a) Entangling a and b. (b) Explicitly including an identity on a and naming the operator systems.

This CX is drawn upside down compared to how I've usually drawn it before, with the control under the target. We saw this previously in Equation 5.48, where I called it CX'. We confirmed there that this works just as we'd hoped, with the target on the lower line controlling the application of the X qugate on the upper line. In this algorithm, I'll call the qugate CX_{ba} to emphasize that qubit b is controlling qubit a. As a reminder, the matrix form of CX_{ba} from Equation 5.48 is given in Equation 7.3.

$$CX_{ba} = \begin{bmatrix} 1 & 0 & 0 & 0 \\ 0 & 0 & 0 & 1 \\ 0 & 0 & 1 & 0 \\ 0 & 1 & 0 & 0 \end{bmatrix} \tag{7.3}$$

Returning to our entanglement step, our experience from Chapter 5 tells us that the output of Figure 7-3 should be $|\beta_s^+\rangle$. We can confirm this by first tensoring together $a = |0\rangle$ and $b = |0\rangle$ to make the starting state $|00\rangle$, then modifying them by the system $I \otimes H$ shown in Figure 7-3(b). This qugate system is written out in Equation 7.4.

$$I \otimes H = \begin{bmatrix} 1 & 0 \\ 0 & 1 \end{bmatrix} \otimes \vee \begin{bmatrix} 1 & 1 \\ 1 & -1 \end{bmatrix} = \vee \begin{bmatrix} 1 & 1 & 0 & 0 \\ 1 & -1 & 0 & 0 \\ 0 & 0 & 1 & 1 \\ 0 & 0 & 1 & -1 \end{bmatrix} \tag{7.4}$$

The second system is CX_{ba}, which we just found. Let's apply both systems to the starting state $|00\rangle$, as shown in Equation 7.5.

$$CX_{ba}(I \otimes H)|00\rangle = CX_{ba}((I \otimes H)|00\rangle) \qquad \text{Apply } I \otimes H \text{ first}$$

$$= CX_{ba}\left(\vee \begin{bmatrix} 1 & 1 & 0 & 0 \\ 1 & -1 & 0 & 0 \\ 0 & 0 & 1 & 1 \\ 0 & 0 & 1 & -1 \end{bmatrix} \begin{bmatrix} 1 \\ 0 \\ 0 \\ 0 \end{bmatrix} \right) \qquad \text{Use } I \otimes H \text{ from Eq. 7.4}$$

$$= CX_{ba} \vee \begin{bmatrix} 1 \\ 1 \\ 0 \\ 0 \end{bmatrix} \qquad \text{Multiply the matrices}$$

$$= \begin{bmatrix} 1 & 0 & 0 & 0 \\ 0 & 0 & 0 & 1 \\ 0 & 0 & 1 & 0 \\ 0 & 1 & 0 & 0 \end{bmatrix} \vee \begin{bmatrix} 1 \\ 1 \\ 0 \\ 0 \end{bmatrix} \qquad \text{Use } CX_{ba} \text{ from Eq. 7.3}$$

$$(7.5)$$

$$= \vee \begin{bmatrix} 1 \\ 0 \\ 0 \\ 1 \end{bmatrix} \qquad \text{Multiply the matrices}$$

$$= \vee(|00\rangle + |11\rangle) = |\beta_s^+\rangle \qquad \text{The Bell state } |\beta_s^+\rangle$$

Great! Figure 7-3 does indeed give us the Bell state $|\beta_s^+\rangle$.

In our thought experiment, Bob now places qubit b in a special bottle and takes it with him to Mars. Alice also places a in a special bottle and puts it somewhere safe in her lab.

A year passes. One day, Alice completes her experiment, resulting in a qubit named s in the state $|\sigma\rangle$. This is the state she wants to send to Bob.

Alice can only modify s and a, the qubits that she has with her on Earth. In order for Alice to cause operations on the qubits s or a to affect Bob's qubit b far away, qubits s and a need to be entangled with b. She can create that entanglement by using s as the control on a CX targeting either a or b. Of these, only a is in the lab with Alice, so she entangles s with a. This second entanglement step is shown in Figure 7-4. I've written CX_{sa} for the CX in the usual orientation, using s as a control on a.

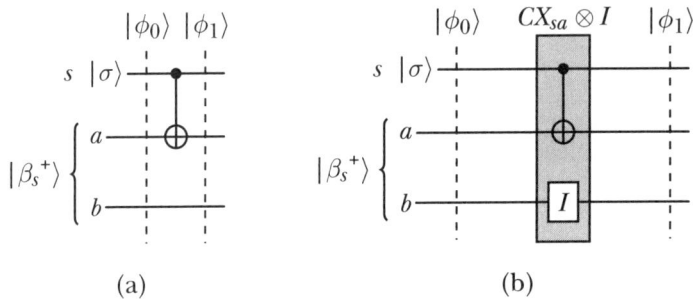

Figure 7-4: (a) Entangling s and a. (b) Including the implied I qugate.

Because we're focusing here on matrix elements, let's write out the components of the qubit system $|\phi_0\rangle$, the system just before Alice entangles s with a. This is $|\sigma\rangle$, the state of Alice's qubit s, tensored with $|\beta_s^+\rangle$, which we derived in Equation 7.5. The result is shown in Equation 7.6.

$$|\phi_0\rangle = |\sigma\rangle \otimes |\beta_s^+\rangle = \begin{bmatrix} \alpha \\ \beta \end{bmatrix} \otimes \vee \begin{bmatrix} 1 \\ 0 \\ 0 \\ 1 \end{bmatrix} = \vee \begin{bmatrix} \alpha \\ 0 \\ 0 \\ \alpha \\ \beta \\ 0 \\ 0 \\ \beta \end{bmatrix} \tag{7.6}$$

Now we'll entangle s with a. Figure 7-4(a) shows using s as a control on a. As usual, it omits the identity qugate I we could place on the b line. But although it's not in the picture, that identity must be in our operator!

Let's write out this matrix, as shown in Equation 7.7. Because CX_{sa} applies a control on the topmost line to a target immediately below it, its matrix is the familiar CX matrix.

$$CX_{sa} \otimes I = \begin{bmatrix} 1 & 0 & 0 & 0 \\ 0 & 1 & 0 & 0 \\ 0 & 0 & 0 & 1 \\ 0 & 0 & 1 & 0 \end{bmatrix} \otimes \begin{bmatrix} 1 & 0 \\ 0 & 1 \end{bmatrix} = \begin{bmatrix} 1 & 0 & 0 & 0 & 0 & 0 & 0 & 0 \\ 0 & 1 & 0 & 0 & 0 & 0 & 0 & 0 \\ 0 & 0 & 1 & 0 & 0 & 0 & 0 & 0 \\ 0 & 0 & 0 & 1 & 0 & 0 & 0 & 0 \\ 0 & 0 & 0 & 0 & 0 & 0 & 1 & 0 \\ 0 & 0 & 0 & 0 & 0 & 0 & 0 & 1 \\ 0 & 0 & 0 & 0 & 1 & 0 & 0 & 0 \\ 0 & 0 & 0 & 0 & 0 & 1 & 0 & 0 \end{bmatrix} \quad (7.7)$$

Now we can apply the system $CX_{sa} \otimes I$ to the state $|\phi_0\rangle$ we found in Equation 7.6 to get $|\phi_1\rangle$, as shown in Equation 7.8.

$$|\phi_1\rangle = (CX_{sa} \otimes I)\,|\phi_0\rangle = \begin{bmatrix} 1 & 0 & 0 & 0 & 0 & 0 & 0 & 0 \\ 0 & 1 & 0 & 0 & 0 & 0 & 0 & 0 \\ 0 & 0 & 1 & 0 & 0 & 0 & 0 & 0 \\ 0 & 0 & 0 & 1 & 0 & 0 & 0 & 0 \\ 0 & 0 & 0 & 0 & 0 & 0 & 1 & 0 \\ 0 & 0 & 0 & 0 & 0 & 0 & 0 & 1 \\ 0 & 0 & 0 & 0 & 1 & 0 & 0 & 0 \\ 0 & 0 & 0 & 0 & 0 & 1 & 0 & 0 \end{bmatrix} \vee \begin{bmatrix} \alpha \\ 0 \\ 0 \\ \alpha \\ \beta \\ 0 \\ 0 \\ \beta \end{bmatrix} = \vee \begin{bmatrix} \alpha \\ 0 \\ 0 \\ \alpha \\ 0 \\ \beta \\ \beta \\ 0 \end{bmatrix} \quad (7.8)$$

I'll be pragmatic now. Looking at $|\phi_1\rangle$ in Equation 7.8, what would it take to turn this into the teleportation state $|\tau\rangle$ in Equation 7.1?

The trick is to write out $|\tau\rangle$ as a single state and compare it to $|\phi_1\rangle$ in Equation 7.8. Then we'll see if we can find a sequence of operations that juggle around the elements of $|\phi_1\rangle$ so that they match $|\tau\rangle$.

I'll find this explicit form of $|\tau\rangle$ in two steps. First, I'll expand each basis state $|00\rangle$ through $|11\rangle$ into its corresponding four-element ket, and then I'll replace each modified version of $|\sigma\rangle$ with the coefficients of its matrix.

As Equation 7.1 shows, we'll need four transformations in all, one for each basis state.

Equation 7.9 summarizes those four transformations for reference. In the fourth row, I applied Z and X operators (in that order) to make the combined operator XZ.

$$I\,|\sigma\rangle = \begin{bmatrix} 1 & 0 \\ 0 & 1 \end{bmatrix}\begin{bmatrix} \alpha \\ \beta \end{bmatrix} = \begin{bmatrix} \alpha \\ \beta \end{bmatrix}$$

$$X\,|\sigma\rangle = \begin{bmatrix} 0 & 1 \\ 1 & 0 \end{bmatrix}\begin{bmatrix} \alpha \\ \beta \end{bmatrix} = \begin{bmatrix} \beta \\ \alpha \end{bmatrix}$$

$$Z\,|\sigma\rangle = \begin{bmatrix} 1 & 0 \\ 0 & -1 \end{bmatrix}\begin{bmatrix} \alpha \\ \beta \end{bmatrix} = \begin{bmatrix} \alpha \\ -\beta \end{bmatrix}$$

$$XZ\,|\sigma\rangle = \begin{bmatrix} 0 & -1 \\ 1 & 0 \end{bmatrix}\begin{bmatrix} \alpha \\ \beta \end{bmatrix} = \begin{bmatrix} -\beta \\ \alpha \end{bmatrix}$$

(7.9)

With these in hand, let's rewrite $|\tau\rangle$. The steps are in Equation 7.10. The first line repeats $|\tau\rangle$ from Equation 7.1. The second line expands the basis states into kets, and the third replaces each modified version of $|\sigma\rangle$ with its state from Equation 7.9. In this last line, the rules of operator precedence tell us to perform the tensor operations before the additions.

$$|\tau\rangle = \frac{1}{2}\left(\,|00\rangle\,I\,|\sigma\rangle + |01\rangle\,X\,|\sigma\rangle + |10\rangle\,Z\,|\sigma\rangle + |11\rangle\,XZ\,|\sigma\rangle\,\right)$$

$$= \frac{1}{2}\left(\begin{bmatrix} 1 \\ 0 \\ 0 \\ 0 \end{bmatrix}I\,|\sigma\rangle + \begin{bmatrix} 0 \\ 1 \\ 0 \\ 0 \end{bmatrix}X\,|\sigma\rangle + \begin{bmatrix} 0 \\ 0 \\ 1 \\ 0 \end{bmatrix}Z\,|\sigma\rangle + \begin{bmatrix} 0 \\ 0 \\ 0 \\ 1 \end{bmatrix}XZ\,|\sigma\rangle\right)$$

(7.10)

$$= \frac{1}{2}\left(\begin{bmatrix} 1 \\ 0 \\ 0 \\ 0 \end{bmatrix}\otimes\begin{bmatrix} \alpha \\ \beta \end{bmatrix} + \begin{bmatrix} 0 \\ 1 \\ 0 \\ 0 \end{bmatrix}\otimes\begin{bmatrix} \beta \\ \alpha \end{bmatrix} + \begin{bmatrix} 0 \\ 0 \\ 1 \\ 0 \end{bmatrix}\otimes\begin{bmatrix} \alpha \\ -\beta \end{bmatrix} + \begin{bmatrix} 0 \\ 0 \\ 0 \\ 1 \end{bmatrix}\otimes\begin{bmatrix} -\beta \\ \alpha \end{bmatrix}\right)$$

Finally, let's explicitly compute the tensor products in the last line of Equation 7.10, giving us Equation 7.11.

$$|\tau\rangle = \frac{1}{2}\left(\begin{bmatrix} \alpha \\ \beta \\ 0 \\ 0 \\ 0 \\ 0 \\ 0 \\ 0 \end{bmatrix} + \begin{bmatrix} 0 \\ 0 \\ \beta \\ \alpha \\ 0 \\ 0 \\ 0 \\ 0 \end{bmatrix} + \begin{bmatrix} 0 \\ 0 \\ 0 \\ 0 \\ \alpha \\ -\beta \\ 0 \\ 0 \end{bmatrix} + \begin{bmatrix} 0 \\ 0 \\ 0 \\ 0 \\ 0 \\ 0 \\ -\beta \\ \alpha \end{bmatrix}\right) = \frac{1}{2}\begin{bmatrix} \alpha \\ \beta \\ \beta \\ \alpha \\ \alpha \\ -\beta \\ -\beta \\ \alpha \end{bmatrix}$$

(7.11)

Great! Now we have an eight-element ket for $|\tau\rangle$. This is the very same τ we originally saw in Equation 7.1, but it's now represented as a single state vector, where all three qubits are entangled together.

This $|\tau\rangle$ is our goal. We want to turn the $|\phi_1\rangle$ in Equation 7.8 into this $|\tau\rangle$. How can we do this? Is there some operator A (or some sequence of operators that we can multiply together to make A) that we can plug into Equation 7.12 to do the trick?

$$
A\,|\phi_1\rangle = |\tau\rangle, \quad \text{or} \quad \vee A
\begin{bmatrix}
\alpha \\ 0 \\ 0 \\ \alpha \\ 0 \\ \beta \\ \beta \\ 0
\end{bmatrix}
= \frac{1}{2}
\begin{bmatrix}
\alpha \\ \beta \\ \beta \\ \alpha \\ \alpha \\ -\beta \\ -\beta \\ \alpha
\end{bmatrix}
\tag{7.12}
$$

Happily, we can indeed build an operator A that does just what we want. If you write down Equation 7.12 with a big empty matrix for A, then you can work through each element and fill in the entries. You'll find some elements must be 1, others −1, and still others must be 0. And to turn the $\vee$ in $|\phi_1\rangle$ into the $1/2$ in $|\tau\rangle$, the matrix will need to include another factor of $\vee$. The resulting matrix is shown in Equation 7.13.

$$
\vee
\begin{bmatrix}
1 & 0 & 0 & 0 & 1 & 0 & 0 & 0 \\
0 & 1 & 0 & 0 & 0 & 1 & 0 & 0 \\
0 & 0 & 1 & 0 & 0 & 0 & 1 & 0 \\
0 & 0 & 0 & 1 & 0 & 0 & 0 & 1 \\
1 & 0 & 0 & 0 & -1 & 0 & 0 & 0 \\
0 & 1 & 0 & 0 & 0 & -1 & 0 & 0 \\
0 & 0 & 1 & 0 & 0 & 0 & -1 & 0 \\
0 & 0 & 0 & 1 & 0 & 0 & 0 & -1
\end{bmatrix}
= \vee
\left[
\begin{array}{c|c}
I_4 & I_4 \\ \hline
I_4 & -I_4
\end{array}
\right]
\tag{7.13}
$$

You can pull this pattern apart, as shown by the block matrix on the right. In the upper-left, upper-right, and lower-left blocks, we have the 4-by-4 identity matrix given by $I_4 = I \otimes I$. In the bottom-right block, we have its negative, $-(I \otimes I)$. This pattern of positives and negatives is just what we get from forming $H \otimes (I \otimes I)$. The parentheses aren't needed, but I've put them there to emphasize that we're thinking of $I \otimes I$ as one matrix that gets tensored with H.

In other words, the matrix in Equation 7.13 that takes us from $|\phi_1\rangle$ to $|\tau\rangle$ is given by $H \otimes I_4$, or $H \otimes I \otimes I$. In the circuit diagram, we draw an H on the top line and usually just imply the I qugates, as in Figure 7-5(a). Figure 7-5(b) shows what it looks like with the I qugates drawn explicitly.

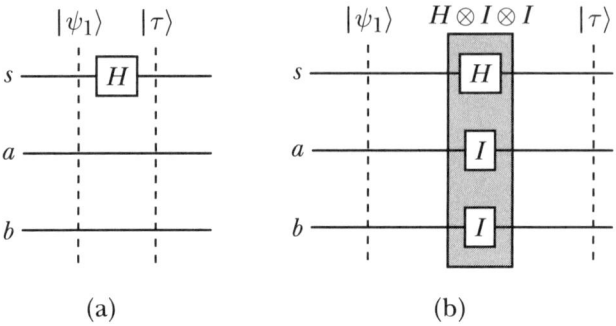

Figure 7-5: (a) Applying the final H to the top qubit, giving us the teleportation state $|\tau\rangle$. (b) Including the implied I qugates.

This process demonstrates that sometimes we build up a circuit from principles, and sometimes we just create something that performs a specific task.

Let's put it all together. Starting with qubits a and b in the state $|0\rangle$, we entangle them with an H and CX_{ba}, then we apply CX_{sa}, and finally we apply an H to s to create the teleportation state $|\tau\rangle$. The process is shown in Figure 7-6.

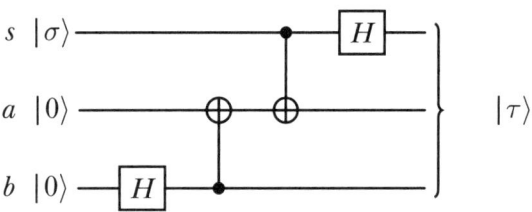

Figure 7-6: The full setup step for teleportation

This is the conceptual peak of the argument: We've created the teleportation state $|\tau\rangle$. The hard work is done! Now we're on the downhill slope from Figure 7-1.

Alice Measures Her Qubits

Now that the teleportation state has been set up, Alice will collapse it to just one state. This will push the state $|\sigma\rangle$ onto Bob's qubit (because it will have nowhere else to go) and simultaneously collapse the qubit s that has held $|\sigma\rangle$ until now.

Alice can measure her two qubits s and a in either order, or even at the same time. Appending this measurement to our existing circuit from Figure 7-6 gives us Figure 7-7. I'm labeling the output bits with the letter m (for measurement) rather than my usual b (for bit) because we're already using b for Bob's qubit.

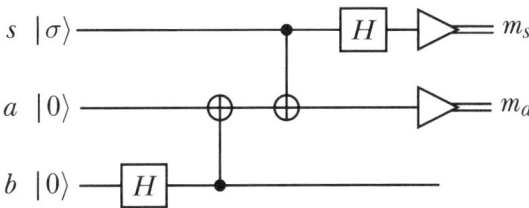

Figure 7-7: After the circuit of Figure 7-6, Alice measures her qubits.

Let's write the measured values from this system as a bitstring $m_s m_a$.

Suppose that Alice measures $m_s = 0$ and $m_a = 1$, or the bitstring 01. The law of partial measurement says that the system state $|\tau\rangle$ must collapse to states that are consistent with this measurement. That is, the system collapses to include only those states that start with $|01\rangle$. There is only one such state in $|\tau\rangle$ from Equation 7.1, and that's $|01\rangle X |\sigma\rangle$. So in this case, after the measurement, Alice's qubits a and s are now $|0\rangle$ and $|1\rangle$ respectively, and Bob's qubit b must be $X |\sigma\rangle$.

Alice Tells Bob the Measurements

When Alice's measurements are complete, our three-qubit system has collapsed to one of the four states in Equation 7.1. That is, depending on what Alice measured, Bob is holding a qubit that is in the state $I |\sigma\rangle$, or in the state $X |\sigma\rangle$, or in the state $Z |\sigma\rangle$, or in the state $XZ |\sigma\rangle$.

If Bob can determine the state of his qubit, he can apply the correct qugates to leave him with $|\sigma\rangle$. So the big question is, how can Bob tell which state he has?

That depends on what Alice measured. So the new question is, how can Bob determine what Alice saw on her meters?

The answer is that he can't! *Alice has to tell him.*

There's no quantum way for Alice to tell Bob anything, because Alice has already collapsed their entangled states.

The only remaining way for Alice to tell Bob what she measured is by using classical means. She can send the values of her two measurements via radio, or she can use a laser to bounce light off the moon, or she can send him a newspaper with the measurements printed somewhere. No matter how she chooses to send this information to Bob, she needs to use classical means, which are limited by the speed of light.

Bob Recovers $|\sigma\rangle$

Once Bob receives the two classical bits that Alice sent, telling him what she measured, he can recover $|\sigma\rangle$ from his qubit. His work is made easier by another remarkable feature of the structure of the teleportation state: Bob will need only two qugates!

The idea is that Bob will use the received classical bits as controls on a controlled-X qugate and a controlled-Z qugate. We've only discussed using quantum bits as controls, but we can use classical bits as well. Think of initializing a pair of qubits using the corresponding bits.

To see what Alice's measurements tell Bob, I've repeated the teleportation state $|\tau\rangle$ from Equation 7.1 here as Equation 7.14.

$$|\tau\rangle = \frac{1}{2}\Big(\,|00\rangle\,I\,|\sigma\rangle + |01\rangle\,X\,|\sigma\rangle + |10\rangle\,Z\,|\sigma\rangle + |11\rangle\,XZ\,|\sigma\rangle\,\Big) \qquad (7.14)$$

Bob's actions to recover $|\sigma\rangle$ are shown in Figure 7-8.

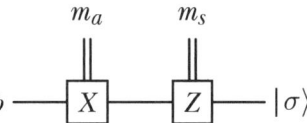

Figure 7-8: Bob decoding his qubit based on Alice's classical bits

Let's look at the four possible pairs of bits that Bob might receive.

If Alice measured 00, then $m_a = m_s = 0$. Equation 7.14 tells us that when Alice measures 00, Bob has $I\,|\sigma\rangle = |\sigma\rangle$. Since both controls are 0, neither qugate is applied. This is just right, because Bob already has $I\,|\sigma\rangle = |\sigma\rangle$, and he's done.

If Alice measured 01, then $m_s = 0$ and $m_a = 1$. Equation 7.14 tells us that Bob's qubit is in state $X\,|\sigma\rangle$. Bob applies the controlled-X qugate, and because X is its own inverse, this gives him $XX\,|\sigma\rangle = |\sigma\rangle$, and he's recovered $|\sigma\rangle$.

If Alice measured 10, then $m_s = 1$ and $m_a = 0$. This tells Bob to apply the controlled-Z qugate. Like the X qugate, Z is its own inverse, so Bob gets $ZZ\,|\sigma\rangle = |\sigma\rangle$.

Finally, If Alice measured 11, then $m_a = m_s = 1$, and Bob knows that his qubit is $XZ\,|\sigma\rangle$. To undo the transformation XZ, Bob applies X and then Z in that order, which is equivalent to applying the single operator ZX (remember to read the operators from right to left). The steps are shown in Equation 7.15.

$$
\begin{aligned}
ZX(XZ)\,|\sigma\rangle &= Z(XX)Z\,|\sigma\rangle && \text{Regroup matrix multiplies} \\
&= ZZ\,|\sigma\rangle && \text{Since } XX = I \\
&= |\sigma\rangle && \text{Since } ZZ = I
\end{aligned}
\qquad (7.15)
$$

It's important to keep the order of the operations at each point in the algorithm clear in your mind. If Alice's qubit is in the state $XZ\,|\sigma\rangle$, then to recover $|\sigma\rangle$ Bob has to apply the inverse of operator XZ, which is ZX.

And we're done. For each of Alice's four possible measurements, Bob has successfully ended up with his qubit in the state $|\sigma\rangle$. We've teleported $|\sigma\rangle$ from Alice to Bob!

Drawing the Teleportation Protocol

Let's now put all the pieces together, giving us Figure 7-9.

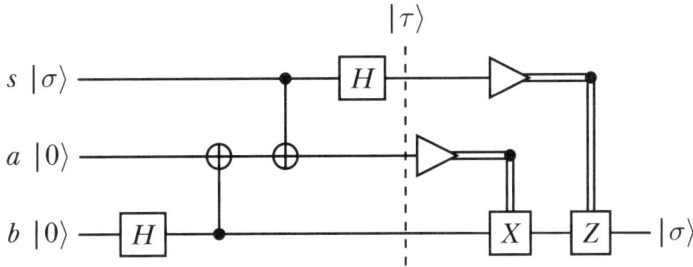

Figure 7-9: The full teleportation circuit as it's normally drawn, with the teleportation state $|\tau\rangle$ marked

Figure 7-9 is the complete, traditional quantum teleportation protocol. This theoretical process has been experimentally confirmed [22].

However, Figure 7-9 may be somewhat misleading, because it shows all three qubits at the far left, suggesting that Alice and Bob have them all in their control at the start of the process. But if Alice already has s in the state $|\sigma\rangle$, and Bob is standing there, she could just hand the s qubit (protected carefully) to Bob, and there would be no need for teleporting anything!

For that reason, I prefer drawing this as in the overall recap of Figure 7-10. The delayed introduction of s clarifies that s isn't yet in the state $|\sigma\rangle$ when Alice and Bob are entangling a and b. Only later does Alice compute s, and then continue the protocol.

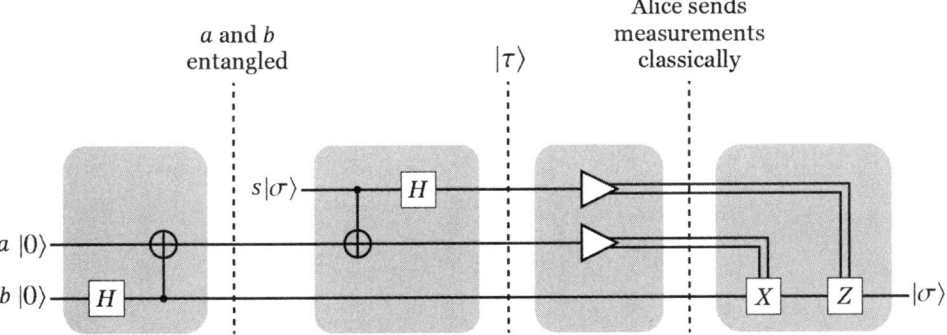

Figure 7-10: A recap of the quantum teleportation algorithm

Teleportation is usually drawn as in Figure 7-9, so keep in mind that in practice, the qubit s is usually not in the state $|\sigma\rangle$ at the start, when Alice and Bob are creating their entangled pair.

Probabilistic Teleportation

We've seen that after Alice has measured her qubits, Bob's qubit is in one of the four states in the superposition $|\tau\rangle$, but he doesn't know which one. Alice has to tell him by sending him two classical bits by classical means.

But let's suppose that for some reason, Alice can't send Bob her bits. Is the situation hopeless for Bob, or is there some way, perhaps with a combination of effort and luck, that he will be able to recover $|\sigma\rangle$ from his qubit?

Let's try another thought experiment.

Suppose that there's been a terrible accident on Mars. The habitat blew up when Bob was out on a mission, leaving Bob the only survivor. The explosion also damaged the rocket he and his colleagues were going to use to return to Earth, and almost all the fuel has leaked out. There's enough fuel to lift off the surface, but there isn't nearly enough to get the rocket back to Earth.

Bob's supplies will run out long before a rescue mission can reach him, so he needs to find a new way home.

Luckily, an earlier Mars mission placed an emergency rescue satellite in Mars orbit. If Bob can reach it, the resources there will not only keep him alive, but he'll also be able to repair his rocket. And there's enough fuel there to fill the rocket's tanks. It's a plan to get home!

Getting his damaged rocket safely up to the rescue satellite will require an elaborate flight plan with multiple steps that he'll have to perform at the right moments. The specific plan for any given day will depend a lot on the local weather.

Unfortunately, Bob doesn't have access to the weather satellites above Mars. But he does have a working radio, and he contacts Alice, who can read the weather satellite data without a problem. They agree that Alice will use that data to work out a flight plan for the next day and send it to Bob. Because these flight plans are complicated, she'll encode the entire plan into a single state, $|\sigma\rangle$. This is a good strategy for them, because before Bob left, he and Alice created lots and lots of entangled pairs to use for teleporting states over the duration of Bob's mission. Bob's half of each pair survived the accident. So once Alice tells him her measurements, Bob can take down the bottle containing the next qubit to be used and apply Alice's bits to put it into the state $|\sigma\rangle$.

On Mars, Bob has cobbled together a decoder to turn a quantum state sent by Alice into a flight plan. He's also written a simulation program that will look at a flight plan and tell him whether it's safe and he'll reach the satellite, or it's unsafe and the rocket will blow up, along with Bob.

The next morning, as Bob prepares to hear from Alice what her measurements were, Bob's radio won't even turn on. It's busted, and he doesn't have the parts to fix it, so now he's lost all touch with Alice. He can't get her measurements, so he can't confidently process his qubit to turn it into $|\sigma\rangle$.

Bob isn't completely without hope, though. He can just plain old guess. Suppose he guesses that Alice measured 00, so his qubit is in the state $I|\sigma\rangle$, and he doesn't have to process it. He has a three in four chance of being wrong, but more optimistically, a one in four chance of being right!

So he feeds his qubit into the decoder, which gives him a flight plan. He then gives that to the safety testing program. If he's lucky, the test will tell him that he guessed correctly and that the flight plan makes sense and is safe. But if he's unlucky, then his guess was wrong (that is, his qubit wasn't in the state $|\sigma\rangle$, but one of the other states in $|\tau\rangle$). In that case, the decoder will have produced a nonsensical flight plan, and the test will tell him that following that plan would end in disaster.

If Alice and Bob were sharing only a single entangled pair, this would be the end of the story. Bob would have no option but to launch the rocket anyway and hope for the best.

Is there some way, *any way*, that Bob can improve his odds?

Recall that Alice and Bob created not one entangled pair before Bob left, but many dozens or hundreds of them. They intended to use them to teleport different quantum states over the course of Bob's mission. But now that Alice knows Bob's radio is out, she'll use them all right away. Alice will run the quantum protocol to teleport the same state $|\sigma\rangle$ over every pair of entangled qubits. Since she can't clone the $|\sigma\rangle$ she's made, she runs her plan-making program many times, creating many distinct qubits that are all $|\sigma\rangle$. She plugs each of these qubits into the teleportation protocol, and even measures the output bits, though she can't send them to Bob. At this point, she's done all she can.

Back on Mars, suppose that Bob's guess for his first qubit resulted in a meaningless and unsafe flight plan. He hopes that Alice is following their backup plan, and gives her some time to compute and entangle $|\sigma\rangle$ on all of their remaining pairs.

After a little while, he'll take down the next of his entangled qubits, guess again, and process the qubit according to his guess of what state the qubit is in (he could just guess it's $|\sigma\rangle$ every time, and apply no qugates to it). Then Bob will decode his qubit and test the resulting flight plan, hoping it will be safe.

The process is shown graphically in Figure 7-11. Alice measures s and a, producing classical bits m_s and m_a, but she does nothing with them. The measurements were just to collapse the states of the qubits. Bob then feeds his qubit b into the decoding algorithm G that turns that qubit into a flight path, represented by the binary number g that comes from measuring the output of G. Because g is a classical binary number, Bob can make as many copies of it as he pleases.

So, Bob makes a copy of g and feeds it into his test, along with whatever other inputs it needs. If the test says the flight plan is safe, then he can follow the steps in g and he's all set for launch!

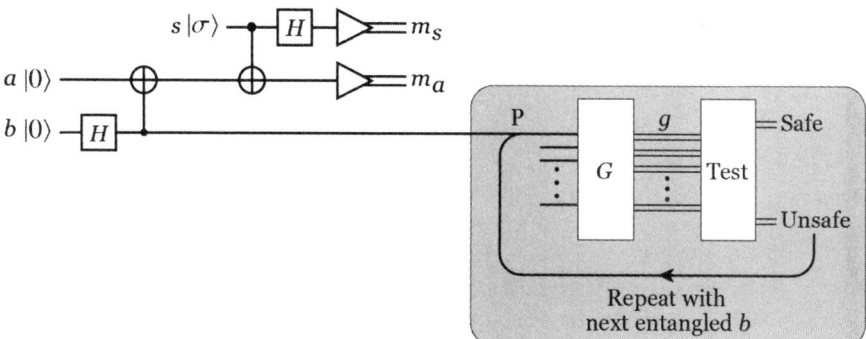

Figure 7-11: After Alice has measured her qubits, Bob guesses that he has $|\sigma\rangle$, computes the binary bitstring g, and then tests that bitstring to see if it's a safe flight plan.

If the test tells him that the flight plan is unsafe, he takes down his next qubit b, makes another guess, and tries again.

Because there's no way for Bob to be sure beforehand that a given qubit actually has the value he's guessing, we might call this process *probabilistic teleportation*.

An important thing to keep in mind is that Bob doesn't really care about the state $|\sigma\rangle$. He'll *use* each qubit, rather than study it, and it's the results of the decoder and test that he cares about.

Once Bob has guessed at the state of any qubit b, he has a one in four chance, or a probability of 0.25, that he'll have guessed correctly and the test will tell him the plan is safe. Those aren't great odds.

What are the chances that when Bob uses this approach, he will ultimately guess right, and thereby get a safe flight plan that could save his life?

To see Bob's chances of success, consider his odds of failure. After one guess (and any processing it might require), there's a three in four chance that b is *not* in the state $|\sigma\rangle$. Thus, Bob has a 0.75 probability of being wrong (and getting an unsafe plan). But this means he has a $1 - 0.75 = 0.25$ probability of being right (and getting a safe plan). After two repeats of the teleportation, his probability of guessing incorrectly both times is $0.75 \times 0.75 = 0.75^2 = 0.5625$. Thus, his probability of being correct at least once is $1 - 0.5625 = 0.4375$. Much better!

After n repeats of the protocol, the probability that Bob has guessed correctly at least once is $1 - 0.75^n$, which I've graphed in Figure 7-12.

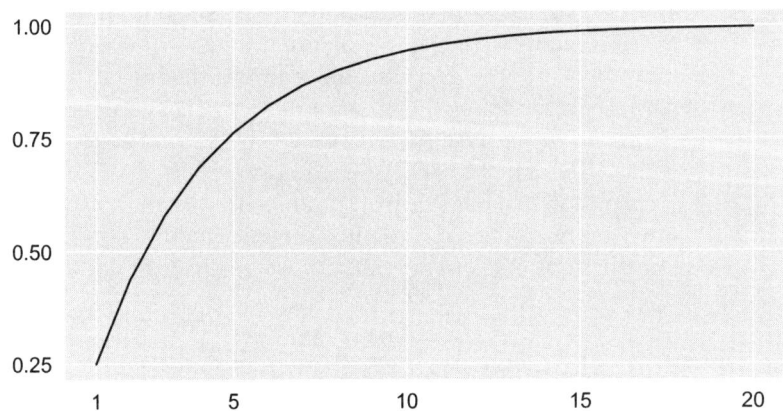

Figure 7-12: A plot of $1 - 0.75^n$ for n from 1 to 20

After 10 repeats, Bob has about a 0.94 probability of having been right at least once. After 20 repeats, his probability of having been right at least once is almost 0.997.

If Alice and Bob are willing to share 20 entangled qubits before Bob leaves, *and* Alice computes her side of the protocol from scratch 20 times, *and* Bob runs his decoder and test 20 times, there's about a 99.7 percent likelihood that Bob will have guessed correctly at some point and obtained a safe flight plan. He only needs one safe flight plan, and he can stop as soon as he has it.

After 50 attempts, Bob's chance of never guessing right, even once, is less than 1 in a million. But even though the odds of Bob guessing correctly go up with each repeat, if he's super unlucky, he might never guess correctly. Worse, Bob has no way to tell Alice if he's guessed correctly or not.

When Alice and Bob are able or willing to share two classical bits, they can run the classical protocol once and teleport the qubit for sure. If they don't want to share those classical bits, or they're unable to, they can hope to teleport the qubit, but they'll have to put in a lot of extra effort, and they might still fail.

To be guaranteed success, Alice needs to send the two classical bits representing her measurements to Bob, using classical means. If they can't do that, they can hope that luck is on Bob's side and he'll get an answer that passes his test before he runs out of qubits.

This is only one of many interesting modifications to the basic teleportation algorithm [4] [158] [289]. You can read up on the references, or try your own ideas. Exploring variations on circuits you already understand is a great way to gain experience with quantum algorithms.

Summary

The teleportation protocol lets us transfer a quantum state from one qubit to another, which can be arbitrarily far away. The process requires that Alice and Bob already share an entangled pair of qubits, and that Alice can transmit two classical bits to Bob. Alice's measurements cause qubit s to collapse. This means that there is never more than one qubit in the state $|\sigma\rangle$.

The big surprise of teleportation is that the state we've transferred contains two complex numbers, each of which is built from two real numbers, for a total of four real numbers. These numbers can require arbitrary numbers of digits if written out, but they will be transferred with perfect precision.

Once Bob has operated on his state, he can measure it. As always, measurement will give him only a 0 or a 1, so there's no way to extract those four real numbers that were transmitted. But before measurement, Bob can use his qubit, now in state $|\sigma\rangle$, in further computations. So Bob can then build on Alice's work, using her result as an input to his own algorithm.

Although the collapse of Bob's qubit is immediate after Alice's measurements, the need to then share classical bits prevents us from using this protocol to send information faster than the speed of light. That's too bad, but it doesn't change the fact that quantum teleportation is still a pretty amazing feat.

If Alice and Bob have lost all classical communication, and they have some additional resources, they can use a probabilistic approach that is likely to give Bob the state $|\sigma\rangle$ eventually. But he'll probably have to try several different instances of b (independently computed by Alice) and then test each result.

In this chapter, we looked at the matrix elements behind Alice's computation of the teleportation state $|\tau\rangle$, and then we unpacked that state to teleport $|\sigma\rangle$ to Bob. We had to deal with a few big matrices, but they were manageable. In general, a system of n qubits will need an operator described by a matrix 2^n elements on a side, which quickly becomes too big to write out and manually compute with. So from now on, we'll focus on the algebraic approach most of the time, rather than writing out the matrices and kets.

Teleportation is an amazing algorithm, and it shows the power of entanglement for sharing information at a great distance. It's pretty incredible that Alice can perform some quantum operations on qubits on Earth and change the states of Bob's qubits on Mars (or even a planet in orbit around a star millions of light years away).

In the next few chapters, we'll look at more quantum algorithms, each of which introduces one or more new concepts into our quantum repertoire.

8

DEUTSCH'S ALGORITHM

Is you is or is you ain't my baby?
Way you're acting lately makes me doubt.
Yous is still my baby, baby.
Seems my flame in your heart's done gone out.
—Louis Jordan, "Is You Is or Is You Ain't My Baby?," 1943 [105]

Now it's time to put the "computing" into quantum computing! In this chapter, we'll meet a quantum circuit that actually computes the solution to a problem.

Don't set your expectations too high, though. The problem we'll solve is a contrived, toy problem. You'll probably never need to solve it in practice, and if you did, you could do so with a one-line program on any classical computer. However, this quantum algorithm will solve the problem in a way that no classical computer ever could.

The big idea is that we're given a small quantum circuit that we're promised belongs to one of two possible categories, but we're not told which. We're allowed to feed an input to the circuit and collect its output, which we call a single *evaluation* of the circuit. Our goal is to determine which category the circuit belongs to by evaluating it the smallest possible number of times. In the classical world, we must evaluate the circuit twice. But Deutsch's algorithm uses quantum computing to get the answer with only one evaluation!

This algorithm is worth our attention because it serves as a terrific introduction to many of the important ideas that enable today's more complex and useful quantum algorithms. We'll look at circuit fragments that get used time and time again, we'll see the ideas behind those fragments, and we'll cover mathematical idioms and analysis techniques that will be useful parts of our standard toolbox when we look at more complex algorithms. Most importantly, we'll see our first example of *quantum parallelism*, which (when fully unleashed) lets us evaluate astronomical numbers of inputs simultaneously. Getting used to these ideas in the context of this small problem is a great way to become familiar with them without distractions.

This chapter uses the problem we want to solve as a running thread, but we'll make several digressions along the way to meet new tools. As you'll see, each tool will move us a little further along in our analysis. By the time we're done, we'll have a complete solution to the problem and a whole lot of useful tools under our belts.

Deutsch's algorithm is historically one of the first demonstrations of the power of quantum algorithms, and it's considered essential knowledge for everyone working in quantum computing.

Deutsch's Problem

I'll start with the problem we want to solve. This problem, and its solution, were originally proposed by David Deutsch, so they are now known as *Deutsch's problem* and *Deutsch's algorithm* [49].

Suppose that we've been given a tiny classical function f that takes a single bit as input and produces a single bit as output. There are only four possible functions that fit this description. I've written them out in Table 8-1 with the names f^{00}, f^{01}, f^{10}, and f^{11} (this frees up the subscript for later use). Note that these superscripts are labels, not exponents. Each column in the table gives the outputs of that function for the inputs $x = 0$ and $x = 1$.

Table 8-1: The four functions for Deutsch's problem

x	$f^{00}(x)$	$f^{01}(x)$	$f^{10}(x)$	$f^{11}(x)$
0	0	0	1	1
1	0	1	0	1
	Constant	Balanced	Balanced	Constant

In Table 8-1, I also gave a "category" name to each function. If the output is the same for both inputs (that is, it's f^{00} or f^{11}), then I've labeled it as a *constant function*. If the output is different for the two inputs (that is, it's f^{01} or f^{10}), then I've labeled it as a *balanced function*. Every one-bit function is either one or the other.

We are not allowed to know which of these four functions we've been given. All we can do is give the mystery function inputs, wait for them to be processed, and then collect its outputs. Each such evaluation of the function is also called a *query* of the function.

The problem is to determine, with the smallest number of queries, whether the function is constant (f^{00} or f^{11}) or balanced (f^{01} or f^{10}). It's important to be clear that we are not trying to find out which function we have, and we're not interested in the particular output for any given input. We *only* want to determine which of these two categories the function belongs to.

I said earlier that this problem could be solved in a single line on a classical computer. Let's see how.

I'll write a one-line routine that takes as its argument a function named f. This function takes an input that's either 0 or 1 and returns either 0 or 1. Our goal is to return True if f is balanced (so one input returns 0 and the other returns 1) and False otherwise. The code, in a Python-like language, is shown in Listing 8-1.

```
def is_balanced(f):
    return f(0) != f(1)
```

Listing 8-1: A classical solution to Deutsch's problem

We have to call the function f twice to determine if it's balanced or not. We say that on a classical computer, Deutsch's problem requires a minimum of two queries.

There seems to be no way around this. We have to evaluate the function once for the input 0 and then a second time for the input 1. Then we can look at the outputs and declare which category the function belongs to. This is just what is_balanced does when it evaluates both f(0) and f(1).

The beauty of Deutsch's algorithm is that it can answer this problem with only *one* evaluation of the function, rather than the two evaluations required by any classical algorithm. That takes half the time!

Oracles

At the heart of Deutsch's algorithm lies the function we want to evaluate. Therefore, we need some way to include the function in the algorithm's quantum circuit. This embedding of an unknown function into a quantum circuit is part of many quantum algorithms, including several we'll see later in this book.

The general approach is to abstract away the function we want to learn about and wrap it up in an *oracle*. This word has multiple meanings, from a person who claims to speak for a deity [139] (such as the famous Oracle of Delphi [35]) to an unspecified mechanism that can solve a computational problem [275].

An oracle is also called a *black box*. Like the word *oracle*, this phrase means different things in different fields, from the flight recorder on an airplane or other vehicle [138] to an early medical instrument [21]. For us, the terms oracle and black box are synonymous.

In quantum computing, we use the term oracle to refer to a fragment of quantum circuitry that we can query, but which we cannot examine. All we can do is give it inputs and receive its outputs.

Often, though not always, we initially define the oracle classically in terms of the classical bits 0 and 1. We then adapt that description to the quantum world by using the states $|0\rangle$ and $|1\rangle$ instead and making any necessary adjustments. I'll follow that approach here and initially discuss this oracle in terms of classical bits, not qubits.

We can draw such an oracle as a box that takes in a query and produces a response. Usually, the query is named x and enters on the left. The response, $f(x)$, leaves on the right, as in Figure 8-1. This figure is not a quantum circuit! It's a block diagram of a classical algorithm, where the single lines carry classical bits and the boxes contain classical functions. The rest of the figures in this section will also be classical diagrams (our figures will return to being quantum circuits at the start of the next section, "Quantum Oracles").

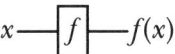

Figure 8-1: A classical one-bit oracle. This is a classical block diagram, not a quantum circuit.

This form of oracle isn't appropriate for quantum computing because it's not *reversible*. Recall that in Chapter 3 we found that all quantum operators (that is, qugates) must be unitary, and all unitary operators are reversible, meaning that we can undo them (any given operator might or might not be its own inverse, but for every operator, there's some set of operations out there that can undo it). In other words, given the output of an operator, we can always recover its input. But the oracle in Figure 8-1 doesn't give us that power. If we don't know the function f, then having only the output bit isn't enough information to say what the input was (in fact, even if we know f, that might still not be enough information).

You can probably imagine several ways to fix this problem. One approach is to propagate the input as a second output, as in Figure 8-2.

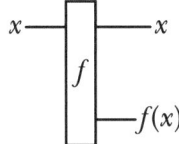

Figure 8-2: Propagating an oracle's input query as an output

This is better, but it's still not reversible. The problem can be seen as a question of conservation of information. If two bits come out, then to be reversible, two bits need to go in.

To achieve this, let's add another input, which we usually call y. So now x and y go in, and x and $f(x)$ come out, as in Figure 8-3.

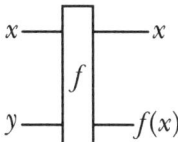

Figure 8-3: A classical oracle with two inputs and two outputs

It's now easy to recover the input x, as we have it at the output, but how do we recover y? A clever solution involves using the *exclusive OR* operation [141].

We still propagate the input x to an output, but instead of presenting $f(x)$ on the second line, we present the exclusive OR of y and $f(x)$. You may recall that the exclusive OR of two bits a and b, written $a \oplus b$, is 0 if the bits are the same and 1 if they aren't (that is, it's 1 if either bit is 1, but not both). The truth table for the exclusive OR (usually written XOR) is shown in Table 8-2.

Table 8-2: The XOR truth table

a	b	$a \oplus b$
0	0	0
0	1	1
1	0	1
1	1	0

Using XOR, an improved version of the oracle with classical bits is shown in Figure 8-4. This approach will generalize nicely when the inputs and outputs are bitstrings rather than single bits. The y input is called a *helper*, *ancilla*, or *auxiliary* input.

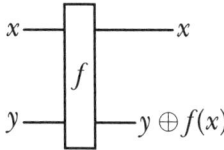

Figure 8-4: The final version of our classical one-bit oracle

Have we made the application of f reversible? If so, might this little block diagram be *self-inverse*? The answer to both questions is yes! Running two copies of the oracle, one after the other, brings us back to our starting point. To see why, consider Figure 8-5.

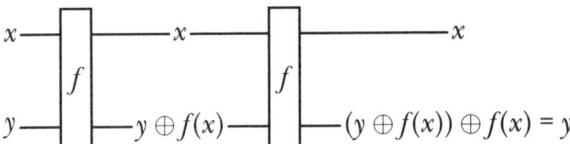

Figure 8-5: Applying the oracle twice recovers the original inputs.

We end up with y at the end of the lower line because the XOR of any bit with itself is 0 and the XOR of any bit with 0 is itself, as summarized in Table 8-3. Here, I've given the three properties names (X1–X3) so we can refer to them later.

Table 8-3: Some properties of the XOR, or $\oplus$, operator

Property	Name
$x \oplus x = 0$	X1
$x \oplus 0 = x$	X2
$(x \oplus y) \oplus z = x \oplus (y \oplus z)$	X3

Equation 8.1 uses these properties on the lower output from Figure 8-5 to show that the oracle is self-inverse.

$$\left(y \oplus f(x) \right) \oplus f(x) = y \oplus \left(f(x) \oplus f(x) \right) \quad \text{Apply property X3}$$
$$= y \oplus 0 \quad \text{Apply property X1} \quad (8.1)$$
$$= y \quad \text{Apply property X2}$$

This confirms that if we run the oracle twice, we recover both inputs x and y.

So far, we've discussed classical oracles that work with bits. Let's now bring oracles into the quantum world.

Quantum Oracles

To turn Figure 8-4 into a quantum operator, we need to somehow turn the classical bits into quantum bits. We usually do this by corresponding each classical 0 with the state $|0\rangle$ and each 1 with $|1\rangle$. Because all the inputs and outputs in Figure 8-4 are bits, this means we can create the quantum output by replacing x with $|x\rangle$, y with $|y\rangle$, and $y \oplus f(x)$ with $|y \oplus f(x)\rangle$.

This correspondence will make our notation easier both here and in the rest of the book. If we know that a particular qubit can be only $|0\rangle$ or $|1\rangle$, then we can also write it as its corresponding classical bit 0 or 1.

A nice example of this philosophy of casually interpreting bits as states and vice versa is represented by the expression $|y \oplus f(x)\rangle$. Because we know that our quantum input x can be only $|0\rangle$ or $|1\rangle$, we interpret it as 0 or 1. Now we can give that bit to the classical function f and get back a classical 0 or 1 as output. Qubit y is also either $|0\rangle$ or $|1\rangle$, so we also interpret it as a bit and form the XOR of y with $f(x)$. That classical result of either 0 or 1 is then promoted back into the quantum realm, where 0 becomes $|0\rangle$ and 1 becomes $|1\rangle$. We do that by wrapping the output value in a ket. So, qubits x and y come in and we treat them classically, perform the whole calculation, and then promote the output $|y \oplus f(x)\rangle$ to a ket. This agreement helps us keep the notation relatively clean.

For the same reason, we often allow qubit names to refer to either bits or qubits depending on the context. For example, suppose that we have a qubit that we've given the name q. The state of the qubit could then be written $|q\rangle$. In the special case where $|q\rangle$ is either $|0\rangle$ or $|1\rangle$, we can treat it as a classical bit and refer to that bit's value as q. Therefore, the symbol q can refer to either the name of the qubit, its quantum state, or its binary equivalent. This overloading of the symbol q definitely has the potential for confusion, but this is a widely used convention, so it's worth getting used to. If there's ever a chance of confusion, it's the author's job to make things clear.

Making the changes we just discussed to Figure 8-4 gives us the two equivalent oracles of Figure 8-6, one with the labels outside the box and one with the labels inside. These are both popular ways to draw this sort of operator. I prefer the version on the left, as the text is usually larger and easier to read. Note that our figures now illustrate quantum circuits again.

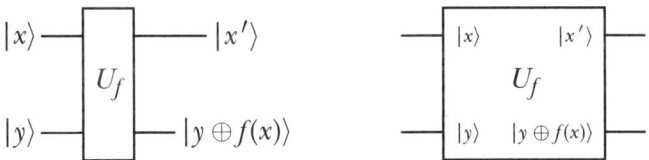

Figure 8-6: A quantum oracle in two equivalent forms; note the use of a prime in the upper output

We sometimes label a quantum gate implementing f with U_f, as I've done here, to remind us that this box now represents a unitary quantum operator rather than a classical function. For simplicity, I'll leave it as f for the rest of this chapter.

In Figure 8-6, I've written the upper output as $|x'\rangle$, even though the classical oracle in Figure 8-4 showed this as x itself. This is a *vital* distinction between the classical and quantum versions of this oracle. Curiously, it's rarely noted explicitly, and figures in the literature often casually label the upper qubit as $|x\rangle$, thereby asserting that it is identical to the upper input. But frequently this won't be the case.

The reason we need to write the upper output as $|x'\rangle$ rather than $|x\rangle$ is because qubits are more complicated than classical bits. While the output will often be $|x\rangle$, sometimes the $|x\rangle$ input is changed inside the oracle, producing a new output on the upper qubit. We'll see that the oracle in Deutsch's algorithm can indeed output an $|x'\rangle$ that is different from its input $|x\rangle$. Labeling the output $|x'\rangle$ reminds us that this output might not be $|x\rangle$ itself.

Some oracles can, in the course of whatever actions they perform, produce entangled outputs. In this case, we should write the output of the oracle using one of the graphic symbols for entanglement in Figure 5-27. But even when entanglement happens, we usually don't mark it explicitly unless that entanglement is essential to understanding the algorithm.

Sometimes an oracle will be drawn as in Figure 8-7. This drawing may seem to suggest that the output of *f* is being split in horizontal and vertical directions, violating the no-cloning theorem, even though there is no cloning involved. I think the pictures in Figure 8-6 are clearer, so I'll use that style in this book.

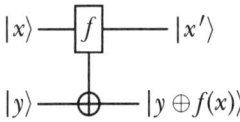

Figure 8-7: An alternative drawing of an oracle

The time and memory requirements, or costs, of any circuit are called its *computational complexity*. When we discuss circuits that use oracles, we usually assume that the complexity of the oracle dominates the overall complexity of the entire circuit. As a result, we often describe the complexity of the entire circuit by counting only the number of queries we make to the oracle while ignoring all the other costs, such as the number of other qugates and their execution times. We call the number of times we evaluate the oracle the *oracle complexity* of the circuit [71]. We usually want the oracle complexity of any circuit to be as small as we can make it, so the circuit uses a minimum of resources, runs quickly, and thereby also minimizes the impact of noise.

Promise Oracles

The oracle that we'll use in Deutsch's algorithm, as well as in many other algorithms, is not a completely arbitrary function. It's a function that we are told, ahead of time, has certain properties. We usually exploit these properties to simplify the calculations, or even make them work at all.

We often say that such an oracle comes with a *promise*, or a guarantee, that it will fulfill certain conditions. We call these *promise oracles*.

In Deutsch's algorithm, the promise is that the oracle is either constant or balanced.

Quantum Parallelism

The surprise of Deutsch's algorithm is that it only needs to evaluate the oracle a single time. Our little classical algorithm in Listing 8-1 called the function f twice, and it's hard to see how to avoid that. But with quantum computers, we can! The essence of the idea is a remarkable feature of quantum gates that we've already seen a few times but now will embrace in earnest.

Suppose we have some system of *n* qubits that's in a superposition of many states. Let's name the superposition $|\psi\rangle$ and assert that it is made up of *s* states. I'll call each such state $|\phi_k\rangle$, where *k* runs from 0 to *s* − 1, and I'll say that each of these states has an amplitude γ_k. Then we can write $|\psi\rangle$ as in Equation 8.2.

$$|\psi\rangle = \sum_{k\in[s]} \gamma_k |\phi_k\rangle, \quad \text{where} \quad \sum_{k\in[s]} |\gamma_k|^2 = 1 \tag{8.2}$$

Now we'll take this system of n qubits, containing s states in superposition, and feed it into some n-qubit qugate that I'll name A. If it were up to us, what would we choose for A to do?

Maybe it should apply itself to just one of the input states. But which one? However we pick, there are probably situations where we would wish we'd chosen differently.

If we could pick any action for A we wanted, I think the most general one would be this: *Disassemble the input superposition into its individual states, apply A to each state, and then assemble those results into a new superposition.* That new superposition would be the output of A.

Let's write this operation in symbols, as shown in Equation 8.3.

$$|\psi'\rangle = A\,|\psi\rangle = \sum_{k\in[s]} \gamma_k\, A\,|\phi_k\rangle, \quad \text{where} \quad \sum_{k\in[s]} |\gamma_k|^2 = 1 \tag{8.3}$$

The corresponding circuit diagram is shown in Figure 8-8.

$$|\psi\rangle\!-\!\boxed{A}\!-\!|\psi'\rangle$$

Figure 8-8: A circuit diagram of Equation 8.3

A conceptual illustration of this idea is shown in Figure 8-9. It illustrates the idea that applying A to $|\psi\rangle$ produces a result *as if* the superposition $|\psi\rangle$ were split apart into its individual states, they were each operated upon by A, and those results were then summed up to create a new output superposition $|\psi'\rangle$.

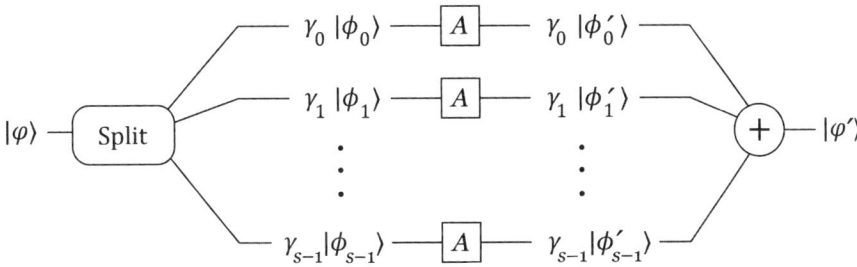

Figure 8-9: A conceptual illustration of quantum parallelism

Whatever the actual underlying mechanism might be, the result corresponds to what we'd get if we applied A to every state in the superposition simultaneously (or in parallel) and then combined all of those results into a new superposition.

We could perform the operation in Figure 8-9 on a classical computer. For instance, suppose we had a list of 10,000 possible sets of initial conditions for a complicated simulation (maybe water waves tossing a boat around in a storm, or the flames produced by a roaring fire), and we wanted to find the set of conditions that caused the simulator to produce the closest match to our desired results (for example, *almost* tipping the boat over three times, but never completely). We might evaluate all 10,000 inputs, one by one, and then give each output a numerical score that describes how well it matched the result we wanted. Then we'd return the starting set with the best score.

If it took m minutes to run the simulation once, it would take $10,000m$ minutes to process every set of inputs.

Now suppose that we did this on a quantum computer that behaved as I described earlier. We'd create a superposition of all 10,000 initial conditions and run them through a quantum version of the simulation, evaluating *all* 10,000 *inputs in parallel*. In the time it takes to process only a single state, our circuit would evaluate every input state in the superposition *simultaneously* and would produce an output superposition containing all the outputs.

This is how quantum gates actually seem to work! When a quantum gate receives a superposition as input, it appears to evaluate every input in the superposition simultaneously and then output a new superposition of those results, as we saw in the conceptual diagram in Figure 8-9. We call this *quantum parallelism*, and it's one of the reasons people are so excited by quantum computers.

Note that I hedged the description a little and said that qugates "seem" to work this way. That's because we don't know if that's exactly how a qugate is accomplishing this feat. Does it really do just what I described, processing all those inputs in parallel? It appears that way. But how could that possibly be happening? It defies common sense. What if the input superposition has billions of states? Where is all that compute power coming from? Where is the energy to drive it coming from? Where is all that information being stored? In short, *by what means could quantum parallelism possibly occur?*

Nobody knows. All we can say is that countless experiments and quantum circuits, including Deutsch's algorithm, are built on the observation that, however it's achieved, the output of a quantum gate appears to be a superposition of its output for every state in its input superposition. Every quantum gate, from I to X to H to CX and all the other qugates we've seen and haven't seen, appear to process input superpositions in this parallel fashion. For this reason, quantum parallelism is a great *functional description* of how quantum gates process superpositions: It tells us the *what* of their behavior but is silent on describing the *how*.

It's also possible that quantum parallelism is a mirage, and what's actually going on is due to a more subtle phenomenon that we're currently unaware of. For now, quantum parallelism seems to be the simplest way to understand how quantum gates process an input in superposition, though there's still an aura of mystery around how they do so.

Quantum parallelism is a central element of most quantum algorithms, including those that use oracles. Because a quantum oracle is built from qugates, it enjoys the same ability to process its inputs in parallel that's offered by the individual qugates. In other words, the entire oracle processes its inputs in parallel and returns a superposition of all the results.

Returning to Deutsch's algorithm, we want to find the outputs of an oracle f given the inputs $|0\rangle$ and $|1\rangle$. Therefore, we create a superposition of those inputs and give them to the oracle, which processes both states *simultaneously* and returns those results in superposition. In symbols, we can write this as in Equation 8.4.

$$f(\vee\,|0\rangle + \vee\,|1\rangle) = \vee f|0\rangle + \vee f|1\rangle \qquad (8.4)$$

In words, applying f just once to a superposition of $|0\rangle$ and $|1\rangle$ produces the same result as if we'd applied f to each state separately and then summed the results to produce a new superposition.

By analogy, suppose that the two input states are walls in a room that you want to paint, and the function f refers to the process of painting them. Then, the left side of Equation 8.4 tells us that we can take the two walls as a single conceptual unit and apply paint to that combined unit. The right side tells us to paint each wall on its own. In practical terms, both viewpoints would give us the same result: two painted walls. The special magic of quantum gates is that using the form on the left, both walls get painted in the time it takes to paint only one.

Wait a minute. We've seen that when H gets the superposition $|+\rangle$ as input, it doesn't return a superposition as output, but instead just $|0\rangle$. Doesn't that contradict my claim that if a superposition goes into a qugate, a superposition comes out?

It's not a contradiction because when the individual outputs of H are summed to create a new superposition, they interfere with each other. The result is still a superposition, but it's a particularly simple one with just one state. We saw this phenomenon in Equation 4.9, and I've shown it in diagram form in Figure 8-10. As claimed, H does process each input individually, in parallel. But when those output states are summed to construct the output superposition, they interfere, causing $|1\rangle$ to have an amplitude of 0 and thus not contribute to the output superposition.

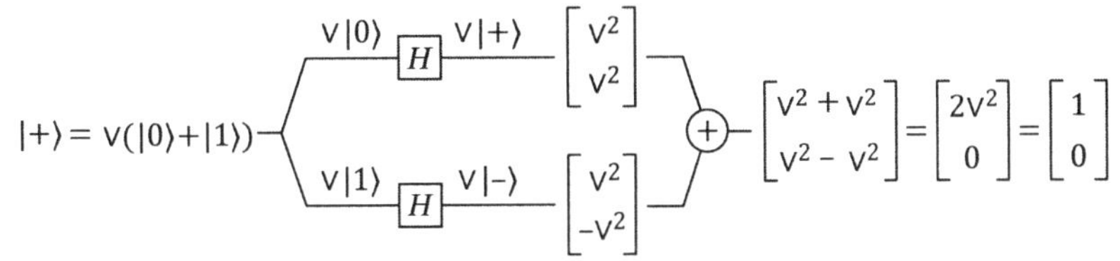

Figure 8-10: A conceptual illustration of how states interfere when computing $H\,|+\rangle$

You might say that quantum parallelism is "just" linear algebra, since we're applying a matrix to every element in a vector. That's fair. But what's special here is that this linear algebra appears to be actually manifested in the real world, at massive scale!

Now that we have quantum parallelism covered, let's see how it's used by Deutsch's algorithm.

The Three Steps of Deutsch's Algorithm

Deutsch's algorithm uses quantum parallelism at the core of a three-step process. The first step is to create a uniform superposition of $|0\rangle$ and $|1\rangle$, the second is to evaluate f (once!) on that superposition to produce a superposition of $f|0\rangle$ and $f|1\rangle$, and the third is to cleverly process this output superposition to determine whether its two results are the same or not.

Using the terms we introduced in previous sections, we say that Deutsch's algorithm is a *promise oracle algorithm* that solves a *promise problem* [182]. The oracle is a quantum circuit that implements one of the four one-bit functions in Table 8-1. As we've seen, its promise is that whatever function it implements, that function is either *constant*, returning the same output for both inputs, or *balanced*, returning a different output for each of the two possible inputs. For a one-bit function, this isn't so much a promise as an unavoidable description, but in Chapter 9 we'll generalize the algorithm to bitstrings of arbitrary length and then the promise will be more general.

Our job will be to query the oracle and somehow determine which category of function it implements. As I mentioned earlier, we aren't trying to determine which of the four functions we have, only which of the two categories it belongs to. One way to look at this is to ask the question, "Is this function balanced?" A final result of 1 will mean yes, and a final result of 0 will mean no (meaning that it must be constant, since it must be one or the other).

As with most oracle algorithms, we'll be given a quantum circuit for the oracle, but we're not allowed to examine that circuit and figure out what it's doing. We can only query it with inputs and then process its outputs.

Let's start with the quantum block diagram of Figure 8-11, which shows the general structure of Deutsch's algorithm.

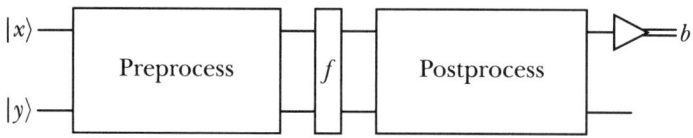

Figure 8-11: A block diagram for Deutsch's algorithm

The preprocessing in the starting block puts both input qubits into superposition. The oracle f evaluates each input in the superposition, producing a superposition of the outputs. The postprocessing block performs interference to refine the oracle's output. Then the measurement collapses the superposition. Figure 8-12 shows a more specific block diagram.

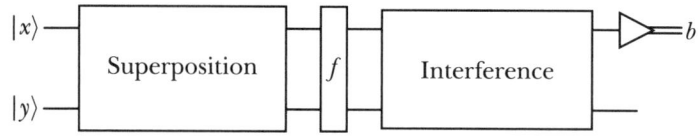

Figure 8-12: An improved block diagram for Deutsch's algorithm

Let's fill in the blocks and get a preview of the whole algorithm, shown in Figure 8-13. This figure is just to show you where we're going, so don't worry if it means little to you right now.

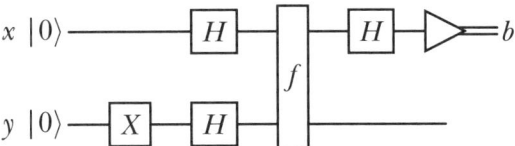

Figure 8-13: Deutsch's algorithm

There's much more going on than meets the eye in this relatively uncomplicated diagram!

An interesting feature of this algorithm is that we ignore the lower output of the oracle. There's definitely a quantum state on that wire (and we'll later see exactly what that state is), but we don't need it to solve our problem. Ultimately, if the bit b at the upper right has the value 0, then the oracle holds one of the constant functions (f^{00} or f^{11}) Otherwise, the bit is 1, and the oracle holds one of the balanced functions (f^{01} or f^{10}).

Analyzing Deutsch's algorithm will take us over a threshold we haven't passed yet. For most of us, what's going on here is more complex and subtle than we are able to immediately grasp purely from an initial look at the circuit. This is when all our effort in setting up the mathematics in Part I will start paying off. As we continue, the math will become indispensable.

Let's work our way through Deutsch's algorithm in the same way the quantum bits do, from left to right [71].

Step 1: Initialization

The first three checkpoints are shown in Figure 8-14.

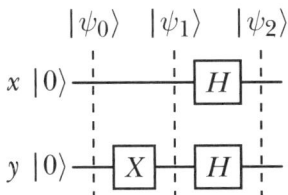

Figure 8-14: The preprocessing step of Deutsch's algorithm

Immediately upon starting, both qubits are initialized to $|0\rangle$, so they form the qubit system state $|00\rangle$. The y qubit goes through an X qugate, giving us the system $|01\rangle$. Then both qubits go through H qugates, transforming x from $|0\rangle$ to $|+\rangle$ and y from $|1\rangle$ to $|-\rangle$. This gives us the new system $|+-\rangle$. For reference, I've written these states down in Equation 8.5.

$$|\psi_0\rangle = |00\rangle$$
$$|\psi_1\rangle = |01\rangle \tag{8.5}$$
$$|\psi_2\rangle = |+-\rangle$$

This wraps up the first step. It might be unclear to you why we produced this particular state, and that's normal when you read almost any quantum algorithm for the first time. The whole circuit is a carefully balanced sequence of qubit manipulations, and it's often not clear how all the pieces go together until you can grasp the whole circuit at once. So let's just keep rolling for now.

Step 2: Querying the Oracle

The next step is to apply $|\psi_2\rangle$ to the oracle, as shown in Figure 8-15. Here, I've labeled the input x and y qubits with their values from $|\psi_2\rangle$.

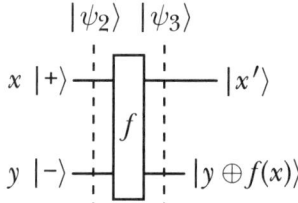

Figure 8-15: Applying the oracle

In Figure 8-6, we saw that the oracle's outputs are $|x'\rangle$, a possibly modified version of the input $|x\rangle$, on the upper wire and $|y \oplus f(x)\rangle$ on the lower wire. Thus, remembering that f stands for a quantum circuit, we can write the action of the oracle on any input $|xy\rangle$ as in Equation 8.6.

$$f|xy\rangle = |x'\rangle\,|y \oplus f(x)\rangle \tag{8.6}$$

We'll have a lot of operations involving $f(x)$ in this chapter, and all of those parentheses around the x's are going to clutter up our equations and make them harder to understand visually. For this reason, I'll write f_x for $f(x)$ for the rest of this chapter.

I'll also use another shortcut. For any bit b, the expression $1 \oplus b$ gives us 0 if b is 1 and 1 if b is 0. That is, it gives us the classical NOT of b. I'll write the NOT of b using a tilde above the label, so $\tilde{b} = 1 \oplus b$. Finally, we've already seen in property X2 in Table 8-3 that the exclusive OR of any bit with 0 is the bit itself, unchanged. I've summarized these statements in symbols in Equation 8.7.

$$f(x) = f_x$$
$$1 \oplus b = \tilde{b} \tag{8.7}$$
$$0 \oplus b = b$$

Let's now apply f to $|\psi_2\rangle$, producing $|\psi_3\rangle$, as shown in Equation 8.8.

$$
\begin{aligned}
|\psi_3\rangle &= f|\psi_2\rangle && \text{Apply the oracle to } |\psi_2\rangle \\
&= f|+-\rangle && \text{Use } |\psi_2\rangle \text{ from Eq. 8.5} && (8.8)\\
&= f(\sqrt{2}(|00\rangle - |01\rangle + |10\rangle - |11\rangle)) && \text{Expand } |+-\rangle \text{ into basis states}
\end{aligned}
$$

Because f is a linear operator, we can apply f to each term, as shown in Equation 8.9.

$$
|\psi_3\rangle = \sqrt{2}(f|00\rangle - f|01\rangle + f|10\rangle - f|11\rangle) \tag{8.9}
$$

It's time for quantum parallelism! Let's apply the oracle in Equation 8.6 to each term, giving us Equation 8.10. The hardware appears to apply the oracle to all four states simultaneously.

$$
\begin{aligned}
|\psi_3\rangle &= \sqrt{2}\Big(\; |0\rangle\,|0\oplus f_0\rangle - |0\rangle\,|1\oplus f_0\rangle + \\
&\qquad |1\rangle\,|0\oplus f_1\rangle - |1\rangle\,|1\oplus f_1\rangle \Big) && \begin{array}{l}\text{Apply Eq. 8.6 to}\\ \text{each term}\end{array} && (8.10)\\
&= \sqrt{2}\Big(\; |0f_0\rangle - |0\widetilde{f_0}\rangle + |1f_1\rangle - |1\widetilde{f_1}\rangle \Big) && \text{Use } 1\oplus b = \tilde{b} \text{ from Eq. 8.7}
\end{aligned}
$$

This is what I meant earlier when I said we were going to rely on the math! I wouldn't have guessed at this final expression just by looking at the illustration in Figure 8-13.

Conceptually, we're just about done, though the mechanics of the next few steps involve a lot of symbol manipulation. I think it's worth going through this process because it helps us become comfortable with this kind of work, which is often just alternating steps of expanding things out and collecting other things back together. When we reach the conclusion and know how the algorithm works, I'll show you a nice shortcut that will allow us to repeat the analysis with less effort and end up at the same conclusion.

Let's dig in! I'll take it slowly. We'll look at the next step twice: once when the oracle is a constant function, and again when it's balanced.

A Constant Oracle

Let's begin by supposing that f is constant. If it is, then $f_0 = f_1$ and $\widetilde{f_0} = \widetilde{f_1}$. So let's replace each f_1 and $\widetilde{f_1}$ in the last line of Equation 8.10 with f_0 and $\widetilde{f_0}$, respectively, giving us $|\psi_{3C}\rangle$ (using C for "constant"), as shown in Equation 8.11.

$$
\begin{aligned}
|\psi_{3C}\rangle &= \sqrt{2}\big[\, |0f_0\rangle - |0\widetilde{f_0}\rangle + |1f_0\rangle - |1\widetilde{f_0}\rangle \, \big] && \begin{array}{l}\text{Replace } f_1, \widetilde{f_1} \text{ in}\\ \text{Eq. 8.10 with } f_0, \widetilde{f_0}\end{array} \\
&= \sqrt{2}\big[(|0\rangle + |1\rangle)\,|f_0\rangle - (|0\rangle + |1\rangle)\,|\widetilde{f_0}\rangle \big] && \begin{array}{l}\text{Gather terms on } |f_0\rangle\\ \text{and } |\widetilde{f_0}\rangle\end{array} && (8.11)\\
&= \sqrt{2}\big[(|0\rangle + |1\rangle)(|f_0\rangle - |\widetilde{f_0}\rangle) \big] && \text{Because } (|0\rangle + |1\rangle) \text{ is shared} \\
&= \sqrt{\,}\big[\, |+\rangle\,(|f_0\rangle - |\widetilde{f_0}\rangle) \big] && \text{Use definition of } |+\rangle
\end{aligned}
$$

Well isn't that interesting! The first qubit, x, has a state of $|+\rangle$, just as it had when it went into the oracle. We know that $H|+\rangle = |0\rangle$, so if f is constant, and we apply an H to the upper qubit of $|\psi_{3C}\rangle$ and measure, we're guaranteed to get back 0.

A Balanced Oracle

Let's repeat the same process, but now assuming that f is a balanced oracle. In this case, $f_1 = \widetilde{f_0}$ and $\widetilde{f_1} = f_0$. Let's make these substitutions into Equation 8.10, as shown in Equation 8.12.

$$|\psi_{3B}\rangle = \surd^2 \big[\, |0f_0\rangle - |0\widetilde{f_0}\rangle + |1\widetilde{f_0}\rangle - |1f_0\rangle \,\big] \qquad \text{Replace } f_1, \widetilde{f_1} \text{ in Eq. 8.10 with } \widetilde{f_1}, f_0$$

$$= \surd^2 \big[(|0\rangle - |1\rangle)\, |f_0\rangle - (|0\rangle - |1\rangle)\, |\widetilde{f_0}\rangle \,\big] \qquad \text{Gather terms on } |f_0\rangle \text{ and } |\widetilde{f_0}\rangle$$

$$= \surd^2 \big[(|0\rangle - |1\rangle)(|f_0\rangle - |\widetilde{f_0}\rangle)\big] \qquad \text{Because } (|0\rangle - |1\rangle) \text{ is shared}$$

$$= \surd\big[\, |-\rangle(|f_0\rangle - |\widetilde{f_0}\rangle)\big] \qquad \text{Use definition of } |-\rangle$$

(8.12)

Another interesting result! The state of the upper qubit, x, was $|+\rangle$ when it entered the oracle but $|-\rangle$ when it left. We know that $H|-\rangle = |1\rangle$, so if f is balanced, and we apply H to the first qubit of $|\psi_{3B}\rangle$ and measure, we will always get a meter reading of 1.

Step 3: Postprocessing and Measurement

Now that we've evaluated the oracle, we'll apply its output to the H qugate I've been mentioning and then measure the upper qubit. The H qugate and the meter are shown in Figure 8-16.

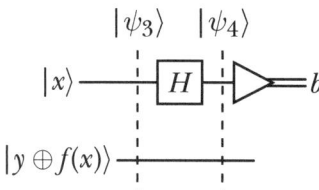

Figure 8-16: The last steps in Deutsch's algorithm

We'll read $b = 0$ when the function in the oracle is constant and $b = 1$ when the oracle is balanced.

Done! Deutsch's algorithm has solved our problem with just one query of the oracle. We've now done something that no classical computer could ever do.

Now you can see why I used the label $|x'\rangle$ for the upper output of the quantum version of the oracle. The input $|x\rangle$ is $|+\rangle$, but the output is either $|+\rangle$ or $|-\rangle$.

You may have noticed that I haven't done anything with the output of the bottom line in Figure 8-16. That's because we have no need of it, since

b holds the output we care about. That seems a little strange: We're using a qubit in the computation, but we don't care about its output value. That qubit is important, though. Let's see why.

Phase Kickback

We went through a lot of math in the previous section to get from $|\psi_2\rangle$, the input to the oracle, to a final state that we could interpret.

But there's an easier way! When the second input to this oracle is $|-\rangle$, we can write its output for any input $|x-\rangle$ in a nicely compact form. The steps are shown in Equation 8.13.

$$
\begin{aligned}
f|x-\rangle &= f|x \vee (|0\rangle - |1\rangle)\rangle & \text{Expand } |-\rangle \\
&= f|\vee(|x0\rangle - |x1\rangle)\rangle & \text{Write states compactly} \\
&= \vee(f|x0\rangle - f|x1\rangle) & \text{Distribute } f \text{ to both states} \quad (8.13) \\
&= \vee\big[\ |x\rangle\,|0 \oplus f_x\rangle - |x\rangle\,|1 \oplus f_x\rangle\ \big] & \text{Apply } f \text{ using Eq. 8.6} \\
&= \vee\,|x\rangle\,(|f_x\rangle - |\widetilde{f_x}\rangle) & \text{Extract common term } |x\rangle
\end{aligned}
$$

Let's look at this result for the two possible values of f_x. If $f_x = |0\rangle$, we get Equation 8.14.

$$
\vee\,|x\rangle\,(|0\rangle - |1\rangle) = \vee\,|x-\rangle \qquad (8.14)
$$

If $f_x = |1\rangle$, we get Equation 8.15.

$$
\vee\,|x\rangle\,(|1\rangle - |0\rangle) = -\vee\,|x-\rangle \qquad (8.15)
$$

These two results are the same except for a global phase of -1 at the start.

We can write both results in one form with a little mathematical trick that we'll use a few times in this book. Recall that $(-1)^0 = 1$ and $(-1)^1 = -1$. The only difference between Equations 8.14 and 8.15 is that when $f_x = 1$, we introduce a factor of -1. So, we can put $(-1)^{f_x}$ in front and write $f|x-\rangle$ as in Equation 8.16.

$$
\boxed{f|x-\rangle = (-1)^{f_x}\,|x-\rangle \qquad (8.16)}
$$

Equation 8.16 tells us that no matter what function is implemented by f, if we give it an input of $|x-\rangle$, that state comes out *unchanged, except for a possible global phase of* -1. We say that the relative phase of -1 in $|-\rangle = \vee(|0\rangle - |1\rangle)$ is *kicked back* as a global phase to the upper qubit. This phenomenon is called *phase kickback*.

It may seem that this is a useless operation, since we know that global phase is unobservable (and thus irrelevant) to the state of a qubit. But that's because the equation only applies when x is one of the individual basis states $|0\rangle$ and $|1\rangle$. It can't be directly applied to a superposition.

The magic appears when we apply this to all the states in a superposition. Because Equation 8.16 is linear, we apply it independently to each term.

Since the resulting terms are summed, the kicked-back global phase, *in a superposition*, becomes a relative phase. And that's what makes phase kickback so useful! It kicks back the global phase in $-|-\rangle$ into the superposition, where the -1 factor changes the relative phases between different terms. Unlike global phases, different relative phases describe different states, which can lead to different results. So while phase kickback appears pointless when applied to basis states, it's powerful when applied to superpositions.

Let's apply this tool to Deutsch's algorithm.

Analyzing Deutsch's Algorithm with Phase Kickback

Because Deutsch's algorithm always presents the oracle with an input of the form $|x-\rangle$, we can apply phase kickback to each term in x to interpret the final step of the circuit more efficiently.

Let's pick up the discussion from $|\psi_2\rangle$, all the way back in Equation 8.5. Applying f to this gives us the form of $|\psi_3\rangle$ shown in Equation 8.17.

$$
\begin{aligned}
|\psi_3\rangle &= f|+-\rangle \\
&= f[\vee(|0\rangle + |1\rangle)\,|-\rangle] && \text{Expand } |+\rangle \\
&= f\left[\vee\,(|0-\rangle + |1-\rangle)\right] && \text{Combine kets} \\
&= \vee\left[f|0-\rangle + f|1-\rangle\right] && \text{Extract } \vee \text{ and apply } f \text{ to each term} \\
&= \vee\left[(-1)^{f_0}\,|0-\rangle + (-1)^{f_1}\,|1-\rangle\right] && \text{Apply phase kickback from Eq. 8.16}
\end{aligned}
\tag{8.17}
$$

As before, let's consider this for both the constant and balanced cases.

If the oracle implements a constant function, then $f_1 = f_0$, and the last line of Equation 8.17 turns into the first line of Equation 8.18.

$$
\begin{aligned}
|\psi_{3C}\rangle &= \vee\left[(-1)^{f_0}\,|0-\rangle + (-1)^{f_0}\,|1-\rangle\right] && \text{Replace } f_1 \text{ in Eq. 8.17 with } f_0 \\
&= \vee(-1)^{f_0}[\,|0-\rangle + |1-\rangle] && \text{Pull out common term } (-1)^{f_0} \\
&= \vee(-1)^{f_0}[(|0\rangle + |1\rangle)\,|-\rangle] && \text{Gather factors on } |-\rangle \\
&= (-1)^{f_0}\,|+-\rangle && \text{Definition of } |+\rangle
\end{aligned}
\tag{8.18}
$$

Because the global phase has no influence on the measurement of a qubit, if we apply H to the top qubit and measure, we'll get back $|0\rangle$.

What if we have a balanced function? We start by replacing f_1 with $\widetilde{f_0}$, as shown in Equation 8.19.

$$
|\psi_{3B}\rangle = \vee\left[(-1)^{f_0}\,|0-\rangle + (-1)^{\widetilde{f_0}}\,|1-\rangle\right]
\tag{8.19}
$$

We can't simplify this, so let's break it into two cases for different values of f_0. If $f_0 = 0$, then we get Equation 8.20.

$$
\vee\left[\,|0-\rangle - |1-\rangle\right] = \vee\left[(|0\rangle - |1\rangle)\,|-\rangle\right] = |--\rangle
\tag{8.20}
$$

If $f_0 = 1$, then the signs are reversed and we get Equation 8.21, producing the term $-|--\rangle$. This might look kind of weird, but it's a compact way of writing $(-1)|--\rangle$.

$$
\vee\left[-|0-\rangle + |1-\rangle\right] = -\vee\left[(|0\rangle - |1\rangle)\,|-\rangle\right] = -|--\rangle
\tag{8.21}
$$

In both cases, running the upper qubit through H and measuring will give us 1.

Using phase kickback, we've reached the same conclusion as before: When the oracle is constant, we measure a 0, and when it's balanced, we measure a 1. But we got there a lot faster.

Let's look at this result in another way. In Table 8-4, I've written down the four functions and the results of applying the phase kickback process from Equation 8.16 to the superposition $|+-\rangle$. I've done this by thinking of that superposition as $(\vee |0\rangle + |1\rangle) |-\rangle$.

Table 8-4: Applying Equation 8.16 to the oracle's input

Function	f_0	f_1	$(-1)^{f_0} |0-\rangle$	$(-1)^{f_1} |1-\rangle$	New first qubit	Matrix and ket
f^{00}	0	0	$|0-\rangle$	$|1-\rangle$	$|0\rangle + |1\rangle$	$\vee \begin{bmatrix} 1 \\ 1 \end{bmatrix} = |+\rangle$
f^{01}	0	1	$|0-\rangle$	$-|1-\rangle$	$|0\rangle - |1\rangle$	$\vee \begin{bmatrix} 1 \\ -1 \end{bmatrix} = |-\rangle$
f^{10}	1	0	$-|0-\rangle$	$|1-\rangle$	$-|0\rangle + |1\rangle$	$\vee \begin{bmatrix} -1 \\ 1 \end{bmatrix} = -|-\rangle$
f^{11}	1	1	$-|0-\rangle$	$-|1-\rangle$	$-|0\rangle - |1\rangle$	$\vee \begin{bmatrix} -1 \\ -1 \end{bmatrix} = -|+\rangle$

The rightmost two columns show the first qubit that results. Note that constant functions produce either $|+\rangle$ or $-|+\rangle$, which become $H|+\rangle$ or $-H|+\rangle$, respectively. Because global phase doesn't affect measurements, measuring either result will produce the bit 0. In the same way, the balanced functions produce either $|-\rangle$ or $-|-\rangle$. These then become $H|-\rangle$ and $-H|-\rangle$, both of which give us a measured value of 1.

Deutsch's Algorithm Revisited

Deutsch's original algorithm, published in 1985 [49], wasn't what we've seen here. It was actually probabilistic and delivered the correct answer only about half the time [258]. The algorithm was extended and simplified by Cleve et al. into the form I've shown here [39]. Nevertheless, it's still known as Deutsch's algorithm in honor of its original inventor.

Let's look at the whole algorithm again, in Figure 8-17.

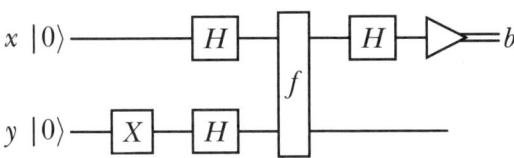

Figure 8-17: A repeat of Deutsch's algorithm

Inside the oracle, the local phase of -1 on the y input gets "kicked back" up to the x superposition. From the classical definition of the oracle, this makes no sense, because there's no such thing as a "negative bit." But in the quantum realm, things are different. Computing f on the combined state $|+-\rangle$ sometimes changes the upper qubit from $|+\rangle$ to $|-\rangle$.

Let's confirm this experimentally. Figure 8-18 shows the four functions from Figure 8-1 in circuit form, ready to be dropped into the f box in Figure 8-17.

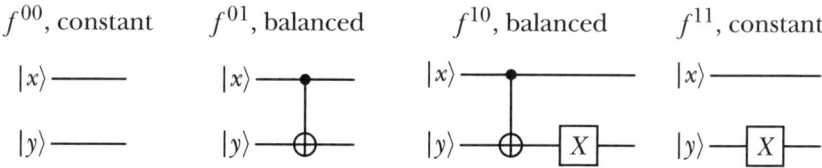

Figure 8-18: Implementations of the four functions from Table 8-1 as the oracle f

Figure 8-19 shows the results of running these circuits for 1,024 shots on a five-qubit quantum computer, and then measuring the upper qubit to get an output bit of either 0 or 1.

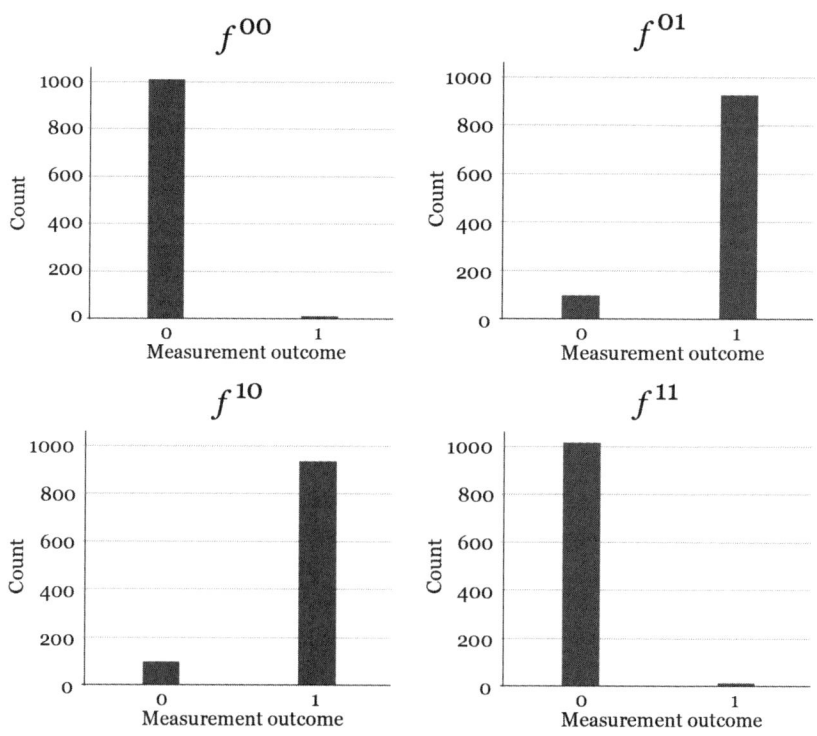

Figure 8-19: The results of evaluating all four oracles in Figure 8-18. Top row: f^{00} and f^{01}. Bottom row: f^{10} and f^{11}.

There's some noise (as usual), but the results generally match our predictions. The two functions f^{00} and f^{11} both almost always return a measured bit of 0, telling us that they are constant. Similarly, the two functions f^{01} and f^{10} both usually return a measured bit of 1, telling us that they are balanced.

Deutsch's algorithm gets the right answer with just a single evaluation of the oracle, which no classical algorithm can do!

CX Terminology

In Figure 8-18, I used a *CX* qugate to build the balanced oracles. The language we use when discussing the *CX* qugate is sometimes a good match to what it's doing, but sometimes it can be misleading. This came up when we discussed this qugate in Chapter 5, where I mentioned that the terms *control* and *target*, and even the name *CX*, were all "somewhat troublesome." Now we can see what I meant by that.

Consider our implementation of f^{01} in Figure 8-18. It's just a single *CX* qugate, apparently using x as a control and y as a target.

But as Equation 8.16 tells us explicitly, when y is in the state $|-\rangle$, it is not changed, no matter what state x is in. Instead, a relative phase of -1 is kicked back up to the superposition. Thus, the control changes and the target doesn't!

This is why I said those names had issues. While "control" and "target" make sense when we use *CX* as a switch, in other situations, those names don't apply and can even be misleading. In this case, a better approach to describing *CX* is to forget about calling the lines control and target and instead just treat this as a two-qubit qugate that operates on two inputs to produce two outputs.

Return to Quantum Parallelism

Because quantum parallelism is so critical to quantum computing, let's revisit the idea once more.

At its core, Deutsch's algorithm can't get around the need to evaluate the oracle for two different inputs, just as we required in the classical case. The magic of the technique is that those two evaluations are performed *simultaneously*.

The input qubit x is put into the superposition $|+\rangle$ before it's given to the oracle. Thanks to linearity, the oracle evaluates both inputs, $|0\rangle$ and $|1\rangle$, and combines both outputs to create a new superposition.

When we increase the number of input qubits to our algorithms, we can operate on hundreds, or hundreds of millions, or almost any number of inputs simultaneously, in the time that it takes to operate on only one qubit.

From a purely mathematical point of view, we might say that this result is nothing special or surprising. After all, the action of a system of qugates on a system of qubits in superposition, as we saw in Equation 8.4, boils down to multiplying matrices, and a conventional computer can do that. All of that is entirely true. Aside from the fact that quantum measurements are truly unpredictable, there's nothing that a quantum computer can compute that a classical computer can't (in theory, anyway). That's why we can run quantum simulators on classical hardware.

But in practice, they can be worlds apart. In the next chapter, we'll generalize Deutsch's algorithm to functions that use an arbitrary number of qubits. As quantum algorithms become more complicated, the sheer number of computations required to solve the problem when there are a large number of inputs (that is, the number of multiplications and additions in the matrix operations) can become astronomical. Depending on the number of inputs and the size of our classical computer, simulating a given quantum circuit on a classical computer could take from minutes to centuries, or even longer. This is why most quantum simulators that run on classical computers are limited to small numbers of qubits.

Computer scientists like to joke that in programming, there are only three integers: 0, 1, and "any." Deutsch's algorithm shows us how to use quantum parallelism to evaluate two inputs in parallel, putting the number of inputs in the "any" category.

Quantum parallelism is one of the key ideas that influences the design of quantum algorithms, enabling us to evaluate enormous numbers of computations simultaneously. Yet to get anything out of the system, we must measure the qubits involved, and we then obtain exactly one result.

Deutsch's algorithm, along with a few others we'll see later, is a rare case where the output is guaranteed. The upper line is ultimately either $|0\rangle$ or $|1\rangle$, so we know with certainty that we'll measure either 0 or 1, respectively.

But the usual case is that the output qubits will be in superpositions. Therefore, we can't be sure what our measurements will show us. The result of measurement will be that we obtain a single state that existed in the superposition with a nonzero probability. If we run the algorithm only once, we'll have no idea what other states were in the superposition, or even if one of them was vastly more probable than the one we measured.

It's not hard to create superpositions of unimaginably vast numbers of states. You need only start with a few hundred qubits in state $|0\rangle$ and run them all through H qugates. Then, any qugates you apply to those qubits seem to affect all the states in the superposition simultaneously. But although all of that information is apparently right there, tantalizingly close, nobody has ever found a way to access it except by measuring the qubits, and thereby losing every state in the superposition except for one. Addressing this phenomenon is one of the most interesting and challenging aspects of designing quantum algorithms.

Revisiting Phase Kickback

Phase kickback is so important that it's worth a closer look. The essential piece of the mechanism using a *CX* qugate is isolated in Figure 8-20.

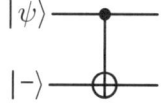

Figure 8-20: Phase kickback with a CX qugate

Suppose first that $|\psi\rangle = |0\rangle$. Then the CX qugate is not applied, and the output state is $|\psi\rangle\,|-\rangle$.

Next, suppose that $|\psi\rangle = |1\rangle$. Then the CX qugate is applied, and the $|-\rangle$ turns into $X\,|-\rangle = -\,|-\rangle$. That minus sign before the lower output is the key thing to keep your eye on.

Finally, suppose that the upper qubit $|\psi\rangle$ is in some superposition $\alpha\,|0\rangle + \beta\,|1\rangle$ (if we choose $|\psi\rangle = |+\rangle$, then $\alpha = \beta = \vee$). Since the control line is in superposition, quantum parallelism kicks in, and we get a superposition of both outputs. This is shown on the first line of Equation 8.22.

$$
\begin{array}{ll}
\alpha\,|0\rangle\,|-\rangle + \beta\,|1\rangle\,(-\,|-\rangle) & \text{Output of Figure 8-20} \\[2mm]
\alpha\,|0\rangle\,|-\rangle + \beta\,|1\rangle\,(-1)\,|-\rangle & \text{Write } -\,|-\rangle \text{ as } (-1)\,|-\rangle \\[2mm]
\Big(\alpha\,|0\rangle + \beta\,|1\rangle\,(-1)\Big)\,|-\rangle & \text{Factor out the } |-\rangle \\[2mm]
\Big(\alpha\,|0\rangle - \beta\,|1\rangle\,\Big)\,|-\rangle & \text{Apply the } (-1) \text{ to } \beta\,|1\rangle
\end{array}
\qquad (8.22)
$$

What's happened is that the relative phase of -1 in the lower qubit has been propagated to the upper qubit.

I think the term "phase kickback" for this phenomenon could be a little misleading for two reasons. First, it's not clear whether it refers to a global or relative phase. Second, a kickback is usually a response to some inciting event, and while we might call "applying a qugate" an inciting event, we do that all the time without this effect. It might be more helpful to think of phase kickback as "propagation of relative phase to a superposition," clarifying the type of phase involved, and that the relative phase on the lower qubit isn't lost, but is passed on, or propagated, to the upper qubit.

This is actually a special case of a more general phenomenon. Let's say that instead of a CX qugate we have a more general controlled-U qugate for some unitary operator U. For any U, there will be some inputs $|\phi\rangle$ for which the output $U\,|\phi\rangle = e^{i\theta}\,|\phi\rangle$, where the θ generally depends on which $|\phi\rangle$ is applied. We call these inputs the *eigenstates* (or *eigenvectors*) of U. Phase kickback will happen for any controlled qugate U as long as the control is in superposition, and the input to U is one of its eigenstates.

We can show this by rewriting the first line of Equation 8.22 as in Equation 8.23, using an input $|\phi\rangle$ that's an eigenstate of U. Factoring out the common term $|\phi\rangle$ has the effect of propagating $e^{i\theta}$ to the upper qubit.

$$
\begin{array}{ll}
\alpha\,|0\rangle\,|\phi\rangle + \beta\,|1\rangle\,(e^{i\theta}\,|\phi\rangle) & \text{Output of system} \\[2mm]
\Big(\alpha\,|0\rangle + e^{i\theta}\beta\,|1\rangle\,\Big)\,|\phi\rangle & \text{Propagate } e^{i\theta} \text{ to upper qubit}
\end{array}
\qquad (8.23)
$$

In particular, the X qugate turns $|-\rangle$ into $-\,|-\rangle = e^{i\theta}\,|-\rangle$ for the value $\theta = \pi$. Therefore, $|-\rangle$ is an eigenstate of X, and the relative phase $e^{i\pi} = -1$ gets propagated to the upper input. In this book we'll always use CX and its eigenstate $|-\rangle$ for our phase kickbacks, but we can produce phase kickback with any controlled qugate as long as its control is in a superposition state, and the input to the qugate is one of its eigenstates.

Summary

In this chapter, we met our first complete quantum algorithm. Deutsch's algorithm is designed to tell us if an unknown one-qubit function does or does not return the same value for both possible inputs.

We characterized the unknown function as an oracle or black box, a piece of quantum circuitry that someone has given us, but which we cannot examine. All we can do is give it inputs and get back its corresponding outputs. We saw that Deutsch's problem could be described as using a promise oracle, where we are assured that the oracle satisfies certain conditions.

If we send a superposition of states into an oracle (or any quantum gate), what seems to happen is that the oracle (or qugate) is simultaneously applied to every state in the superposition, and then all the results are combined into a single new superposition. This happens in parallel, taking no more time than that required to evaluate a single input. We call this quantum parallelism. In the final combination, the coefficients on the states can interfere, changing their amplitudes.

We often describe the cost of running an algorithm that involves an oracle in terms of oracle complexity, or the number of times the oracle must be called. Deutsch's algorithm manages to describe which kind of oracle we have by calling the oracle only once.

Deutsch's algorithm uses quantum parallelism to characterize the oracle f by giving it a superposition of $|0\rangle$ and $|1\rangle$, and thus getting back as output a superposition of $f|0\rangle$ and $f|1\rangle$. It then applies an H qugate, leading to a measured output of 0 if the oracle always returns the same value and 1 otherwise.

We can analyze Deutsch's algorithm more efficiently by using the idea of phase kickback. We also looked at this phenomenon of propagation of relative phase, for any type of controlled qugate.

In later chapters, the ideas of an oracle, quantum parallelism, and phase kickback will come up repeatedly. They are essential building blocks of many quantum algorithms.

This is a great time to pause and reflect. We've seen a fundamentally new way to solve a problem: quantum parallelism. Though it was a simple problem, we used a whole new suite of tools and ideas. The final algorithm is a carefully interlocked sequence of steps that interact in ways that can be surprising.

You now know how to solve a problem using some of the unique capabilities of quantum computers. In the following chapters, we'll see how to solve bigger and more interesting problems.

9

DEUTSCH–JOZSA'S ALGORITHM

If something is there, you can only see it with your eyes open, but if it isn't there, you can see it just as well with your eyes closed. That's why imaginary things are often easier to see than real ones.
—Norton Juster, *The Phantom Tollbooth*, 1961 [108]

One qubit is a fine thing. Deutsch's algorithm in Chapter 8 uses a one-qubit oracle, and we've seen that it's a great example of quantum computing.

But now we'll move on to multiple qubits, by generalizing Deutsch's oracle to one that operates on a bitstring of multiple bits. The result is called the *Deutsch–Jozsa algorithm*, after its creators, David Deutsch and Richard Jozsa [258].

Analyzing the Deutsch–Jozsa algorithm will introduce us to a few new pieces of notation and some new mathematical idioms that we'll find useful for many other quantum algorithms, including those that we'll see later in this book.

Introducing Deutsch–Jozsa

As in Deutsch's algorithm, we start with a function that someone has given us, implemented as a promise oracle.

The input to the oracle, in classical terms, is a bitstring of n bits. Together, the different combinations of 0 and 1 among these bits represent 2^n possible inputs to the oracle.

For each input, the oracle produces a single bit of output. We are promised that the output of the oracle obeys one of three conditions: either the output is always 0 for every input, the output is always 1 for every input, or the output is 0 for exactly half of the inputs, and 1 for the other half. The oracle definitely follows one of these three rules.

Table 9-1 shows four such functions, named f_0 through f_3.

Table 9-1: Four possible oracles

x	$f_0(x)$	$f_1(x)$	$f_2(x)$	$f_3(x)$
000	0	1	0	1
001	0	1	0	1
010	0	1	0	0
011	0	1	0	1
100	0	1	1	0
101	0	1	1	0
110	0	1	1	0
111	0	1	1	1

If the oracle always produces either 0 or 1, we say it's *constant*. Otherwise, its output is 0 for exactly half the inputs and 1 for the other half, and we say it's *balanced*. In Table 9-1, functions f_0 and f_1 are constant and functions f_2 and f_3 are balanced. Since 2^n is always even, such a split of the outputs into two equal-sized collections is always possible.

As with Deutsch's algorithm, we'd like to determine whether the function inside the oracle is constant or balanced using the smallest possible number of queries to the oracle.

In the classical case, we'd need to test the inputs one by one, though we can try them in any order. If we're lucky, then the first two inputs we try will give us different outputs: a 0 for one and a 1 for the other. The only way this can happen is if the oracle is balanced, so we have our answer and we can stop testing.

But if both outputs are the same, we have to keep trying new inputs. Suppose that we've tested half of the total number of possible inputs, and every one has given us the same number as output. It sure looks constant, but it might be balanced, and we just happened to guess all the inputs that gave us the same output. Function f_2 in Table 9-1 would give us this behavior. We have to try one more input. If it gives us the same output again, the oracle is constant; otherwise, it's balanced.

The conclusion is that in the worst case, we'd have to try one more than half the total number of possible inputs to be sure which category the function belongs to. So, classically, we could need as many as $(2^n / 2) + 1$ queries of the oracle.

As you might be anticipating, we can characterize our oracle in *just one query* using the Deutsch–Jozsa algorithm. And we'll do it using a similar strategy to Deutsch's algorithm. This time, we'll create a superposition of all 2^n possible inputs and evaluate them all at once with quantum parallelism. Then we'll use interference to process the results. If the function is constant and the outputs are all the same, the outputs will interfere so that the measured output will be all 0s. But if the function is balanced, the oracle's outputs will interfere so that the measured output will be something other than all 0s (we don't need to characterize that output any more precisely, since the conditions "all 0s" and "not all 0s" are all we need to distinguish the two possible classes). I'll write a bitstring of n bits of value 0 as 0^n, and similarly write 1^n for a bitstring of n bits that are all 1.

The block diagram in Figure 9-1 looks a lot like Deutsch's algorithm (Figure 8-12), only we now have $n + 1$ input qubits: the x qubits named x_{n-1} to x_0, and the single y input. The output bits are named b_{n-1} to b_0.

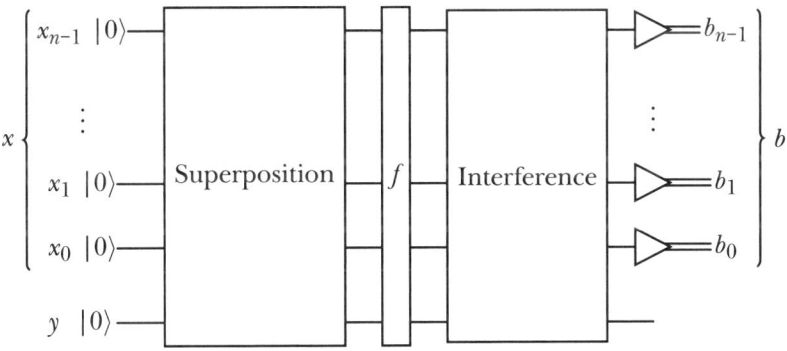

Figure 9-1: An abstract block diagram for the Deutsch–Jozsa algorithm

In Figure 9-1, all qubits start in the state $|0\rangle$. In this diagram, the big braces at the start and end don't indicate entanglement, but rather clusters of qubits or bits that we can think of as a group (this is a common overloading of the brace symbol, relying on context to distinguish entanglement from grouping).

Let's open up the boxes and see the whole algorithm, shown in Figure 9-2. This also looks similar to the representation of Deutsch's algorithm in Figure 8-13, but with the upper line repeated n times.

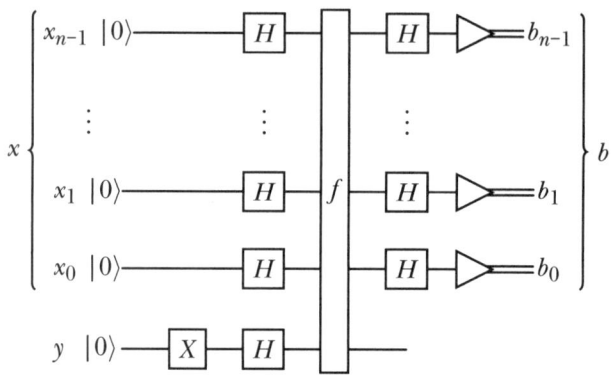

Figure 9-2: A block diagram for the Deutsch–Jozsa algorithm

There's a more compact way to draw Figure 9-2, shown in Figure 9-3. Note that at the end, the letter b refers to the entire output bitstring of n bits.

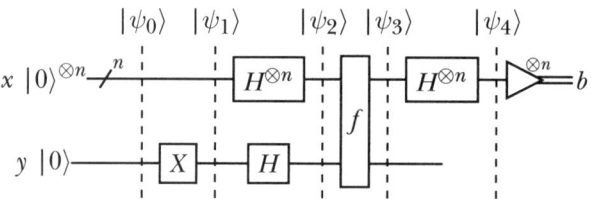

Figure 9-3: A compact form of Figure 9-2

There are two new conventions in this diagram. First, to represent a *bundle* of qubits we use a single wire but place a slash on it, along with the number of wires (or qubits) in the bundle. Here, the upper-left horizontal line represents a bundle of n qubits. This declaration continues for that wire through the rest of the circuit, unless it's explicitly changed. Therefore, in Figure 9-3, the upper line coming out of f still represents a bundle of n qubits. That bundle then goes into a bundle of n meters. The collective output of the meters is a bitstring of n classical bits.

Treating multiple qubits as a group in this way often makes our diagrams simpler and easier to interpret (compare Figures 9-2 and 9-3). It also helps us conceptually, since now we only need to think about the group of qubits rather than each one individually. We call each such bundle of qubits a *register*.

The second new notation is the $\otimes n$ exponent. This indicates that a state, qugate, or meter is tensored with itself n times. For example, we can write $|0\rangle^{\otimes 3} = |0\rangle \otimes |0\rangle \otimes |0\rangle = |000\rangle$, and $H^{\otimes 3} = H \otimes H \otimes H$. Similarly, a meter with $\otimes 3$ above and to its right represents a stack of three meters, applied to each of three qubits.

In this case, the input $|0\rangle^{\otimes n}$ tells us that x is a system created by tensoring together n qubits, each in state $|0\rangle$. This bundle is then run through $H^{\otimes n}$, a vertical system of qugates composed of n copies of H stacked up and tensored together. They all go through f, along with y. The qubits of x are then fed into another system of n copies of H, and then each qubit is measured independently, creating the n-bit output bitstring b. Figures 9-2 and 9-3 describe the same circuit.

You may see $|0\rangle^{\otimes n}$ written as $|0^n\rangle$ or $|0\rangle^n$, and similarly (but more rarely) sometimes people write $H^{\otimes n}$ as H^n. I won't use those notations in this book.

The Three Steps of Deutsch–Jozsa's Algorithm

As with Deutsch's algorithm, we'll go through Deutsch–Jozsa in three steps. These steps correspond to the elements of the block diagram in Figure 9-1.

Step 1: Initialization

Like with Deutsch's algorithm, I'll begin with the initialization step. Our qubits all start out as $|0\rangle$. The y qubit is inverted by an X qugate, and then each qubit in both x and y goes into an H qugate, as shown in Figure 9-4.

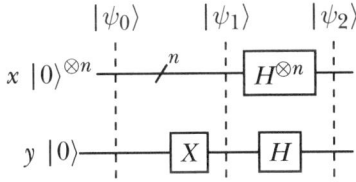

Figure 9-4: The preprocessing step of Deutsch–Jozsa's algorithm

As we proceed, it will be convenient to keep the x register distinct from the y qubit. Therefore, I'll write many of the system states of the Deutsch–Jozsa algorithm with the x register first, tensored with the y register. These first states are shown in Equation 9.1

$$|\psi_0\rangle = |0\rangle^{\otimes n} |0\rangle$$
$$|\psi_1\rangle = |0\rangle^{\otimes n} |1\rangle \qquad\qquad (9.1)$$
$$|\psi_2\rangle = |+\rangle^{\otimes n} |-\rangle$$

Note that the y register is $|-\rangle$. If you're thinking that this is to set up phase kickback in the oracle, you're right!

The x register in $|\psi_2\rangle$ is a special state in quantum computing: It's the *equal superposition of all n-qubit states*, also called a *uniform superposition*.

Recall that $H|0\rangle = \vee |0\rangle + \vee |1\rangle$, which is an equal superposition of the two one-qubit computational basis states. If we apply $(H \otimes H)$ to $(|0\rangle \otimes |0\rangle)$, which we can write as $H^{\otimes 2} |00\rangle$, we get $\vee^2(|00\rangle + |01\rangle + |10\rangle + |11\rangle)$. If we apply $H^{\otimes n}$ to $|0\rangle^{\otimes n}$, we get back all 2^n possible states of n qubits, scaled by $\vee^n$ so that the state has a magnitude of 1. We often write this state as $|+\rangle^{\otimes n}$.

This superposition is how Deutsch–Jozsa will query the oracle for all 2^n inputs simultaneously. It's a common starting point for many quantum algorithms. Putting the register x into the state $|+\rangle^{\otimes n}$ puts the qubits of that register in a state that is an equal superposition of *every possible combination* of n qubits in the basis states (that is, either $|0\rangle$ or $|1\rangle$).

I've written this in mathematical form in Equation 9.2.

$$|+\rangle^{\otimes n} = \surd^n \sum_{k \in \mathbb{B}^n} |k\rangle \qquad (9.2)$$

In this notation, writing $k \in \mathbb{B}^n$ is a way to say that k takes on every one of the 2^n bitstrings of n bits. It's equivalent to $k \in [2^n]$, but it emphasizes that we're thinking of these values as bitstrings. So, Equation 9.2 describes a loop in which k takes on every n-bit binary number, one at a time, and then all of the states $|k\rangle$ are summed together.

For example, when $n = 3$, the expression $|+\rangle^{\otimes 3}$ can be interpreted as any of the forms in Equation 9.3.

$$|+\rangle^{\otimes 3} = \surd^3 \sum_{k \in \mathbb{B}^3} |k\rangle = \surd^3 \sum_{k \in [8]} |k\rangle = \surd^3 \sum_{k=0}^{7} |k\rangle$$

$$= \surd^3 \Big(|000\rangle + |001\rangle + |010\rangle + |011\rangle + |100\rangle + |101\rangle + |110\rangle + |111\rangle \Big)$$

$$= \surd^3 \Big(|0\rangle + |1\rangle + |2\rangle + |3\rangle + |4\rangle + |5\rangle + |6\rangle + |7\rangle \Big)$$

$$= \surd^3 \left(\begin{bmatrix}1\\0\\0\\0\\0\\0\\0\\0\end{bmatrix} + \begin{bmatrix}0\\1\\0\\0\\0\\0\\0\\0\end{bmatrix} + \begin{bmatrix}0\\0\\1\\0\\0\\0\\0\\0\end{bmatrix} + \begin{bmatrix}0\\0\\0\\1\\0\\0\\0\\0\end{bmatrix} + \begin{bmatrix}0\\0\\0\\0\\1\\0\\0\\0\end{bmatrix} + \begin{bmatrix}0\\0\\0\\0\\0\\1\\0\\0\end{bmatrix} + \begin{bmatrix}0\\0\\0\\0\\0\\0\\1\\0\end{bmatrix} + \begin{bmatrix}0\\0\\0\\0\\0\\0\\0\\1\end{bmatrix} \right) = \surd^3 \begin{bmatrix}1\\1\\1\\1\\1\\1\\1\\1\end{bmatrix} \qquad (9.3)$$

Thanks to linearity and quantum parallelism, applying one or more qugates to this superposition means that those qugates will be applied to every one of these states simultaneously.

Let that soak in. Suppose $n = 30$. Then $2^n = 2^{30} = 1{,}073{,}741{,}824$, and applying a qugate to $|+\rangle^{\otimes 30}$ applies it to every one of these 1 billion states simultaneously, in the time it takes to apply the qugate to a single state. Try doing *that* with a classical computer!

All the states in $|+\rangle^{\otimes n}$ start with the same amplitude, $\surd^n$. Typically, the algorithm's goal will be to change the amplitudes and/or the phases on some states, ultimately giving the state (or states) that we want to measure a larger amplitude when we make our measurements. Ideally, if there's just one solution to our problem, that state will end up with an amplitude of 1 (and thus a probability of 1), while all others will have an amplitude of 0 (and thus a probability of 0), so that we measure the output we want with certainty. Usually, however, we have to settle for making the probabilities close to these values and running the circuit many times to find our answer, which will be the most frequently produced result.

Let's use our new notation to write $|\psi_2\rangle$ from Equation 9.1 as shown in Equation 9.4.

$$|\psi_2\rangle = |+\rangle^{\otimes n}\,|-\rangle = \vee^n \sum_{k\in\mathbb{B}^n} |k\rangle\,|-\rangle \tag{9.4}$$

It doesn't matter if we sum up the terms $|k\rangle\,|-\rangle$ or sum up only the $|k\rangle$ terms and tensor on the $|-\rangle$ to that sum afterward, since the $|-\rangle$ term is the final part of every state. This is illustrated in Equation 9.5.

$$\vee^n \left(\sum_{k\in\mathbb{B}^n} |k\rangle \right)|-\rangle = \vee^n \sum_{k\in\mathbb{B}^n} \left(|k\rangle\,|-\rangle \right) \tag{9.5}$$

Because both forms are the same, we don't need the parentheses. But when I write these kinds of expressions, I usually have the left-hand version of Equation 9.5 in the back of my mind.

Step 2: Querying the Oracle

Now that we have placed x into an equal superposition of all possible inputs for n qubits and placed y into $|-\rangle$, we're ready to give those inputs to the oracle. The diagram for this operation is shown in Figure 9-5.

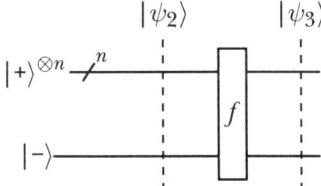

Figure 9-5: Applying f to $|\psi_2\rangle$ to get $|\psi_3\rangle$

As we know, thanks to quantum parallelism, the oracle will process each of the states in the superposition $|\psi_2\rangle$ at the same time and produce a new superposition $|\psi_3\rangle$ as output. Each state in the output superposition is the result of querying the oracle with the corresponding state in the input superposition.

Since every input to the oracle is in the form $|k\rangle\,|-\rangle$, we can find the output of the oracle by applying phase kickback from Equation 8.16. This gives us $|\psi_3\rangle$, as shown in Equation 9.6.

$$|\psi_3\rangle = f|\psi_2\rangle$$

$$= f\vee^n \sum_{k\in\mathbb{B}^n} |k\rangle\,|-\rangle \qquad \text{Replace } |\psi_2\rangle \text{ with Eq. 9.4}$$

$$= \vee^n \sum_{k\in\mathbb{B}^n} f|k-\rangle \qquad \text{Use linearity property L2} \qquad (9.6)$$

$$= \vee^n \sum_{k\in\mathbb{B}^n} (-1)^{f_k}|k-\rangle \qquad \text{Apply phase kickback from Eq. 8.16}$$

We've now got the result of applying the oracle, *in one step*, to every possible input of n qubits. This is a new superposition that contains the output for every input. Each such output has the same form: the input state $|k\rangle$ tensored with the y input $|-\rangle$, multiplied by either 1 or -1, depending on the oracle's response to that input.

Step 3: Postprocessing and Measurement

Our next job is to turn the oracle's output into something that we can measure, enabling us to determine whether the outputs are all the same (corresponding to a constant oracle) or half 0 and half 1 (corresponding to a balanced oracle).

The superposition $|\psi_3\rangle$ holds all the information we need, though we need to manipulate it into a form that's appropriate for measuring in the standard basis.

Because we've seen Figure 9-3, we know the next step is to apply an H qugate to every qubit in register x, as in Figure 9-6.

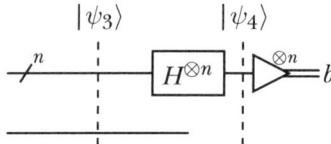

Figure 9-6: Applying $H^{\otimes n}$ to the upper register after f

But imagine you didn't know that. Suppose you were developing your own algorithm to solve the problem of distinguishing a balanced function from a constant one. As you worked on the problem, trying different approaches, at one point you produced $|\psi_3\rangle$, as in Equation 9.6.

Would you know that you were almost done? Would it be clear to you that you basically had the problem solved? Would you know what to do next?

These questions reveal one of the great challenges of quantum algorithm design. Even when you're on the verge of a solution, you might not know it. Every step of the algorithm can be subtle and complicated and hard to predict, and each step interlocks with the rest of the algorithm, like a big jigsaw puzzle. We have the advantage here of following along in the footsteps of

Deutsch and Jozsa, who worked out what this step should be, and its analysis by Sevag Gharibian [71].

Let's first look at where we're headed. We'll take the oracle's output $|\psi_3\rangle$ from Equation 9.6 and run all the qubits in the upper register through H qugates. From this point on, we can ignore the y qubit. Its purpose was to create the conditions for phase kickback. That's done now, so we have no further use for this qubit.

Let's temporarily write $|\psi_3\rangle_x$ for the x register (that is, the first n qubits of $|\psi_3\rangle$, excluding the $|-\rangle$ at the end). Applying H to each of these gives us Equation 9.7.

$$
\begin{aligned}
|\psi_4\rangle &= H^{\otimes n} |\psi_3\rangle_x & \text{From Fig. 9-6} \\
&= H^{\otimes n} \vee^n \sum_{k \in \mathbb{B}^n} (-1)^{f_k} |k\rangle & \text{Use } |\psi_3\rangle \text{ from Eq. 9.6} \\
&= \vee^n \sum_{k \in \mathbb{B}^n} (-1)^{f_k} H^{\otimes n} |k\rangle & \text{Because } H \text{ is linear}
\end{aligned}
\tag{9.7}
$$

The last line of Equation 9.7 is correct, because it's a transformation of the first line, but what does it mean? It's pretty complicated, and it would be hard to express in just a few words. I would really like to simplify it.

We'll get a more sensible version of that expression if we find a simpler expression for each $H^{\otimes n} |k\rangle$. The most efficient way to do that, ironically, is to write the Hadamard transform in a more complicated way!

Another Way to Write H

We're about to embark on another digression, but it's an important one, because along the way we'll encounter three new tools. First, we'll see a new, flexible way to write $H|0\rangle$ and $H|1\rangle$. Second, we'll meet the bitstring dot product, which will come in handy in many algorithms. Third, we'll develop a new way to represent applying H to every qubit in a register. These three tools are just what we need to see why Equation 9.7 is already holding the answer to our problem of categorizing the function inside the oracle.

To get started, let's write $H|0\rangle$ in a more complicated way. I know, that sounds like a bad idea, but it will turn out that this complicated form will let us simplify our analysis.

Remember that, by definition, $(-1)^0 = 1$ and $(-1)^1 = -1$. Equation 9.8 shows a new, expanded expression for $H|0\rangle$.

$$
H|0\rangle = \vee\left(1\,|0\rangle + 1\,|1\rangle\right) = \vee\left((-1)^0\,|0\rangle + (-1)^0\,|1\rangle\right)
\tag{9.8}
$$

Writing $H|1\rangle$ is similar, only now we'll multiply $|1\rangle$ by -1, or $(-1)^1$. Equation 9.9 shows the result.

$$
H|1\rangle = \vee\left(1\,|0\rangle - 1\,|1\rangle\right) = \vee\left((-1)^0\,|0\rangle + (-1)^1\,|1\rangle\right)
\tag{9.9}
$$

Let's summarize these by looking at $H|s\rangle$, where s can be either 0 or 1. Then we can write $H|s\rangle$ as in Equation 9.10.

$$
H|s\rangle = \vee\left((-1)^0\,|0\rangle + (-1)^s\,|1\rangle\right)
\tag{9.10}
$$

Let's write the right-hand side of Equation 9.10 as a tiny loop of only two values. I'll use an index k that takes on the values 0 and 1. The result is shown in Equation 9.11. Here, the expression sk means we're multiplying s and k.

$$H\,|s\rangle = \vee \sum_{k\in\mathbb{B}} (-1)^{sk}\,|k\rangle \qquad (9.11)$$

As I promised, this looks complicated, so let's check it. First, I'll write everything out for $s = 0$. In Equation 9.12, I've used the $\times$ sign for multiplication.

$$
\begin{aligned}
H\,|0\rangle &= \vee \sum_{k\in\mathbb{B}} (-1)^{0\times k}\,|k\rangle && \text{Eq. 9.11 for } s = 0 \\
&= \vee\left((-1)^{0\times 0}\,|0\rangle + (-1)^{0\times 1}\,|1\rangle \right) && \text{Expand the loop} \\
&= \vee(1\,|0\rangle + 1\,|1\rangle) && \text{Apply exponents to } -1 \text{ terms} \\
&= \vee(|0\rangle + |1\rangle) && \text{Simplify} \\
&= |+\rangle && \text{Definition of } |+\rangle
\end{aligned}
\qquad (9.12)
$$

This confirms that Equation 9.11 gives us $|+\rangle$ for $H\,|0\rangle$. Let's try it again for $s = 1$, in Equation 9.13.

$$
\begin{aligned}
H\,|1\rangle &= \vee \sum_{k\in\mathbb{B}} (-1)^{1\times k}\,|k\rangle && \text{Eq. 9.11 for } s = 1 \\
&= \vee\left((-1)^{1\times 0}\,|0\rangle + (-1)^{1\times 1}\,|1\rangle \right) && \text{Expand the loop} \\
&= \vee(1\,|0\rangle + (-1)\,|1\rangle) && \text{Apply exponents to } -1 \text{ terms} \\
&= \vee(|0\rangle - |1\rangle) && \text{Simplify} \\
&= |-\rangle && \text{Definition of } |-\rangle
\end{aligned}
\qquad (9.13)
$$

As we'd hope, Equation 9.11 gives us $|-\rangle$ for $H\,|1\rangle$.

Great, we've come up with a way to write the Hadamard transform of a single qubit that's both more complicated and harder to read. But the value here is that we can generalize this result to just what Equation 9.7 needs.

I think the easiest way to see this generalization is to write it down for increasing numbers of qubits and then observe the pattern. Let's start with a system of two qubits, apply H to both using Equation 9.11, and see what comes out. The intermediate steps get messy, but then we'll clean things up.

Suppose we have a quantum state $|g\rangle$. For our starting system of two qubits, I'll write $|g\rangle$ in terms of its qubits, as $|g_1 g_0\rangle$ or $|g_1\rangle \otimes |g_0\rangle$.

Let's apply Equation 9.11 to this system.

$$
\begin{aligned}
H^{\otimes 2}\,|g\rangle &= H(|g_1\rangle \otimes |g_0\rangle) && \text{Expand } |g\rangle \\
&= H\,|g_1\rangle \otimes H\,|g_0\rangle && \text{Because } H \text{ is linear} \\
&= \vee \sum_{k\in\mathbb{B}} (-1)^{kg_1}\,|k\rangle \otimes \vee \sum_{k\in\mathbb{B}} (-1)^{kg_0}\,|k\rangle && \text{Apply Eq. 9.11 to each term}
\end{aligned}
\qquad (9.14)
$$

Expanding the loops in the last line of Equation 9.14 gives us Equation 9.15.

$$H^{\otimes 2} \, |g\rangle = \sqrt{2} \left(\left((-1)^{0g_1} \, |0\rangle + (-1)^{1g_1} \, |1\rangle \right) \otimes \left((-1)^{0g_0} \, |0\rangle + (-1)^{1g_0} \, |1\rangle \right) \right) \quad (9.15)$$

Carrying out the tensor products in Equation 9.15 gives us the four terms in Equation 9.16.

$$H^{\otimes 2} \, |g\rangle = \sqrt{2} \Big((-1)^{0g_1}(-1)^{0g_0} \, |00\rangle + (-1)^{0g_1}(-1)^{1g_0} \, |01\rangle +$$
$$(-1)^{1g_1}(-1)^{0g_0} \, |10\rangle + (-1)^{1g_1}(-1)^{1g_0} \, |11\rangle \Big) \quad (9.16)$$

That's a whole lot of stuff! Let's simplify it a little by gathering together the exponents on (-1) for each state, giving us Equation 9.17.

$$H^{\otimes 2} \, |g\rangle = \sqrt{2} \Big((-1)^{0g_1+0g_0} \, |00\rangle + (-1)^{0g_1+1g_0} \, |01\rangle +$$
$$(-1)^{1g_1+0g_0} \, |10\rangle + (-1)^{1g_1+1g_0} \, |11\rangle \Big) \quad (9.17)$$

We can write this as a summation. Suppose we write an index k that loops over all four bitstrings of two bits. Writing each bitstring k as $k_1 k_0$, where each of k_1 and k_0 refer to a single bit that's 0 or 1, we can greatly simplify Equation 9.17 as Equation 9.18.

$$H^{\otimes 2} \, |g\rangle = \sqrt{2} \sum_{k \in \mathbb{B}^2} (-1)^{k_1 g_1 + k_0 g_0} \, |k\rangle \quad (9.18)$$

In this expression, $k_1 g_1$ is the product of the bits k_1 and g_1, and $k_0 g_0$ is the product of the bits k_0 and g_0. Remember that the expression $k \in \mathbb{B}^2$ under the summation means that k takes on the four values $00, 01, 10$, and 11 as the loop executes.

We can make Equation 9.18 even more compact with a new piece of notation. I'll define a new operation on two bitstrings. I'll use the centered dot symbol $\cdot$ for this and call it the *bitstring dot product*, because structurally it's similar to the dot product we saw in Equation 2.37.

In words, we start with two bitstrings, a and b, each with the same number of bits (let's say there are d bits). We take the leftmost bit of each bitstring, a_{d-1} and b_{d-1}, and multiply them together. This gives us 1 if both bits are 1; otherwise, it's 0. Then we take the next pair of bits, a_{d-2} and b_{d-2}, multiply them together, and so on, all the way down to a_0 and b_0. Finally, we add up all of these products and then find the result modulo 2 (written *mod 2*). This gives us back the remainder after dividing the sum by 2. In other words, it gives us 0 if the sum is even or 1 if the sum is odd. We write this as in Equation 9.19 (some authors don't include the mod 2 in this definition, but include it later when the dot product is used).

$$a \cdot b \stackrel{\triangle}{=} \left(\sum_{k \in [d]} a_k b_k \right) \bmod 2 \qquad (9.19)$$

The sum in parentheses in Equation 9.19 tells us how many pairs of bits are both 1. By taking the result modulo 2, we get back 0 if the number of these pairs is even, and 1 if the number of such pairs is odd.

Another way to write the bitstring inner product is to compute the XOR of the product of each pair of bits, as in Equation 9.20.

$$a \cdot b = a_0 b_0 \oplus a_1 b_1 \oplus \ldots \oplus a_{d-1} b_{d-1} \qquad (9.20)$$

As before, if there are an even number of products where $a_k b_k = 1$, this XOR will be 0, and otherwise it will be 1.

You'll sometimes see Equation 9.20 written using a big $\oplus$ sign, as in Equation 9.21. This is an iterator, like a big Σ sign, telling us to run a loop and compute the elements on the right for each value of k in $[d]$, then combine those elements using the $\oplus$ operator.

$$a \cdot b = \bigoplus_{k \in [d]} a_k b_k \qquad (9.21)$$

Equations 9.19, 9.20, and 9.21 are all equivalent descriptions of the bitstring dot product, so any time we see something like $a \cdot b$ for bitstrings a and b and we want to find its value, we're free to use whichever of these versions is most convenient in that situation.

With the bitstring dot product, we can simplify Equation 9.18. I'll replace the exponent on -1 with the dot product of the bits of k and the bits of g, as in Equation 9.22.

$$k_1 g_1 + k_0 g_0 = k \cdot g \qquad (9.22)$$

This gives us a compact form for the result of applying the H qugate to each of the bits in a two-bit bitstring g, shown in Equation 9.23.

$$H^{\otimes 2} |g\rangle = \sqrt{}^2 \sum_{k \in \mathbb{B}^2} (-1)^{k \cdot g} |k\rangle \qquad (9.23)$$

That's what we get from applying H to both qubits of a two-qubit system. It's kind of a long-winded way to write that result, but now we can easily generalize it to larger numbers of qubits and see the pattern that emerges.

Repeating this analysis for three qubits gives us Equation 9.24.

$$H^{\otimes 3} |g\rangle = \sqrt{}^3 \sum_{k \in \mathbb{B}^3} (-1)^{k \cdot g} |k\rangle \qquad (9.24)$$

All of these results would work out similarly if there were 4 qubits, or 4 million. The general result for n qubits requires only replacing each 2 in Equation 9.23, or each 3 in Equation 9.24, with n. This gives us the final form in Equation 9.25.

$$H^{\otimes n} \, |g\rangle = \sqrt{}^n \sum_{k \in \mathbb{B}^n} (-1)^{k \cdot g} \, |k\rangle \qquad (9.25)$$

This is what we wanted! It's an explicit representation of what we get if we apply $H^{\otimes n}$ to a bitstring g of n bits.

Equation 9.25 is the third of the three tools I mentioned at the start of this section. Recall that most quantum algorithms begin with all qubits in the state $|0\rangle$. The first step of many algorithms, including Deutsch–Jozsa's, is to put all of those qubits into the equal superposition $|+\rangle$ by running each of them through an H qugate, giving us the n-qubit system $|+\rangle^{\otimes n}$.

Equation 9.25 is a way to write the result of applying H to each specific state g in the input superposition. The result is already normalized by $\sqrt{}^n$ for n qubits, so we can just apply Equation 9.25 to each state $|g\rangle$ in $|+\rangle^{\otimes n}$ and add up the results, giving us $H^{\otimes n} \, |+\rangle^{\otimes n}$.

We'll see that having explicit access to the result of applying H to each state in the superposition will allow us to combine those output states in useful ways.

In this digression, we've seen three results that will be useful not only as we wrap up Deutsch–Jozsa but also in many other algorithms, including those we'll see in later chapters.

Back to Step 3

Let's put our new tools to use. I'll take the expression for $|\psi_4\rangle$ from the last line of Equation 9.7 (the output of the oracle, processed by H qugates) and substitute Equation 9.25, for each state, as shown in Equation 9.26. We're going to end up with something that probably won't look like an improvement, but it will be! The variable z introduced in the second line is an arbitrary letter that takes on the values of every bitstring from $0^{\otimes n}$ to $1^{\otimes n}$.

$$
\begin{aligned}
|\psi_4\rangle &= \sqrt{}^n \sum_{k \in \mathbb{B}^n} (-1)^{f_k} H^{\otimes n} \, |k\rangle && \text{Eq. 9.7} \\[2mm]
&= \sqrt{}^n \sum_{k \in \mathbb{B}^n} (-1)^{f_k} \left(\sqrt{}^n \sum_{z \in \mathbb{B}^n} (-1)^{z \cdot k} \, |z\rangle \right) && \begin{array}{l} \text{Replace } H^{\otimes n} \, |k\rangle \text{ with} \\ \text{Eq. 9.25} \end{array} \\[2mm]
&= \sqrt{}^{2n} \sum_{k \in \mathbb{B}^n} \sum_{z \in \mathbb{B}^n} (-1)^{f_k} (-1)^{z \cdot k} \, |z\rangle && \text{Group terms together} \\[2mm]
&= \sqrt{}^{2n} \sum_{k \in \mathbb{B}^n} \sum_{z \in \mathbb{B}^n} (-1)^{f_k + z \cdot k} \, |z\rangle && \text{Combine exponents}
\end{aligned}
\qquad (9.26)
$$

As I mentioned earlier, it's often hard to know when you're near the end of a quantum algorithm. Though it probably won't appear this way, the last line of Equation 9.26 is what we've been aiming for all along. We'll see why in the next section, where that expression will help us find the solution to our problem.

Results in Constant and Balanced Cases

Now we can follow the same pattern that we followed in Chapter 8 for Deutsch's algorithm and look at Equation 9.26 for each category of oracle.

First, we'll look at the last line of Equation 9.26 under the assumption that our oracle contains a constant function. Using that assumption, we can simplify the expression until it tells us that we will always measure an output that is all 0s or all 1s.

Then we'll return to the last line of Equation 9.26, but this time we'll assume that the oracle is balanced. We'll see that the result of the simplifications due to this second assumption is that we will always measure an output that is *not* all 0s.

Thus, if we measure all 0s, we'll know the oracle is constant, and otherwise it's balanced. Let's see how we draw that conclusion by looking at each category of oracle.

A Constant Function

Let's assume that the function is constant. Then every f_k is the same (either all 0s or all 1s). That means we can pull the $(-1)^{f_k}$ part out of the last line of Equation 9.26. Because in this case f_k is the same for every choice of k, I'll arbitrarily use f_0. I'll also swap the order of the two big sigma signs, so we can group things as in Equation 9.27.

$$|\psi_4\rangle = (-1)^{f_0} \sum_{z \in \mathbb{B}^n} \left(\sqrt{}^{2n} \sum_{k \in \mathbb{B}^n} (-1)^{z \cdot k} \right) |z\rangle \tag{9.27}$$

Now comes a trick. Well, not really a trick, but a step that might have been tricky to predict. I'll guess that if we measure all 0s from this state, it might tell us something interesting. So what's the amplitude (and thus the probability) that we'll measure an output bitstring of all 0s? This is the amplitude of the state $|0\rangle^{\otimes n}$ in $|\psi_4\rangle$, given by Equation 9.27.

If we're interested only in the amplitude of $|0\rangle^{\otimes n}$, then that's the only value of z in Equation 9.27 that we want to look at. Therefore, the amplitude of this state can be found by setting z to the bitstring of n elements that are all 0. We can also lose the summation over z, since we're interested only in the single case when z is all 0. These observations give us the amplitude of $|0\rangle$ in Equation 9.28. Note that in the next-to-last line, the denominator is $\sqrt{2}$ raised to $2n$, not 2^n. This is equivalent to 2^n.

$$(-1)^{f_0} \vee^{2n} \sum_{k \in \mathbb{B}^n} (-1)^{0 \cdot k} \qquad \text{Eq. 9.27 with } z = 0$$

$$= (-1)^{f_0} \vee^{2n} \sum_{k \in \mathbb{B}^n} (-1)^0 \qquad \text{Since } 0 \cdot k = 0$$

$$= (-1)^{f_0} \vee^{2n} \sum_{k \in \mathbb{B}^n} 1 \qquad \text{Because } (-1)^0 = 1 \qquad (9.28)$$

$$= (-1)^{f_0} \vee^{2n} 2^n \qquad \text{Sum up } 2^n \text{ copies of } 1$$

$$= (-1)^{f_0} \frac{1}{(\sqrt{2})^{2n}} 2^n \qquad \text{Use the definition of } \vee$$

$$= (-1)^{f_0} \qquad \text{Since } (1/2^n)2^n = 1$$

Because f_0 is a bit, the amplitude of $|0\rangle^{\otimes n}$ given by Equation 9.28 is either -1 or 1. Squaring this tells us that the probability of measuring the state of all 0s is 1.

Since all probabilities in a superposition must add up to 1, we can conclude that if the function is constant, *all of the probabilities for all 2^n states are on state* $|0\rangle^{\otimes n}$. The probability for every other state is 0. Because there's a probability of 1 that our measurement will produce the output string 0^n, we know *with certainty* that if we measure this state at the end of the algorithm, we'll get an *n*-bit bitstring of all 0 elements.

In short, if the function is constant, we are guaranteed to measure a 0 for every bit in the output.

A Balanced Function

If the function is balanced, we can't factor out the $(-1)^{f_k}$ term. So let's go back to $|\psi_4\rangle$ from the last line of Equation 9.26, and again we'll set $z = 0$ to see the probability of measuring all 0s at the output. The steps are in Equation 9.29.

$$\vee^{2n} \sum_{k \in \mathbb{B}^n} \sum_{z \in \mathbb{B}^n} (-1)^{f_k + z \cdot k} \qquad \text{Eq. 9.26}$$

$$= \vee^{2n} \sum_{k \in \mathbb{B}^n} (-1)^{f_k + 0 \cdot k} \qquad \text{Set } z = 0 \qquad (9.29)$$

$$= \vee^{2n} \sum_{k \in \mathbb{B}^n} (-1)^{f_k}$$

What is the last summation? If f is balanced, then we're *promised* that it has exactly as many 0 outputs as 1 outputs. So for every $(-1)^0 = 1$ value, we will have a matching $(-1)^1 = -1$ value, with the result that we are adding equal numbers of 1s and -1s. Thus, *everything cancels out*, and the value of the summation is 0. In other words, Equation 9.29 tells us that if the function is balanced, the amplitude of the state $|0\rangle^{\otimes n}$ is 0. That is, *there is no possibility of measuring all 0s at the output*, since that state has an amplitude (and therefore a probability) of 0, and doesn't even exist in the superposition.

So, any measurement *other* than all 0s at the output means that the function is balanced. We don't care what the output value is, just that it's not all 0s.

Actual Results of Deutsch–Jozsa's Algorithm

Let's try this out. I ran the Deutsch–Jozsa circuit on a three-qubit quantum computer. The results from a constant oracle are on the left side of Figure 9-7, and the results from a balanced oracle are on the right side. In this graph, and others like it to come, only the states that were measured at least once are shown, so any state that doesn't have an associated bar was never measured at the output. This helps keep the images legible when there are many possible output states but only a few were observed.

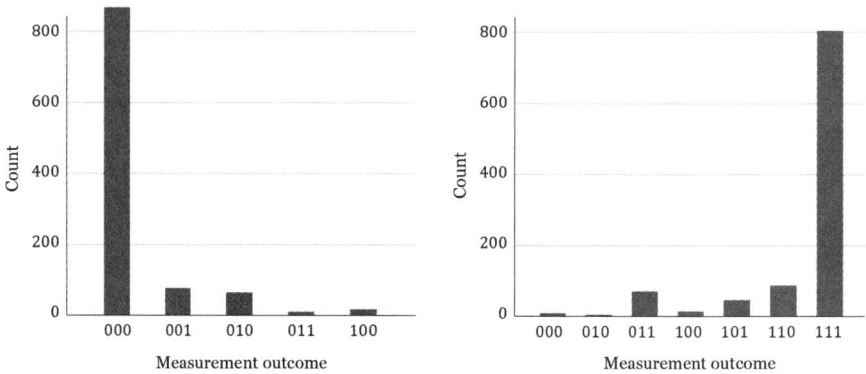

Figure 9-7: Left: The results from a constant oracle. Right: The results from a balanced oracle.

As we'd expect, most of the constant results were all 0s, with a little noise for other outputs, and most of the balanced results were *not* all 0s. We've got our solution: If the measured output is all 0s, then we have a constant function; otherwise, it's balanced.

Once again, the *y* input, which we needed for phase kickback in the oracle, is not measured. It's like an enzyme in chemistry, which helps reactions proceed but isn't itself altered by the process.

Why Does the Math Work So Well?

I hope at this point it's becoming clear to you why we spent so much time on the math in Part I. It's because understanding all but the simplest quantum algorithms requires analyzing the actions algebraically, rather than visually or intuitively.

The failure of our intuition to handle quantum algorithms (at least, until we've had a lot of experience with them) is disappointing, but not surprising. Our intuition is tuned for what happens at the scales of time and space we can directly perceive. Operations at the quantum scale, and their logic, are entirely unfamiliar and therefore outside our experience.

The range of sensory phenomena that we can directly perceive is sometimes called our *umwelt* (pronounced OOM-velt). The limits of our umwelt are all around us. For example, birds show each other beautiful plumage in the ultraviolet range, which other birds can see but most people can't [17]. Sharks sense magnetic fields to help them navigate [226], as do some plants [133], cattle [33], birds [25], and even monarch butterflies [1]. Most people's senses seem completely unable to perceive that those fields even exist.

Mountains grow and change on the scale of thousands of years [80], and subatomic objects blink into and out of existence faster than any cell in our body can respond to their presence [126]. Without direct perception of these phenomena, and evolution-honed physical and mental tools to interpret them, they are simply beyond our immediate reach and experience.

Quantum phenomena are outside of our everyday perceptions in many ways: They're too small, too fast, and too different. If we throw a ball into the air we have a sense of what will happen, but for even moderately complicated quantum interactions, our untrained experience and intuition are unreliable guides.

For reasons nobody exactly knows, the body of symbol manipulation tools that we call mathematics is apparently able to capture and describe not only an abundance of natural phenomena, but also the behavior of objects at the quantum scale. These equations tell us how things work and let us predict new results. Nobody can say why they are so successful at this, but they are. Our math is an augmentation of our own capacities. It's a wheelchair, a microscope, or a rover on Mars. We don't really know why it's so capable at describing and predicting quantum phenomena, though there is no shortage of theories. What we do know is that it works so extraordinarily well that we are able to use it to build everything from GPS satellites and cell phones to medical imaging scanners and quantum computers.

In short, until we have a better understanding of the universe, we may not know why our abstract symbols, obeying specific rules, are so amazingly effective at describing reality at the quantum scale. But they are, and thanks to that astonishing fact, we can build quantum computers and design quantum algorithms to run on them.

Summary

In this chapter, we explored Deutsch–Jozsa's algorithm, which generalizes Deutsch's algorithm for functions of many qubits.

As with Deutsch's algorithm, this function is given to us as a promise oracle, which promises that either it returns the same output for every possible input or the outputs are evenly split between two values.

We developed three important new ideas in our analysis. First, we saw a new expression for the Hadamard qugate. Second, we met the bitstring dot product, which takes two bitstrings, counts the number of times both bitstrings have a 1 in the same place, and tells us if that count is even or odd. Third, we introduced an explicit way to write the result of applying an H qugate to every one of n input qubits.

With our expanded quantum toolkit, we're prepared to tackle more sophisticated algorithms!

10

BERNSTEIN–VAZIRANI'S ALGORITHM

Anything will give up its secrets if you love it enough.
—George Washington Carver (apocryphal), approx. 1934 [32]

You know what would be really cool? Solving a brand-new quantum algorithm with a circuit we've already seen. Guess what we'll do in this chapter?

Not only will we look at a new kind of problem, but we'll also see a new way of looking at some oracle problems. Our goal this time is to guess a secret bitstring that's hidden inside the oracle.

As we've done before, we'll start by looking at our oracle in terms of classical bits.

Suppose the oracle contains a secret bitstring s made up of n bits. If we give the oracle an input x that also has n bits and a helper bit y, the oracle promises to compute $s \cdot x$, the bitstring dot product of s and x from Equation 9.19. Remember that the bitstring dot product counts the number of locations where both bitstrings are 1 and then finds that result modulo 2, so the output is 0 if there are an even number of such locations; otherwise, the output is 1.

The oracle takes $s \cdot x$ (the bitstring dot product of the secret bitstring s and the input x), forms the exclusive OR of that with y, and returns that bit in the output y register. The x input is passed to the output unchanged.

Figure 10-1 shows the idea. As in Figure 9-1, the braces here indicate grouping, not entanglement.

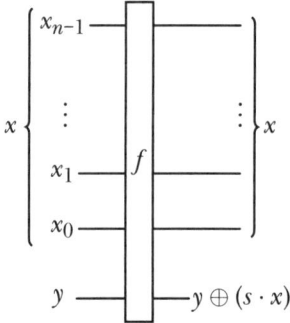

Figure 10-1: The Bernstein–Vazirani oracle for classical bits

Equation 10.1 describes the quantum version of the oracle.

$$f \,|xy\rangle = |x'\rangle \, |y \oplus (s \cdot x)\rangle \tag{10.1}$$

We'd like to discover the secret bitstring s while making the fewest number of queries to the oracle.

The Classical Solution

To solve this classically, one approach is to query the oracle d times, once for each bit in the secret bitstring s. Each query will be all 0s except for a single 1. We won't repeat queries, so since there are d bits to test, we'll have d inputs.

Because $1 \oplus 1 = 0$, if the output of any of these queries is 0, then we know that the bit in the secret bitstring at the position given by the 1 in our query must also be 1. If the output is not 0, then because $1 \oplus 0 = 1$, we know the bit at that position is 0. However we proceed, we need n queries to be sure of getting the answer [293].

The Bernstein–Vazirani Circuit

The Bernstein–Vazirani circuit uses superposition and interference to find the secret bitstring s in a single query of the oracle. It's named for its inventors, Ethan Bernstein and Umesh Vazirani [251] [18].

A compact version of this circuit is shown in Figure 10-2.

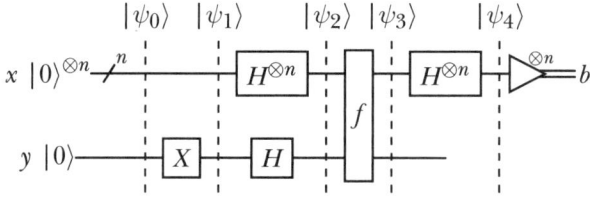

Figure 10-2: The Bernstein–Vazirani algorithm

I introduced the conventions for this compact form in Chapter 9. If it still seems somewhat terse to you, compare it to the expanded version in Figure 10-3.

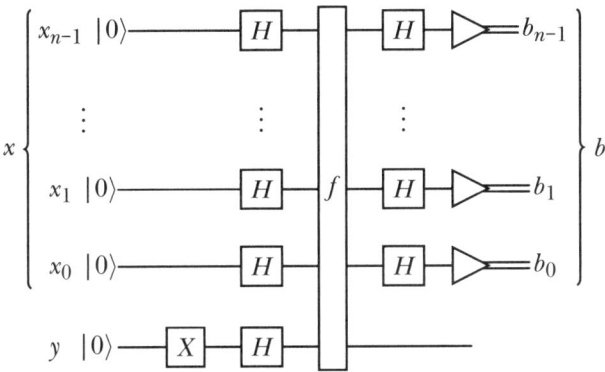

Figure 10-3: An expanded diagram of Figure 10-2

Is something looking familiar? These diagrams are identical to Figures 9-2 and 9-3 for the Deutsch–Jozsa algorithm. The same circuit solves both problems! The only differences are that we're using a different oracle and we interpret the output differently.

We'll perform the same sequence of creating a superposition, evaluating the oracle, and then interfering its output, only this time the output we measure, b, is the secret bitstring s. Let's see how that works.

Circuit Analysis After Deutsch–Jozsa

Since the Bernstein–Vazirani circuit in Figure 10-3 is identical to the Deutsch–Jozsa circuit in Figure 9-2, we can pick up from the end of our analysis of that circuit. If you look back at Chapter 9, we're going to start now just after Equation 9.26, where we found an expression for the state at $|\psi_4\rangle$. As we can see from Figure 10-2, this is the final state of the system immediately before measurement.

We can do this because nothing in our analysis of Deutsch–Jozsa up to that point used any knowledge of the function in the oracle. This means that we can start with $|\psi_4\rangle$, only this time our next step will be to process that expression using features from our new oracle function in Equation 10.1.

Equation 9.26 is repeated here as Equation 10.2. As before, we're looking only at the n qubits in the x register, and we're ignoring that final $|-\rangle$ in the auxiliary qubit.

$$|\psi_4\rangle = \sqrt{2^n} \sum_{k\in\mathbb{B}^n} \sum_{z\in\mathbb{B}^n} (-1)^{f_k + z\cdot k} |z\rangle \tag{10.2}$$

Later, it will be useful for us to have the summations in the opposite order. This makes no change to the computation. Therefore, going forward, I'll use the version of $|\psi_4\rangle$ in Equation 10.3.

$$|\psi_4\rangle = \sqrt{2^n} \sum_{z\in\mathbb{B}^n} \sum_{k\in\mathbb{B}^n} (-1)^{f_k+z\cdot k} |z\rangle \tag{10.3}$$

Now let's plug in our oracle. For an input bitstring k, the oracle evaluates the function $f(k) = s \cdot k$, or the bitstring dot product of the input x and the secret string s. Let's put that into our expression for $|\psi_4\rangle$, as in Equation 10.4.

$$
\begin{aligned}
|\psi_4\rangle &= \sqrt{2^n} \sum_{z\in\mathbb{B}^n} \sum_{k\in\mathbb{B}^n} (-1)^{(s\cdot k)+(z\cdot k)} |z\rangle \quad && \text{Eq. 10.3 with our} \\
&&& \text{new oracle for } f_k \\
&= \sqrt{2^n} \sum_{z\in\mathbb{B}^n} \sum_{k\in\mathbb{B}^n} (-1)^{(s+z)\cdot k} |z\rangle \quad && \text{Collect common factor } k
\end{aligned} \tag{10.4}
$$

Suppose we want to find the amplitude (and thus, the probability) of some specific state z_0. Then we can drop the summation over all the states z and use z_0 instead, giving us Equation 10.5.

$$\text{Amplitude of } |z_0\rangle = \sqrt{2^n} \sum_{k\in\mathbb{B}^n} (-1)^{(s+z_0)\cdot k} \tag{10.5}$$

The state we're actually hoping to measure is $|s\rangle$, the secret bitstring. To find its probability, let's set $|z_0\rangle$ to $|s\rangle$, giving us Equation 10.6.

$$\text{Amplitude of } |s\rangle = \sqrt{2^n} \sum_{k\in\mathbb{B}^n} (-1)^{(s+s)\cdot k} \tag{10.6}$$

Whatever the value of s, we know from Table 10-1 that because s is a bit, $(s + s)$ is 0 because bit addition is modulo 2 (since bits can be only 0 or 1).

Table 10-1: Comparing $(a + b) \bmod 2$ and $a \oplus b$

a	b	$(a + b) \bmod 2$	$a \oplus b$
0	0	$(0 + 0) \bmod 2 = 0$	0
0	1	$(0 + 1) \bmod 2 = 1$	1
1	0	$(1 + 0) \bmod 2 = 1$	1
1	1	$(1 + 1) \bmod 2 = 0$	0

Since both $(0 + 0) \bmod 2$ and $(1 + 1) \bmod 2$ are 0, the $s + s$ term in the exponent on -1 in Equation 10.6 is always 0. That means the whole exponent is 0, and we end up with Equation 10.7.

$$\text{Amplitude of } |s\rangle = \sqrt{2^n} \sum_{k\in\mathbb{B}^n} 1 = \frac{1}{\sqrt{2}^{2n}} 2^n = \frac{1}{2^n} 2^n = 1 \tag{10.7}$$

This tells us something remarkable! The amplitude of state $|s\rangle$ in the superposition is 1. This means the amplitudes of all the other states must be 0. We are therefore *certain* to measure the secret string s at the output!

Running the Algorithm

Once again, we've solved a problem using a circuit in three stages: Create an input superposition of all input states, evaluate them all at once in the oracle (that is, make use of quantum superposition), and then interfere the results together to produce our output.

Let's try it out. I made a circuit of three qubits and set the secret string to 011, as shown in Figure 10-4. The top qubit (corresponding to the bit 0) is unaffected by the oracle, while the two lower qubits (corresponding to the bits 1 and 1) each go through a step of phase kickback.

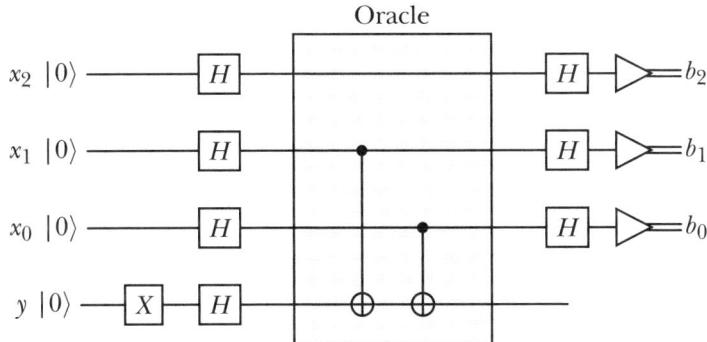

Figure 10-4: The Bernstein–Vazirani algorithm for the secret string 011

I ran this circuit on a five-qubit quantum computer, getting the results shown in Figure 10-5. As usual, this plot omits the states that were never measured.

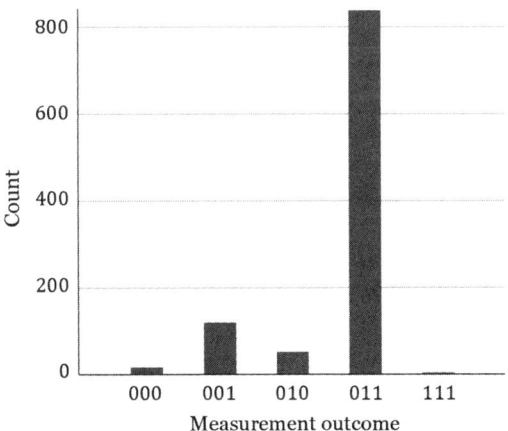

Figure 10-5: Output from running the Bernstein–Vazirani circuit with a secret string of 011

Aside from the usual noise, the result 011 stands out. As always, the theory worked!

Simplifying with a Quantum Identity

Everything we've done so far is fine, but there's another way to look at this oracle. Surprisingly, this point of view has almost no "quantum-like" feeling to it [137] [293].

A *quantum identity* is a way to write one qugate using other qugates. Sometimes this makes things easier to understand, or leads to simplifications. In a moment, I'll use a *CX* qugate mirrored vertically from the orientation we've been using so far, placing the control on the lower line and the target on the upper line. It will operate the same way as a *CX* in the more usual orientation after we surround it with *H* qugates. This equivalence, or identity, is shown in Figure 10-6.

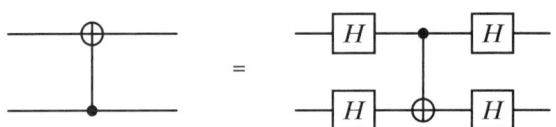

Figure 10-6: The *CX'* qugate with the control on the lower line is equivalent to a *CX* with the control on the upper line, sandwiched between *H* qugates.

To confirm the identity in Figure 10-6, I wrote out the matrix for $CX' = H^{\otimes 2} CX H^{\otimes 2}$ in Equation 10.8.

$$CX' = H^{\otimes 2} CX H^{\otimes 2}$$

$$= \surd^2 \begin{bmatrix} 1 & 1 & 1 & 1 \\ 1 & -1 & 1 & -1 \\ 1 & 1 & -1 & -1 \\ 1 & -1 & -1 & 1 \end{bmatrix} \begin{bmatrix} 1 & 0 & 0 & 0 \\ 0 & 1 & 0 & 0 \\ 0 & 0 & 0 & 1 \\ 0 & 0 & 1 & 0 \end{bmatrix} \surd^2 \begin{bmatrix} 1 & 1 & 1 & 1 \\ 1 & -1 & 1 & -1 \\ 1 & 1 & -1 & -1 \\ 1 & -1 & -1 & 1 \end{bmatrix}$$

$$= \surd^4 \begin{bmatrix} 4 & 0 & 0 & 0 \\ 0 & 0 & 0 & 4 \\ 0 & 0 & 4 & 0 \\ 0 & 4 & 0 & 0 \end{bmatrix} = \begin{bmatrix} 1 & 0 & 0 & 0 \\ 0 & 0 & 0 & 1 \\ 0 & 0 & 1 & 0 \\ 0 & 1 & 0 & 0 \end{bmatrix} \qquad (10.8)$$

The matrix in Equation 10.8 matches the matrix given in Equation 5.48 where the control is on the lower line, so we've confirmed the identity in Figure 10-6.

If we widen the block marked "Oracle" in Figure 10-4 to embrace the *H* qugates on each side of it, we get Figure 10-7, where I gave this new block the label "Block A."

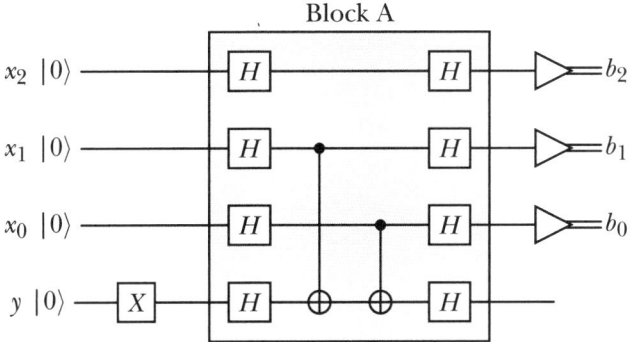

Figure 10-7: Enlarging the oracle block in Figure 10-4

Combining Block A with the identity in Figure 10-6 lets us lose all the H qugates, giving us the simpler Figure 10-8. In this figure, Block B performs the identical function as Block A in Figure 10-7.

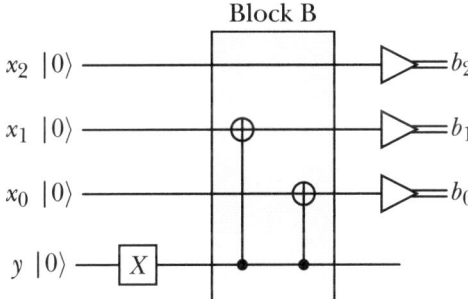

Figure 10-8: Redrawing Figure 10-7 using the identity of Figure 10-6

This circuit is equivalent to the full Bernstein–Vazirani algorithm for the secret string 011! Looking at Figure 10-8, it's no surprise that this works as the oracle. The bottom qubit enters the block in the state $|1\rangle$, the others are $|0\rangle$, and the CX qugates flip x_1 and x_0. The output is then certainly 011. No superpositions, no interference!

Sometimes our problems can vanish when we look at them in a new way.

Summary

In this chapter, we analyzed the Bernstein–Vazirani algorithm, another promise oracle algorithm. This time our oracle holds a secret bitstring, and for any multi-qubit input, it returns the bitstring dot product of that input (interpreted as a bitstring) and the secret string. Our goal was to find the secret bitstring.

Except for the oracle, the Bernstein–Vazirani circuit is identical to the Deutsch–Jozsa circuit, so we had little work to do in our analysis except at the very end of the algorithm. It's amazing that the same set of quantum gates can solve such different problems.

We also saw that there was a way to construct the oracle so that we didn't need even a single superposition. This suggests that, with some work and luck, we might be able to find surprisingly simple forms for other quantum algorithms, including our own.

11

SIMON'S ALGORITHM

JIM: *A good assistant knows what their superior is thinking before they even think it. Meredith, what number am I thinking of right now?*
MEREDITH: *Uh, two.*
JIM: *Nine hundred and eighty-five trillion . . . seventeen.*
—Greg Daniels, Brent Forrester, and Ricky Gervais, *The Office*, "A.A.R.M. Part 1" [43]

In this chapter, I'll broaden our survey of oracle problems to Simon's algorithm. This will introduce us to not just one, or two, but *three* new features.

First, unlike the algorithms we've seen so far, Simon's algorithm doesn't give us a single answer at the end that solves a problem. Rather, each time we run it, we usually measure a unique output. After running the algorithm several times, we'll have accumulated multiple different outputs. Then we'll combine these outputs (and we'll see just how this is done) to find the solution to our problem.

Second, this step of combining and processing the outputs of the quantum algorithm is performed by a classical computer running a classical algorithm. We say that taken together, the two processes form a *hybrid algorithm* comprising a quantum step and a classical step.

The third feature is actually the most exciting. The previous algorithms we've seen can produce results more quickly than classical computers, but the increase in speed wasn't huge. That changes now.

Exponential Growth

Simon's algorithm provides us with an *exponential speedup* over classical algorithms. It's important to be clear about what this means.

In popular use, the word *exponential* has taken on a new life, distinct from its technical origins, to become an adverb that is roughly a synonym for "much" or "a lot." It's common to hear people say things like, "Today is exponentially hotter than yesterday."

The original (though more nerdy) meaning of the word refers to a specific type of mathematical function that raises a variable to an exponent [236]. Generally, an *exponential function* is given by $f(x) = ab^{cx+d}$. The heart of the function is the b^x part, which tells us that the complexity, or cost, of our problem grows by a factor of b raised to x, the size of the problem. This is the meaning of *exponential* in this book.

When people speak of *the* exponential function in the sciences, they mean the function e^x, also written $\exp(x)$, where e is called *Euler's number* or *Napier's number*. It has a value of about 2.718.

Exponential functions grow *quickly*, which is why people have adopted the word to imply staggering growth. When referring to a computation, exponential growth means that the problem eventually starts demanding overwhelming amounts of resources such as memory or running time. The best classical algorithms that we know of to solve Simon's problem are exponential [292]. Because of their dramatically increasing demands, such algorithms quickly become impractical for large problems.

Simon's algorithm, by contrast, is *not* exponential. Even though we have to execute the algorithm multiple times to produce a solution, the number of repeats is expressed *linearly*, or in the form ax for a fixed a and x related to the size of the task. As we use Simon's algorithm to solve larger and larger problems, its computational requirements increase far more slowly than an exponential algorithm's. Because it replaces an exponential algorithm with a linear one, we say that Simon's algorithm provides an *exponential speedup* over classical approaches.

Simon's algorithm was the first to show that a quantum computer could provide exponential speedups to problems, giving us answers beyond the practical reach of today's classical computers.

Like the problems we've seen in previous chapters, the problem we'll solve here is just a toy that you'll probably never encounter in practice. By keeping things simple, we'll be able to see how the new ideas work without the extra burden of a complicated problem to solve.

Before we get to the algorithm itself, let's note a couple of things about exponential algorithms. The explosive growth of exponential functions often catches people off guard, as we're not used to this kind of effect at our usual human scales of time and space.

For problems that involve small inputs (the precise meaning of *small* depends on the algorithm), an algorithm that scales exponentially can still be attractive and practical. The problem with these algorithms is that as the size of the problem increases, the rate at which the algorithm consumes time, memory, energy, or other resources increases at an astounding pace.

Getting a feeling for this phenomenon can be challenging. The physicist Albert Bartlett noted that the principle of growth so revered by capitalists, economists, and investors is rooted in exponential behavior. Such growth is certain to lead to calamity because in the world we inhabit, resources are finite. The unending pursuit of growth will consume all resources until none are left, causing widespread suffering. Bartlett lamented this distressing observation compactly, asserting that "the greatest shortcoming of the human race is man's inability to understand the exponential function" [15].

Algorithms that demand exponential resources such as time and space are usually not considered great solutions to large, real-world problems.

In previous chapters, we ran each algorithm once, and we knew that our measurement at the end would hold the solution to our problem. With practical quantum algorithms, that ideal situation is rare. Usually, we have to run the algorithm, measure an output, and then *check* the result (often with a classical computer) to see if it solves our problem. If not, we run the quantum algorithm again, check the result again, and repeat the loop until we measure an output that solves our problem. This is often efficient, because checking an answer can be far easier than computing it.

One way that computer scientists classify programs is in terms of how long the computer might take to produce an answer. This doesn't mean that every run of the program *will* take that long to produce an answer, but only that it *could*. Problems like the one we're looking at in this chapter belong to a category called the *NP complexity class* [274]. This means that their answers can be verified in an amount of time given by a polynomial equation. In practice, this usually means that even if it's hard to find an answer, it's quick to confirm whether the answer is correct. For example, solving fiendishly hard Sudoku games of arbitrary size is in the NP complexity class [91]. It might take you a long time to complete a grid, but you can check your solution quickly.

This kind of problem is like finding a pair of comfortable shoes. You can read all the reviews you like, but there's no substitute for going to the shoe store and trying on one pair after another until you find shoes that feel good on your feet. Finding the best pair of shoes can be time-consuming (even with expert help), but you can usually tell right away if the pair you're wearing is comfortable or not.

There are many other classes of classical complexity, and there's a corresponding theory of quantum complexity classes [47, §12]. Analyzing the complexity of an algorithm can be difficult, because we have to be sure to take into account every possible step in its computation. In this book, I won't do any complexity analysis except in the broadest terms.

Simon's Oracle

This chapter's focus is *Simon's algorithm*, named for its inventor, Daniel Simon [200]. It's centered around a promise oracle that takes a bitstring as input and produces a new bitstring as output.

The promise made by Simon's oracle is a little strange. It seems unlike anything that would come up in practice, and that's because it didn't! In fact, according to Simon, the algorithm came about as the result of a thought problem. Casting about for a quantum algorithm that would show an exponential speedup, he first created the circuit and then looked for an oracle to fit it [182].

As usual, let's meet the oracle as a classical system that uses bits rather than qubits, and then move it to the quantum world.

Like the Bernstein–Vazirani oracle in Chapter 10, Simon's oracle contains a secret bitstring. The number of bits in the bitstring is its dimensionality. Usually I'd give that the name d, but in this algorithm the number of bits in the secret string will also be the number of qubits we're working with, and we usually use n for that quantity. We'll be spending a lot of time with the qubits, so I'll use n for the number that describes both the number of bits in the oracle and the number of qubits in the algorithm.

All the bitstrings we'll work with in this chapter will have the same dimensionality, or number of bits, n.

The oracle doesn't make any promises about its specific outputs; it only promises that its outputs satisfy a specific (though weird) property. Many outputs will satisfy this property, so there's no single result we're hoping to measure. That is, we're not going to directly measure the secret bitstring s at the output of the algorithm. Instead, we'll run the algorithm multiple times and amass lots of measurements, and then we'll combine those measurements classically to uncover the secret bitstring.

The oracle's definition is based on the XOR operation, whose truth table we saw in Table 8-2. Keep in mind going forward that the XOR operation is symmetrical, or commutative, so $a \oplus b = b \oplus a$.

Simon's oracle generalizes this operation to work with bitstrings, using the *bitstring XOR* operation. Given two bitstrings a and b of the same number of bits, writing $a \oplus b$ means we apply XOR to each corresponding pair of bits to produce a new bitstring of the same number of bits. Equation 11.1 shows the idea.

$$a \oplus b = a_{n-1} \oplus b_{n-1}, a_{n-2} \oplus b_{n-2}, \ldots, a_0 \oplus b_0 \tag{11.1}$$

For example, suppose we have two bitstrings, each with three bits. We could write a in terms of its bits as $a_2 a_1 a_0$ and b similarly as $b_2 b_1 b_0$. The bitstring XOR applies $a \oplus b$ to each corresponding pair of bits, producing the new three-bit bitstring $(a_2 \oplus b_2), (a_1 \oplus b_1), (a_0 \oplus b_0)$.

Simon's oracle promises to use a secret bitstring s such that the output of $f(x)$ (for any x) and $f(x \oplus s)$ will be the same, and different from any other outputs.

This is a strange function, without an obvious use (though it might be helpful in some form of cryptography). In fact, as I mentioned, Simon invented this function by starting with a simple quantum circuit and looking for something, *anything*, it could do better than a classical circuit, and contrived this function to fill that role [182].

Another way to express the oracle's promise is to say that for bitstrings x and y, if $y = x \oplus s$, then $f(x) = f(y)$. Equation 11.2 shows that the situation is symmetrical, so we could also XOR the secret string with y to get x.

$$
\begin{aligned}
x \oplus s &= y && \text{Starting relationship} \\
(x \oplus s) \oplus s &= y \oplus s && \text{XOR both sides with } s \\
x \oplus (s \oplus s) &= y \oplus s && \text{Since XOR is associative} \\
x &= y \oplus s && \text{Because } s \oplus s = 0 \text{ and } x \oplus 0 = x
\end{aligned}
\tag{11.2}
$$

This tells us that Simon's algorithm produces outputs that come in *pairs*. For any input x, the output $f(x)$ will also be the output for $f(y)$ if $y = x \oplus s$, and this shared output will not appear for any other input. If x and y are different, then we call this a *two-to-one function*, as it sends pairs of inputs to the same output. I'll give this the shorthand name of an *F2* function.

Let's get a feeling for an *F2* function by building one. Remember that the oracle doesn't have to produce any specific outputs for these inputs, as long as they obey the rule that $f(x) = f(x \oplus s)$.

I'll pick a secret string s with the value 101. A possible function for this secret string is shown in Table 11-1. The leftmost column shows the eight possible inputs, and the column to its right shows the output $f(x)$ for each input.

Table 11-1: An *F2* function for $s = 101$

x	$f(x)$	$x \oplus 101$	$f(x \oplus 101)$
000	001	101	001
001	110	100	110
010	011	111	011
011	100	110	100
100	$110 = f(001)$	001	110
101	$001 = f(000)$	000	001
110	$100 = f(011)$	011	100
111	$011 = f(010)$	010	011

To better understand what's going on here, let's rebuild this function one step at a time. We'll start by imagining that the table is empty except for the inputs x in the leftmost column and the XOR of those inputs with the secret string x in the third column, labeled $x \oplus 101$. We'll fill in the second and fourth columns by assigning outputs to inputs, working our way from top to bottom.

The first input is $x = 000$. We can give it any three-bit bitstring we like as an output. I'll arbitrarily assign it the output 001 and place that in the $f(x)$ column. The other input is found from $y = x \oplus s = 000 \oplus 101$. The third column, labeled $x \oplus 101$, tells us that this value is 101. Therefore, when the input is 101, the function will produce 001, just as it did for the input 000. So, we also assign 001 to the output of input 101. I've placed 001 in the first row of the rightmost column as well, because we've just found that it's also the output of $x \oplus 101$.

The fragment of the table we've just filled in is in Table 11-2.

Table 11-2: Filling in Table 11-1

x	$f(x)$	$x \oplus 101$	$f(x \oplus 101)$
000	001	101	001
001		100	
010		111	
011		110	
100		001	
101	$001 = f(000)$	000	001
110		011	
111		010	

That's the first pair done. Now let's move down the table. Input 001 doesn't have an assigned output, so we can give it any unused output value. I'll arbitrarily assign it 110. Following the same process as before, we find $y = 001 \oplus 101 = 100$, so we assign the output 110 to the input 100 as well.

Continuing down, input 010 also has no output, so I'll arbitrarily assign it the output 011. Once again, we find y by computing $010 \oplus 101 = 111$ and assign the same output, 011, to the input 111.

Finally, I'll give input 011 the output 100 and assign the same input to the other input element of this pair, $y = 011 \oplus 101 = 110$.

And that's a perfectly fine definition of a Simon's oracle of type $F2$. We can assign the outputs any way we like so long as each $f(x) = f(x \oplus s)$ and each output belongs to exactly one of the pairs. Because the outputs come in pairs, half of the possible outputs never appear.

That would be the end of our discussion of Simon's oracle, except for one snag: When $s = 0$ (that is, s is a bitstring of the same length as x, where every bit is 0). In this special case, every y is $y = x \oplus 0 = x$, so every "pair" of inputs degenerates to a single input, as x and y have the same value. In other words, every input has a unique output. For this special case of $s = 0$, we say that Simon's oracle is a *one-to-one function*, as each input produces a unique output. In other words, every bitstring of n bits appears as an output once. In shorthand, I'll call this an $F1$ function.

An example of an $F1$ function is shown in Table 11-3. Because the function is $F1$, there are eight outputs, which we can assign to the inputs any way we like. Essentially, the output strings are a shuffled version of the input strings. This also defines a perfectly fine Simon's oracle, since $f(x) = f(x \oplus s)$ for every x. Because $s = 0$, for every input x we have $x \oplus s = x$, so every output is unique.

Table 11-3: An $F1$ function when $s = 0$

x	$f(x)$
000	110
001	010
010	001
011	111
100	011
101	101
110	000
111	100

Our goal will be to find the secret string s, whether it's 0 or not.

As usual, we'll promote Simon's oracle to the quantum domain by treating every bitstring as a quantum state formed by the tensor product of $|0\rangle$ and $|1\rangle$ states, matching the bits in the bitstring. So in Table 11-3, the input bitstring 100 corresponds to the input state $|100\rangle$, and its output bitstring 011 corresponds to the output state $|011\rangle$.

The Classical Solution

How much computing power do we need from a classical computer to determine the secret string? We'll continue to use oracle complexity, or the number of queries to the oracle, as our measure [71].

Finding the oracle complexity rigorously for Simon's problem is challenging [30] [38]. The problem is that when algorithms become complicated, there's a world of difference between arguing that any specific solution *sounds* like the best one and actually *proving* that no other solution, from anyone, ever, could be more efficient. Just because nobody's imagined a better solution yet isn't proof that such a solution isn't out there.

In this case, a quick and informal analysis gives us a result that is close to the proven answer, but with much less work, so I'll take this casual route.

If we have n bits, there are 2^n possible input bitstrings to the oracle. We need to find two inputs x_0 and x_1 for which $f(x_0) = f(x_1)$. Taking our 2^n strings one by one and plugging them into f, in the worst case of a two-to-one function we could try half of the inputs and never see a repeated output. So, we need to try one more input. If its output duplicates a previously seen result, then we know we have a two-to-one function; otherwise, it's a one-to-one function.

For example, in Table 11-1, suppose we tried values starting at the top and worked our way down. There are $2^n = 2^3 = 8$ inputs to try. For the first half of the inputs, none of the outputs was a duplicate. But the next input, 100, produced the output 110, which matched one of the earlier outputs, from input 001. It's only at this moment, after trying one more than half of the inputs, that we can say that the secret string in Table 11-1 must be given by $100 \oplus 001 = 101$.

The conclusion is that we may have to evaluate $(2^n / 2) + 1 = 2^{n-1} + 1$ inputs to find the secret string. We say that the cost of the algorithm, in this case the number of queries we have to make to the oracle, is of the *order* 2^n. We drop the other terms because as n grows, the cost of everything else (including the division by 2) becomes insignificant in comparison to 2^n. Since the cost of the algorithm is of the form b^x (for $b = 2$), we say that the classical version of Simon's problem has exponential cost.

This analysis is the same as the one we use for a famous problem called the *birthday problem*. It asks how many people you need to assemble in a room to have more than some degree of confidence that two of them share a birthday. For example, if you want to be about 50 percent sure that there's a shared birthday, you need only 23 people [12] [232].

The birthday problem (and its many variations) has been carefully studied [63], bringing the cost down to the expression $k2^{n/2}$ for some number k that depends on the specifics of the problem [26].

Whether the exponent on 2 is $n-1$ or $n/2$, the cost is still 2 to the something related to n, which makes it an exponential function. Figure 11-1(a) shows the function $2^{n/2}$, and Figure 11-1(b) shows the function 2^{n-1}.

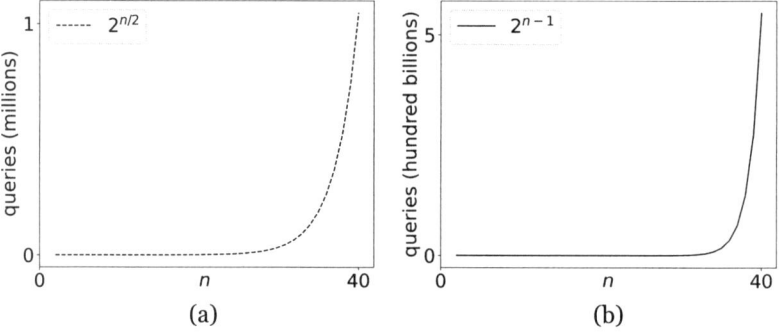

Figure 11-1: (a) The function $2^{n/2}$. The vertical scale is in millions. (b) The function 2^{n-1}. The vertical scale is in hundreds of billions.

As Figure 11-1 shows, the value of the exponent in an exponential function makes a big difference, but speaking generally, both shapes have the sudden, dramatic increase that is characteristic of exponential growth. Due to these costs, any exponential classical algorithm quickly becomes impractical. For example, if you wanted to classify a function in Simon's problem of 50 bits, you would need to call the oracle roughly 2^{50}, or over a *quadrillion*, times to see a cycle. That's about the number of atoms making up the Earth [61]!

Suppose that for some reason you actually wanted to solve Simon's problem for a large number of bits. We've just seen that the time required quickly becomes ridiculously impractical. And that's our cue to look for a quantum solution!

The Quantum Part of the Algorithm

Simon's algorithm isn't something you're ever likely to run. It solves a made-up problem that Simon invented to show off what a quantum computer can do.

One of Simon's innovations was to design an algorithm in two parts, one quantum and one classical. The two processes work sequentially, with the quantum algorithm going first. The classical bitstrings we measure from multiple runs of the quantum algorithm become the input values for the classical algorithm.

This is because each output contains some information about the secret bitstring s. The classical algorithm is designed to combine all this partial information to determine s.

Simon's algorithm is in two pieces that are connected by an abstract mathematical idea. This can make it challenging to understand either piece in isolation. We'll look at the quantum circuit first, and then we'll see how its outputs will be combined in the classical program [182].

Let's get started by considering a compact version of the quantum circuit, shown in Figure 11-2.

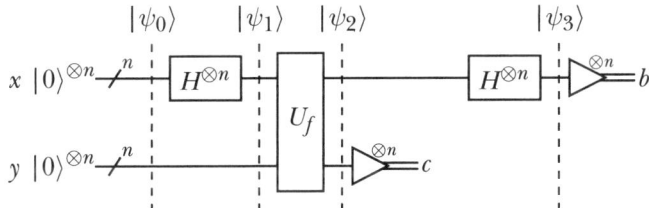

Figure 11-2: The quantum part of Simon's algorithm

This is reminiscent of the Deutsch–Jozsa and Bernstein–Vazirani algorithms in Figures 9-3 and 10-3, but it's not quite the same.

Specifically, the lower input marked y is now also a register of n qubits, and we measure this register. We'll see later that we don't ever use the measured result c, so we don't actually *have* to make these measurements, but they're a helpful conceptual step for understanding how the circuit works. We also don't put any of the input qubits into the $|-\rangle$ state as we usually did before, because this algorithm doesn't use phase kickback.

We can write down the first two steps right away, as shown in Equation 11.3. As usual, I'll keep the x and y registers distinct.

$$|\psi_0\rangle = |0\rangle^{\otimes n} |0\rangle^{\otimes n}$$
$$|\psi_1\rangle = |+\rangle^{\otimes n} |0\rangle^{\otimes n}$$

(11.3)

Again using f_x as a shorthand for $f(x)$, we can write the output of the oracle for any specific input x as $|x\rangle\,|y \oplus f_x\rangle$. Since this oracle doesn't modify the x register, its upper output is x itself. And because every input y is 0, $|y \oplus f_x\rangle$ is just $|f_x\rangle$. With these observations, we can write the output of the oracle, $|\psi_2\rangle$, as shown in Equation 11.4.

$$|\psi_2\rangle = \vee^n \sum_{x \in \mathbb{B}^n} |x\rangle\,|f_x\rangle \tag{11.4}$$

We're going to break with tradition now. In the algorithms we've seen so far, the next step was to apply H qugates to the x register and then measure the results. Instead, we'll measure the y register first.

Why would we want to measure the y register, particularly since we're not going to use the result? It's to exploit the extended partial measurement rule from Chapter 6. This tells us that upon measuring any specific output for y, the remaining qubits (in register x) will collapse to the state (or states) that are consistent with that measurement. Thus, measuring y is a way to filter the superposition in x so it's a smaller, specific set of states.

But do we really need to measure y before we measure x? The principle of deferred measurement, also discussed in Chapter 6, tells us that we don't necessarily need to measure y at any specific moment. We can measure it at any time we like, perhaps just after the oracle, or maybe next year, or in 1,000 years. We can even make this measurement *after* we measure the x register. No matter when we measure it, the x register will collapse to the states in the superposition that are consistent with the measurement of y. Since we can hold off on that measurement as long as we like, in practice we never need to actually measure y at all!

Once again, quantum computing doesn't always work the way we're used to things working.

For this discussion, I'll assume we measure y right after the oracle, when it's part of the state $|\psi_2\rangle$.

Here's the big picture of the next few steps.

Think of the output of the oracle, $|\psi_2\rangle$, as a single system of $n + n = 2n$ qubits. The fact that we chose to think of this system as made up of two pieces with different names in Equation 11.4 doesn't matter to the hardware. To reflect that, I'll rewrite Equation 11.4 as Equation 11.5, which expresses the qubit system as a superposition of states $|x\,f_x\rangle$.

$$|\psi_2\rangle = \vee^n \sum_{x \in \mathbb{B}^n} |x\,f_x\rangle \tag{11.5}$$

If the function is of the class $F1$, getting a specific measurement y_0 means that the x register must hold the corresponding state $|x_0\rangle$, and not a superposition. This is because the only state that is consistent with an $F1$ function giving us a measurement of y_0 on the output is for the x register to be x_0. That is, $f(x_0) = y_0$.

In short, x_0 is the only input that could have produced y_0, so we'll measure x_0 with certainty.

However, if the function is of type $F2$, then measuring y_0 tells us that the x register is necessarily *either* of the x inputs that produce y. Let's call them $|x_0\rangle$ and $|x_1\rangle$. When we measure a specific y_0, the superposition $|+\rangle^{\otimes n}$ in register x collapses to $\vee(|x_0\rangle + |x_1\rangle)$, because all the other states in the input superposition are inconsistent with having measured y_0. In other words, our measurement filtered out all but two input states.

As a result of this collapse of x, we can write the system as $|\psi_2\rangle'$, shown in Equation 11.6.

$$|\psi_2\rangle' = \vee\left(|x_0\rangle + |x_1\rangle \right) |y_0\rangle \tag{11.6}$$

Measuring the upper register will give us either x_0 or x_1, but we can't say which.

Because we're not sure what output we'll measure, it's not clear how we might process that output to find the secret string we're after. But if we work through the math, we'll see that if we run the algorithm enough times, we'll gather enough information for a classical algorithm to extract s from our measurements.

Rewriting x and Measuring

Our last step in analyzing the quantum part of Simon's algorithm will be to characterize $|\psi_3\rangle$, the final state that we're going to measure in the upper register x.

Let's focus first on the $F1$ case. To find $|\psi_3\rangle$, we apply H qugates to the upper register of Equation 11.5. I'll call this result $|\psi_3\rangle_1$. Remembering that we've measured some specific output y_0 from the y register, the x register must be $|x_0\rangle$. Equation 9.25 lets us apply H to this state, giving us $|\psi_3\rangle_1$ in Equation 11.7 (since we've now measured the x register, I'll leave out the y register for the rest of this discussion). I'll use the letter z for our summation, for consistency with the steps to come.

$$
\begin{aligned}
|\psi_3\rangle_1 &= H^{\otimes n} |\psi_2\rangle_x && \text{Apply } H^{\otimes n} \text{ to the } x \text{ part} \\
&&& \text{of } |\psi_2\rangle \\
&= H^{\otimes n} \vee^n \sum_{z \in \mathbb{B}^n} |z\rangle && \text{Use Eq. 11.5 for the } x \text{ part} \\
&&& \text{of } |\psi_2\rangle \\
&= \vee^n \sum_{z \in \mathbb{B}^n} (-1)^{x_0 \cdot z} |z\rangle && \text{Rewrite using Eq. 9.25}
\end{aligned}
\tag{11.7}
$$

As usual, the bitstring dot product $x_0 \cdot z$ is the sum of the products of each pair of values, following the definition in Equation 9.19.

Now let's assume the oracle is of type $F2$. To find the expression when the function is two-to-one, we note that measuring y_0 for the y register means that the x register must be either $|x_0\rangle$ or $|x_1\rangle$. We can use this to rewrite Equation 11.6, giving us $|\psi_3\rangle_2$ in Equation 11.8.

$$|\psi_3\rangle_2 = H^{\otimes n}\,|\psi_2\rangle_x \qquad\qquad \text{Apply } H^{\otimes n} \text{ to the } x \text{ part of } |\psi_2\rangle$$

$$= H^{\otimes n} \vee (|x_0\rangle + |x_1\rangle) \qquad\qquad \text{Use Eq. 11.6 for the } x \text{ part of } |\psi_2\rangle$$

$$= \vee\left(H^{\otimes n}\,|x_0\rangle + H^{\otimes n}\,|x_1\rangle\right) \qquad\qquad \text{Distribute } H^{\otimes n} \qquad (11.8)$$

$$= \vee^{n+1}\left(\sum_{z\in\mathbb{B}^n}(-1)^{x_0\cdot z}\,|z\rangle + \sum_{z\in\mathbb{B}^n}(-1)^{x_1\cdot z}\,|z\rangle\right) \qquad\qquad \text{Rewrite using Eq. 9.25}$$

$$= \vee^{n+1}\sum_{z\in\mathbb{B}^n}\left((-1)^{x_0\cdot z} + (-1)^{x_1\cdot z}\right)|z\rangle \qquad\qquad \text{Collect terms}$$

Whew! All this work is going to pay off, though, because now we're ready to measure the x register!

Suppose that the oracle is of type $F1$. Then the last line of Equation 11.7 tells us the state of the system just before measurement. Let's say that measurement produces a bitstring z. We know by the definition of the oracle that $s \cdot z = 0$.

Alternatively, suppose that the oracle is of type $F2$. Then the last line of Equation 11.8 tells us that we'll measure a bitstring z that corresponds to either $|x_0\rangle$ or $|x_1\rangle$. I'll say that x_0 and x_1 are the bitstrings used by the oracle associated with these two states.

The trick to proceeding comes from noticing that $x_0 \cdot z$ and $x_1 \cdot z$ must be equal. That's because if they're not, the amplitudes in Equation 11.8 cancel one another, and we'll never measure the resulting state because it has an amplitude of 0. To see this, let's look at just the coefficient $(-1)^{x_0\cdot z} + (-1)^{x_1\cdot z}$. In Equation 11.9, I've written this sum for all four possible values of $x_0 \cdot z$ and $x_1 \cdot z$.

$x_0 \cdot z$	$x_1 \cdot z$	$(-1)^{x_0\cdot z}$	$(-1)^{x_1\cdot z}$	Sum
0	0	1	1	2
0	1	1	−1	0
1	0	−1	0	0
1	1	1	1	2

(11.9)

This tells us that if $x_0 \cdot z \neq x_1 \cdot z$, the resulting state has an amplitude of 0. Since the only states we can measure are those with nonzero amplitudes, we can conclude that $x_0 \cdot z = x_1 \cdot z$. Let's write that equivalence down and simplify it as shown in Equation 11.10.

$$x_0 \cdot z = x_1 \cdot z \qquad\qquad \text{Must be equal}$$
$$x_0 \cdot z = (x_0 \oplus s) \cdot z \qquad\qquad \text{Because } x_1 = x_0 \oplus s$$
$$x_0 \cdot z = (x_0 \cdot z) \oplus (s \cdot z) \qquad\qquad \text{Distribute the dot product} \qquad (11.10)$$
$$0 = s \cdot z \qquad\qquad \text{Remove the common term } x_0 \cdot z$$

Combining what we've just seen for the $F1$ and $F2$ cases, we've found that any state z we measure from register x in Simon's algorithm satisfies $s \cdot z = 0$.

Although this relationship by itself doesn't tell us much about s, if we run the algorithm repeatedly and build up a collection of measured z outputs, they can be combined to unlock the secret string s. We'll do that last step of processing on a classical computer.

Combining the Quantum Outputs

Imagine that we've run and measured the quantum circuit many times, producing a bunch of measured outputs. Now we get the payoff: a classical algorithm that combines those outputs to produce the secret string.

Suppose we run Simon's algorithm r times. Let's call our measurements $z^{(k)}$ for $k \in [r]$. Writing k as a superscript leaves the subscript available for us to use in a moment. Putting parentheses around the k reminds us that this is an index, not an exponent.

We've seen that for both $F1$ and $F2$ oracles, finding the bitstring dot product of these measured states $z^{(k)}$ with the secret string s will always produce a result of 0. Let's write this for r measured results, as shown in Equation 11.11.

$$0 = s \cdot z^{(0)}$$
$$0 = s \cdot z^{(1)}$$
$$\vdots \qquad\qquad (11.11)$$
$$0 = s \cdot z^{(r-1)}$$

As all of our bitstrings have n elements, both s and every $z^{(k)}$ have n elements, each 0 or 1. We call Equation 11.11 a *system of r equations in n unknowns*.

Suppose $n = 1$, so registers x and y are each a single qubit. Then one measurement tells us everything we need. Since $s \cdot z^{(0)} = 0$, we immediately have $s = z^{(0)}$.

We found the secret bitstring! Well, we found the secret bit.

What if $n = 2$? To find two values for our two-element bitstring, we'd need two different values for z. Remember that $z^{(k)}$ is the result of the kth measurement. Now we'll use the subscript we kept available earlier to refer to the bits of each measurement. Numbering the bits from the right and starting with 0, measurement $z^{(0)}$ has bits $z_1^{(0)}$ and $z_0^{(0)}$, and measurement $z^{(1)}$ has bits $z_1^{(1)}$ and $z_0^{(1)}$. Similarly, we can write s in binary as $s_1 s_0$.

Our results are then shown in Equation 11.12 (remember that we're adding bits, so $1 + 1 = 0$).

$$0 = z_1^{(0)} s_1 + z_0^{(0)} s_0$$
$$\qquad\qquad (11.12)$$
$$0 = z_1^{(1)} s_1 + z_0^{(1)} s_0$$

We know the bits of $z^{(0)}$ and $z^{(1)}$, so Equation 11.12 has two equations with two unknowns (s_0 and s_1).

Generalizing this for r runs gives us the messy monster Equation 11.13, where again all additions are modulo 2.

$$0 = z_{n-1}^{(0)} s_{n-1} + z_{n-2}^{(0)} s_{n-2} + \cdots z_1^{(0)} s_1 + z_0^{(0)} s_0$$

$$0 = z_{n-1}^{(1)} s_{n-1} + z_{n-2}^{(1)} s_{n-2} + \cdots z_1^{(1)} s_1 + z_0^{(1)} s_0 \qquad (11.13)$$

$$\vdots$$

$$0 = z_{n-1}^{(r-1)} s_{n-1} + z_{n-2}^{(r-1)} s_{n-2} + \cdots z_1^{(r-1)} s_1 + z_0^{(r-1)} s_0$$

This is the same as Equation 11.11, but with all the bitstring dot products written out.

There are lots of classical algorithms that take Equation 11.13 as input, along with the values of the z terms, and efficiently solve for the bits of s. Most are based on a technique called *Gaussian elimination* [238] named for Carl Friedrich Gauss (though special cases for this process were known at least 1,600 years before his birth [264]). Plug the equations into Gaussian elimination, and out pops the secret string s.

As long as $r > n - 1$ (that is, the number of times that we've run the algorithm is greater than the number of qubits in x minus 1), we have a chance of finding each of the n bits in s. But it's not a sure thing. To be sure we can get all n bits, Gaussian elimination requires that the equations be *linearly independent*. This means that we cannot multiply all the terms in any equation by a single number and thereby match any of the other equations. If a set of equations is not linearly independent, we say it is *linearly dependent*.

If the equations in Equation 11.13 are not linearly independent, then the Gaussian elimination algorithm will get stuck, and we won't get an answer. This isn't a fault of Gaussian elimination, because any other system would also get stuck. We just don't have enough information yet.

Because $s = 0$ is always a solution, we want $n - 1$ linearly independent equations (rather than n) [14]. Each run of Simon's algorithm gives us one equation, but we could get unlucky and get back some measurements that lead to equations that are multiples of equations we already have. How many times do we have to run Simon's algorithm to be sure that we have $n - 1$ linearly independent equations?

Because quantum measurement is probabilistic, we can never be completely sure. It's possible that we could measure the same result over and over again, no matter how many times we run the algorithm (the odds of this happening plummet with each additional run, and can become astronomically small, but they never reach 0).

One way to go is to run Simon's algorithm $n - 1$ times and then *try* to apply Gaussian elimination. The algorithm will fail if there aren't enough linearly independent equations (in practice, this usually means the library routine reports an error). Then we can go back, run the algorithm a few more times, append the new equations onto our list from Equation 11.11, and repeat the process until Gaussian elimination succeeds and we get back s.

If we collect a great many outputs and Gaussian elimination still fails, we usually just declare that the secret string is all 0s, though there can always be a shred of doubt.

I won't go into the details, but you can prove that the number of times you need to run Simon's algorithm to almost surely obtain $n - 1$ independent equations and thus discover s is an, or some constant a times the number of bits n [182] [14]. In other words, the growth of Simon's algorithm is linear in n, not exponential.

As I mentioned before, because we never use the measurement of y, the principle of partial measurement (which we discussed in Chapter 6) tells us that we can put that measurement off for a few minutes, or a few millennia. Since we get all the information we need from measuring x, we don't actually *ever* need to measure y.

This is wild! The mere fact that we *could* measure y forces x to be consistent with that measurement, even if we don't actually measure y. Quantum computing is full of surprises.

A version of Figure 11-2 without this measurement is given in Figure 11-3.

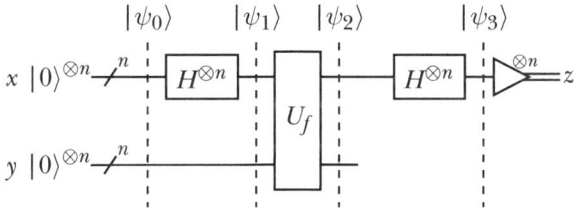

Figure 11-3: Simon's algorithm without measurement of y

This is a strange situation. We need to measure the y register to get the x register into the state we want. But since we know that we *can* measure y, we don't *have* to, because we can always measure it later. That is, if we measure y, then we know x will collapse to a superposition consistent with that measurement of y, regardless of when we make it. So if we agree that we'll measure it, say, in a hundred million years, we can in practical terms just drop that measurement from the circuit, and proceed *as if* we had made it.

Wow.

An Example of Simon's Algorithm

Let's run an example. Designing an oracle for Simon's algorithm that uses an arbitrary secret string can require some delicate logical thinking [111], so let's pick a small, specific situation. I'll say that our input is three qubits, and the secret bitstring is $s = 010$. The complete circuit, including the oracle, is shown in Figure 11-4. I've left off the measurement of the y register, because it's not needed.

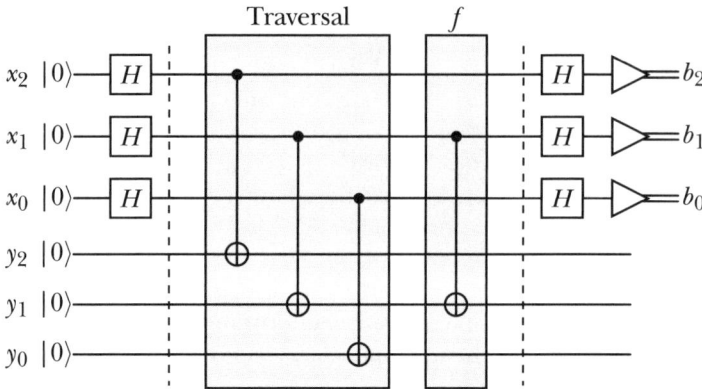

Figure 11-4: A circuit for Simon's algorithm with the secret string $s = 010$

The oracle is between the dashed lines. I've split it into two pieces, shaded in gray. First, I placed a CX qugate from each qubit of the x register to its corresponding qubit in the y register. That is, x_2 is the control for its target, y_2. Similarly, x_1 is the control for its target y_1, and x_0 is the control for its target y_0. This pattern of qugates is sometimes called a *traversal*. Then I applied the oracle for the function f, which performs $x \oplus s = x \oplus 010$.

Note that the f operation, involving a single CX qugate, is the same as the second CX in the traversal, so they undo one another. We could therefore simplify this diagram (and the demands on the hardware that executes it) by removing both of these qugates. I left them both in place here to illustrate the use of a traversal.

A summary of the inputs and the result of applying f to each one is shown in Table 11-4.

Table 11-4: Values involved in Simon's algorithm for the secret string $s = 010$

x	y	$x \oplus s$	$x \cdot s$
000	000	010	0
001	001	011	0
010	000	000	1
011	001	001	1
100	100	110	0
101	101	111	0
110	100	100	1
111	101	101	1

In Table 11-4, there are four values of x for which $x \cdot s = 0$: 000, 001, 100, and 101. So when we run this algorithm many times, we should expect to get each of these four results with equal frequency, and no others.

I ran this for 1,024 shots on a seven-qubit quantum computer, and got back the results in Figure 11-5.

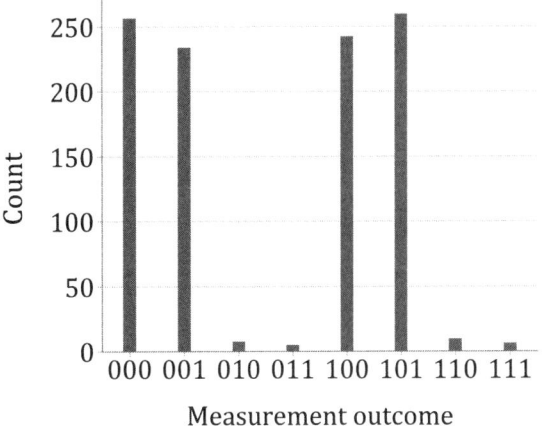

Figure 11-5: Running Simon's algorithm with s = 010 using the circuit of Figure 11-4 on a seven-qubit quantum computer

Pretty nice! Those short bars are due to the noise we can't really avoid these days, but the dominant results show that we definitely received the results we predicted.

The Balancing Act

We've seen the steps of Simon's algorithm, but you might be left wondering how it all came together.

If so, don't let it get to you. This system is complicated, despite its deceptively simple circuit diagram.

And remember that Simon didn't set out to solve this problem. As I mentioned, he started with a circuit containing an oracle, then searched for some interesting function the oracle could compute. He eventually assembled a scheme where the oracle implemented a strange and unfamiliar function. Because he knew about solving systems of linear equations, Simon engineered this function so that it could produce a set of $n - 1$ linearly independent equations, which is just what Gaussian elimination (or any other solver of linear systems) needs as input. The result is a set of equations that Gaussian elimination can solve to produce the bits that make up that secret string.

It's all a balancing act. The whole mechanism is unlikely and artificially constructed, but it does do *something*. And it does it exponentially faster than you can do it on a conventional computer, which is a great demonstration (though mostly a theoretical one) of the power of quantum computing.

The performance gain of Simon's hybrid algorithm versus a classical algorithm, where cost is measured in the number of queries of the oracle, is shown in Figure 11-6.

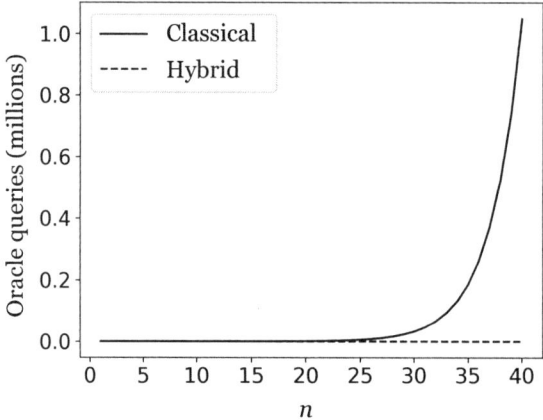

Figure 11-6: The costs of the classical solution to Simon's problem and the hybrid quantum–classical solution. The vertical scale is in millions.

No matter what range we pick for n, the curves always have this form: The cost of Simon's algorithm hardly grows, while the classical cost explodes exponentially.

Simon's algorithm was a breakthrough in quantum computing because it was the first quantum algorithm that demonstrated an exponential speedup, confirming many people's expectations that quantum algorithms could dramatically outperform their classical counterparts [182].

Simon's algorithm is a great example of the subtlety and sophistication of quantum algorithms. Its elegant but complex technique is belied by its apparently simple quantum circuit.

Summary

In this chapter, we looked at Simon's algorithm, which expanded our quantum repertoire in three ways.

First, each run produces an output that doesn't solve the problem on its own, but contributes one piece of information to the final result. Second, it's a hybrid algorithm, because its measured outputs are processed by a classical computer to extract an answer from them. Third, the algorithm shows an exponential speedup relative to the classical solution.

In Simon's problem, we are given a function that contains a secret bitstring that it uses to create a two-to-one or one-to-one mapping of its inputs, and we want to learn that bitstring.

By collecting multiple outputs, we can build a system of linear equations. Then, a classical computer can use a traditional method such as Gaussian elimination to find the secret bitstring.

Simon's algorithm was the first to demonstrate an exponential speedup over the best classical algorithm. Though it doesn't solve a problem of much practical interest, it stimulated new work in quantum computing, and the important ideas it introduced are now a part of every quantum programmer's toolkit.

12

GROVER'S ALGORITHM

Yeah, I've been searchin'
A-searchin'
Oh, yeah, searchin' every which a-way . . .
—Jerry Leiber and Mike Stoller, "Searchin,'" 1957 [122]

So far, we've focused on contrived, or toy, problems. Now we'll move past those and see our first real, useful algorithm! Its purpose is to help us locate objects in an unorganized, or *unstructured*, database.

The algorithm we'll look at in this chapter is called *Grover's algorithm*, for Lov Grover, who published it in 1996 [82] [83].

Our discussion of Grover's algorithm will follow the same general flow as in the previous chapters: We'll discuss what the algorithm is for, see how to construct and execute it, look at results, and finally wrap up with some observations.

Let's get a general feeling for the task that Grover's algorithm addresses with an analogy. Suppose that you would like to put up a poster for your favorite movie in your home. You visit a movie memorabilia store, and the staff point you to a bin full of many different, unorganized posters. You have no choice but to flip through each poster in the bin, checking to see if it's a poster for your favorite film. This bin of posters is analogous to an *unstructured database*, meaning that it holds information (movie posters) but has not been organized (or structured) in a way that helps your search.

Suppose that, instead, the storekeeper always put posters into the bin in alphabetical order by movie title. This would make it much faster to find the one you want, because that order introduces *structure* into the database.

In practice, many difficult database problems are made much easier by creating some kind of structure for the database, like putting the posters in alphabetical order or ordering them by their release date. Building this structure can take some time, but if there are lots of queries to the database, the one-time cost of working out the structure is recovered many times over by the time saved in each query.

Grover's algorithm is helpful for those problems where our database is unstructured, so it has no organization we can exploit. What we do have is an oracle that can tell us, for any given element in the database, whether it's the entry (or one of the entries) that we want.

Extending our movie poster analogy, suppose that you're with a friend, and you're looking for your friend's favorite movie. But they're moody today and won't tell you what that movie is. Instead, you have to show them each poster one by one; they'll say "no" every time until you happen to show them the poster for their favorite film, and then they'll say "yes." If this sounds terribly inefficient to you, then you know why people prefer structured databases!

But sometimes we're given an unstructured database, and we need to work with it.

Grover's algorithm is designed for this kind of situation. Like your friend at the movie memorabilia store, Grover's oracle returns one result for every input element that fits your search criteria and a different result for all others.

I'm going to assume in this chapter that we're looking for only one entry in the database. This is just to keep the notation a little clearer and the discussion a little shorter. If we have a situation where there are multiple elements that fit our criteria, the oracle we'll see here will identify all of them, and the overall algorithm will return one such element, chosen at random.

We say that the oracle *marks*, or labels, our desired entry. Suppose that we query the oracle with a string of n qubits, so we can make 2^n possible queries. If the database has no structure, there's no alternative in the classical case except to run through every possible element until we happen to try the entry that the oracle marks. At worst, the marked entry corresponds to our final query, so it can take up to $2^n - 1$ tests to get our answer (the -1 is there because if we're certain there's at least one entry to be marked, we can skip the final query).

As you may expect, Grover's algorithm will save us time! As usual, we'll measure the cost of the algorithm using oracle complexity, or the number of times we need to call the oracle. It usually won't be just once, but it will be far less than 2^n.

An Overview of Grover's Algorithm

Before we dig into details, let's get an overall picture of Grover's algorithm. The general flow is like what we've seen before: We start with a register x of multiple qubits all in the state $|0\rangle$, and a register y containing a single qubit in the same state. We apply an X qugate to y, and then we put all the qubits into superpositions with H qugates. Next, we run an oracle, though in this case we apply many copies of the same oracle in a row. There's no final post-processing; we just take the output of the final oracle and measure it.

Figure 12-1 shows this process in circuit form.

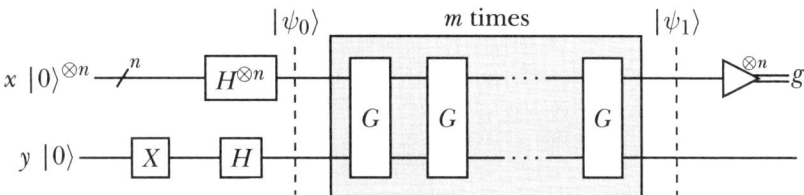

Figure 12-1: Grover's algorithm

The opening stages probably look familiar by now. We start with an upper register named x containing n qubits, all initialized to $|0\rangle$. Each of these qubits goes into an H qugate, producing an equal superposition $|+\rangle^{\otimes n}$ of all 2^n states from $|0\rangle$ to $|2^n - 1\rangle$. The register y contains a single qubit, also initialized to $|0\rangle$. An X qugate turns that into $|1\rangle$, and then an H qugate turns it into $|-\rangle$. Together, we have $|\psi_0\rangle$ in Equation 12.1.

$$|\psi_0\rangle = |+\rangle^{\otimes n} |-\rangle \tag{12.1}$$

We use $|\psi_0\rangle$ as the starting input to a chain, or sequence, of identical oracles, each of which I've named G. They're all the same, and there are m of them. We'll see later in this chapter how to determine m. We'll also see that each G is actually two operations in sequence.

After this chain of operations, we get an output state $|\psi_1\rangle$. We measure the n qubits of this state to get a final bitstring g of n bits.

Since there's no looping in quantum circuits today, this approach requires us to build a new quantum circuit containing m copies of G for each value of m we want to use.

Table 12-1 summarizes the symbols we'll use most often in this chapter. Recall that all of the other notation I use in the book is summarized in the Appendix.

Table 12-1: Symbols used in this chapter

Name	Meaning
n	Number of qubits
N	2^n, the number of basis states
t	Bitstring for the marked entry
$|\tau\rangle$	State corresponding to bitstring t
$|\sigma\rangle$	$|+\rangle^{\otimes n} - |\tau\rangle$, or the superposition of all basis states *except* $|\tau\rangle$

The steps inside of each *G* qugate are two versions of the same operation, called a *reflection*. Let's look at this idea in a general form first. Then we'll return to Grover's algorithm and see how we use reflection to find our marked entry.

Reflections

By three methods we may learn wisdom: first, by reflection, which is noblest; second, by imitation, which is easiest; and third, by experience, which is the bitterest.

—Confucius (apocryphal), approx. −500 [221]

An appealing quality of Grover's algorithm is that we can draw the oracle's operations using traditional 2D geometry. In fact, all we need are two types of reflections of arrows.

It's worth getting familiar with these operations now, so that when we get into Grover's algorithm we can just pull them out and use them, and not have to stop to see how they work.

Let's first visualize reflections using arrows in a plane. I'll name those arrows **v** and **w**. Because we'll ultimately turn the arrows into quantum states, I'll assume all the starting arrows in this section have a magnitude of 1.

I want to derive a formula that starts with the arrows **v** and **w** and reflects **v** around **w**. I'll temporarily call the new arrow $\mathbf{v}_\frown \mathbf{w}$. We'll later turn this formula into an operator that we'll use in Grover's oracle.

The geometry for reflecting **v** around **w** is illustrated in Figure 12-2.

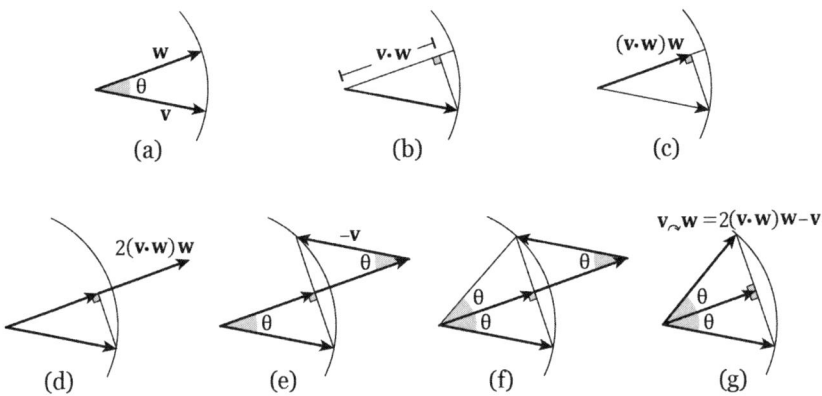

Figure 12-2: Reflecting **v** around **w**

Step (a) shows our two arrows, with an angle θ between them. We want to reflect **v** through the line that is defined by **w**. In step (b), I've drawn a line from the tip of **v** to **w**, perpendicular to **w**. As discussed in Chapter 2, the distance from the origin to this intersection point is given by the dot product **v** · **w**. Step (c) scales **w** by this distance to give us the arrow (**v** · **w**)**w**. In step (d), I added this arrow to itself. In step (e), I subtracted **v** (that is, I added −**v**) to this arrow).

The angle formed in the upper right is again θ, because **v** and $-$**v** are parallel. If we flip this triangle around its short edge, as in step (f), we find that the hypotenuse of this new triangle is exactly the arrow we want. In fact, all three triangles in step (f) are the same right triangle, just flipped into different orientations.

Step (g) shows the result of this whole operation, the arrow **v**$_\frown$**w**. The formula is summarized in Equation 12.2.

$$\mathbf{v}_\frown\mathbf{w} = 2(\mathbf{v} \cdot \mathbf{w})\mathbf{w} - \mathbf{v} \qquad (12.2)$$

Let's promote this to quantum states. I'll rename **v** to $|\nu\rangle$ and **w** to $|\omega\rangle$. We need to be careful when replacing the dot product with a braket, because while the order of the arguments doesn't matter for the dot product, it does matter for quantum states. Recall from Equation 2.56 that we project the *from* state onto the *onto* state by writing the *onto* state first. So, to project $|\nu\rangle$ onto $|\omega\rangle$, we replace the dot product with the braket, or inner product, $\langle\omega|\nu\rangle$.

This gives us Equation 12.3, where I named the reflection of $|\nu\rangle$ around $|\omega\rangle$ with $\text{Ref}(|\nu\rangle, |\omega\rangle)$.

$$\text{Ref}(|\nu\rangle, |\omega\rangle) = 2\langle\omega|\nu\rangle |\omega\rangle - |\nu\rangle \qquad (12.3)$$

The other reflection we'll need for Grover's oracle is a reflection of **v** around the line *perpendicular* to the vector **w**. The geometry is shown in Figure 12-3, where I temporarily named the result **v**$_\frown\perp$**w**.

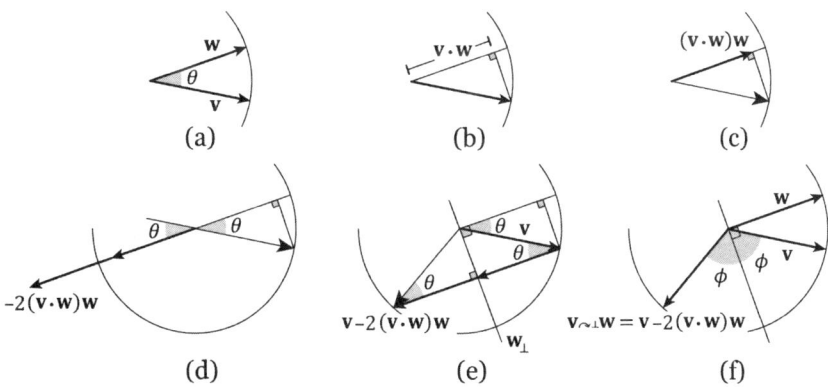

Figure 12-3: Reflecting **v** around the perpendicular to **w**

Steps (a) through (c) are the same as before. But this time, in step (d), I reversed the arrow $2(\mathbf{v} \cdot \mathbf{w})\mathbf{w}$ and drew it starting at the center of the circle. In step (e), I added **v** and the vector $-2(\mathbf{v} \cdot \mathbf{w})\mathbf{w}$ that we just built to get a new point on the unit circle. This gives us two new right triangles. Because $-2(\mathbf{v} \cdot \mathbf{w})\mathbf{w}$ is parallel to **w**, all three triangles in step (e) are the same right triangle, so the smaller angle of each is the same θ we started with. I've removed the construction lines in step (f). The arrow pointing down and to the left is the reflection of **v** through the line perpendicular to **w**. The formula is shown in Equation 12.4.

$$\mathbf{v}_{\curvearrowright \perp \mathbf{w}} = \mathbf{v} - 2(\mathbf{v} \cdot \mathbf{w})\mathbf{w} \tag{12.4}$$

As before, I'll promote these arrow operations to states by turning **v** into $|\nu\rangle$ and **w** into $|\omega\rangle$. This gives us Equation 12.5, where $|\omega\rangle_\perp$ refers to the state corresponding to the vector perpendicular to **w**.

$$\text{Ref}(|\nu\rangle, |\omega\rangle_\perp) = |\nu\rangle - 2\langle\omega|\nu\rangle\,|\omega\rangle \tag{12.5}$$

The arrow forms of these two operations, shown in Equations 12.3 and 12.5, are the same except for their signs, as summarized in Equation 12.6.

$$\text{Ref}(|\nu\rangle, |\omega\rangle) = -\text{Ref}(|\nu\rangle, |\omega\rangle_\perp) \tag{12.6}$$

Equations 12.2 and 12.4 combine the same two arrows, only with a reversed order of subtraction. That tells us that the two arrows point in exactly opposite directions! That's pretty cool. Figure 12-4 shows this relationship.

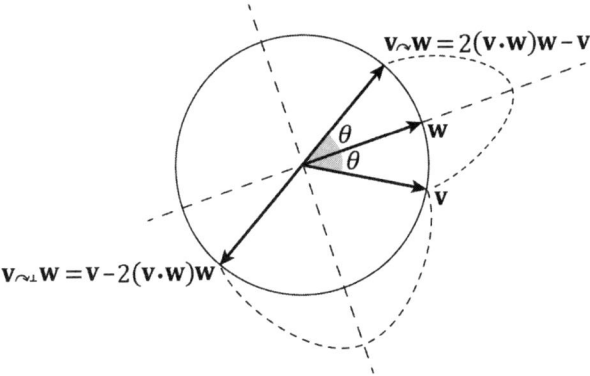

Figure 12-4: New arrows that point in exactly opposite directions

It's worth keeping one more nice piece of geometry in the back of our minds. Though I won't go through the steps here, we can prove that any two reflections performed in a row can always be written as a single rotation, and vice versa [132].

The Stages of Grover's Oracle

In the overview of Grover's algorithm in Figure 12-1, I showed the oracle as a series of boxes marked G. In fact, each of those boxes is made up of a pair of operations, applied in sequence. Let's call them G_1 and G_2.

G_1 is called the *marking* stage, because it identifies which states in the superposition we're looking for. G_2 is called the *diffusion* stage, referring to the mechanism it uses to increase the probability that we'll ultimately measure one of the desired, or marked, states in the superposition.

We first apply G_1 to mark the state we want, then G_2 to increase its probability of being measured. Unfortunately, that step also removes the mark, so we mark it again with G_1, increase the probability with G_2, and so on, until we've repeated this pair of operations a total of m times.

Let's look at these two stages in turn.

G_1: Marking

We'll start with G_1. As usual, we'll begin with a definition of the function it implements using bitstrings for inputs and outputs. I'll use the name t for the "target" bitstring we want to mark (that is, the one we're searching for). The value of t is known to the oracle, but not to us. Recalling our example involving movie posters, the G_1 stage plays a similar role to our moody friend who won't tell us their favorite movie's name, but will tell us when we show them its poster, corresponding to t.

For any input x, the oracle responds with 0 if x is not our marked entry t and 1 if it is t. I've written this in Equation 12.7, where I've again used f_x as a shorthand for $f(x)$.

$$f_x = \begin{cases} 0, & x \neq t \\ 1, & x = t \end{cases} \tag{12.7}$$

The quantum version of the oracle uses this function to perform phase kickback on the input, thanks to the $y = |-\rangle$ coming in on the lower register in Figure 12-1. So, we can write this function as G_1 in Equation 12.8. I'll focus here just on the upper register, x. When the input is the marked bitstring t, I'll say that its corresponding state is $|\tau\rangle$, so the output of G_1 will be $-|\tau\rangle$.

$$G_1 = (-1)^{f_x} |x\rangle = \begin{cases} |x\rangle, & x \neq t \\ -|\tau\rangle, & x = t \end{cases} \tag{12.8}$$

We sometimes call this a *phase oracle*, because its entire job is to "mark" the entry we're looking for by applying a phase of $e^{i\pi} = -1$ to it.

I promised earlier that each of the steps in G is a reflection. We can see why this is true for G_1 by grouping all the input states into two sets: all those that are $|\tau\rangle$ (that's a set with one element), and all those that are *not* $|\tau\rangle$ (that's a bigger set, with $2^n - 1$ elements). I'll write this latter group collectively as $|\sigma\rangle$, a superposition containing all basis states *except* $|\tau\rangle$. With these definitions, we can write the x register of $|\psi_0\rangle$ as a single superposition made up of two parts, $|\sigma\rangle + |\tau\rangle$.

Because all the states in the superposition $|\psi_0\rangle$ are basis states, they're all orthogonal to one another. This includes $|\tau\rangle$, so we can say that $|\tau\rangle$ is orthogonal to the whole set $|\sigma\rangle$.

Let's draw these two sets of states in a Cartesian diagram. This will be a conceptual aid, and it doesn't illustrate the states in any direct way. I'll place $|\tau\rangle$ on the vertical axis and the collection of all the other states $|\sigma\rangle$ on the horizontal axis, as shown in Figure 12-5(a). The axes $|\tau\rangle$ and $|\sigma\rangle$ are perpendicular because the states are orthogonal.

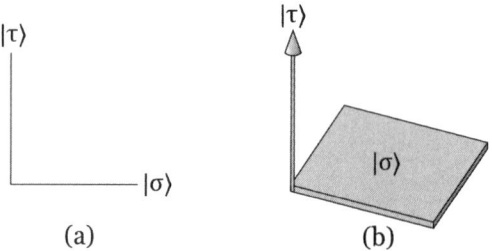

Figure 12-5: We can break up all inputs into the state $|\tau\rangle$ and all other states, $|\sigma\rangle$. (a) When there is only one state in $|\sigma\rangle$. (b) When there are two states in $|\sigma\rangle$.

You can think of the horizontal axis as all the states that aren't $|\tau\rangle$ somehow combined together. Taken individually or as a group, they're all orthogonal to $|\tau\rangle$. If there were two states in $|\sigma\rangle$, we could draw them as a plane, as in part (b) of Figure 12-5. Part (a) would be this 3D diagram viewed from a point on the plane, so the plane appears as a line. We can imagine a corresponding operation for any number of states in $|\sigma\rangle$. For simplicity, I'll stick to the 2D version in part (a).

Let's look at the mechanics of G_1. Suppose that the x input register is in the input state $|\psi\rangle$. We can write that as a sum of the $N = 2^n$ states in the input superposition, each scaled by some number. I'll call the scaling factors γ_k. I've written out the terms in Equation 12.9.

$$|\psi\rangle = \sum_{k\in[N]} \gamma_k |k\rangle = \gamma_0 |0\rangle + \gamma_1 |1\rangle + \cdots + \gamma_\tau |\tau\rangle + \cdots + \gamma_{N-1} |N-1\rangle \quad (12.9)$$

I've written this with our desired answer, $|\tau\rangle$, somewhere in the middle of the superposition. As the order of the states in a superposition doesn't matter, we can always make sure it's away from the ends, which lets us avoid notational clutter to represent those special cases.

Let's group this into two superpositions $|\tau\rangle$ and $|\sigma\rangle$, as in Equation 12.10. I'll list all the states in $|\sigma\rangle$ first, skipping over $|\tau\rangle$, and then add $|\tau\rangle$ back in at the end.

$$|\psi\rangle = \gamma_0 |0\rangle + \gamma_1 |1\rangle + \cdots + \overbrace{\gamma_{t-1} |t-1\rangle + \gamma_{t+1} |t+1\rangle}^{\text{skip } \gamma_\tau |\tau\rangle} + \cdots + \gamma_{N-1} |N-1\rangle + \gamma_\tau |\tau\rangle$$

$$= \left(\sum_{\substack{k\in[N] \\ k\neq t}} \gamma_k |k\rangle \right) + \gamma_\tau |\tau\rangle \quad (12.10)$$

We can simplify this a little. At the start of Grover's algorithm, all of the states will be in an equal superposition. If there are n qubits, then we know each state will have amplitude $\vee^n$. Since each γ_k in Equation 12.10 will be $\vee^n$, we can just use a single value of γ for all of them. I'll write this as γ_σ, giving us Equation 12.11.

$$|\psi\rangle = \gamma_\sigma\,|\sigma\rangle + \gamma_t\,|\tau\rangle \tag{12.11}$$

This will be convenient because as Grover's algorithm proceeds, the value of γ_τ will increase, while all the other amplitudes will decrease together. In other words, every state in $|\sigma\rangle$ will always have the same amplitude.

We can think of Equation 12.11 as telling us that $|\psi\rangle$ is made up of two components: some $|\sigma\rangle$ and some $|\tau\rangle$. It's just like a point in the 2D Euclidean plane, which is some amount of the **x** basis vector and some amount of **y**.

Because all the $N = 2^n$ states in the initial superposition have the same amplitude, that common amplitude is $\sqrt{1/N} = 1/\sqrt{N}$. Therefore, we can write $|\psi\rangle$ as in Equation 12.12.

$$|\psi\rangle = \frac{1}{\sqrt{N}}\,|\tau\rangle + \frac{N-1}{\sqrt{N}}\,|\sigma\rangle \tag{12.12}$$

We can draw this state conceptually as in Figure 12-6. I've labeled the angle between $|\psi\rangle$ and the $|\sigma\rangle$ axis as θ. At the start of Grover's algorithm, θ will be extremely small, so I've exaggerated its value for this figure.

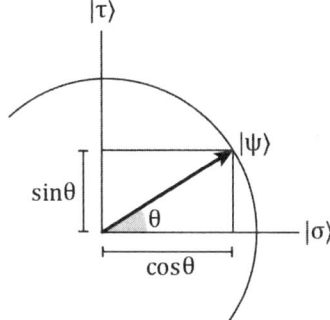

Figure 12-6: We can write the amplitude for $|\psi\rangle$ in terms of the angle θ.

We can rewrite Equation 12.12 using this right triangle as in Equation 12.13.

$$|\psi\rangle = \sin\theta\,|\tau\rangle + \cos\theta\,|\sigma\rangle \tag{12.13}$$

Comparing Equations 12.11, 12.12, and 12.13 tells us that γ_τ is given by two equivalent expressions: $\gamma_\tau = \sin\theta = 1/\sqrt{N}$.

This means that the probability of measuring $|\tau\rangle$ is $\sin^2\theta = 1/N$ (the notation $\sin^2\theta$ is equivalent to $(\sin\theta)^2$ without the parentheses). When we have a large database, N is large, so the amplitude $\sin\theta = 1/\sqrt{N}$ will be nearly 0. The value of the probability $\sin^2\theta = 1/N$ drops to 0 as well, but even faster, as shown in Figure 12-7.

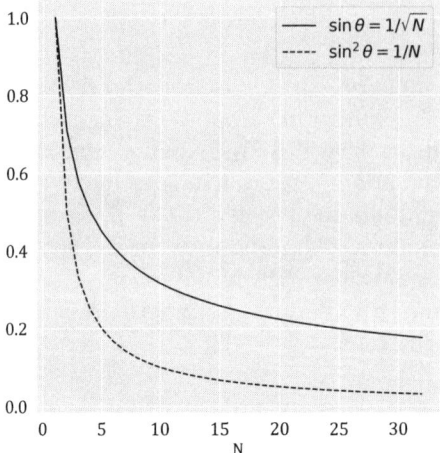

Figure 12-7: As N increases, the amplitude $1/\sqrt{N}$ approaches 0, while the probability $\sin^2 \theta = 1/N$ goes to 0 even more quickly.

This is not encouraging news. Figure 12-7 tells us that there's only a small probability $1/N$ that, if we measured $|\psi\rangle$ right now, it would collapse to state $|\tau\rangle$ and we'd observe the desired bitstring t. If N is large, we'll rarely measure t at the output. This is pretty disappointing.

However, the good news is that we can make $|\tau\rangle$ more likely to be measured by increasing its amplitude. Visually, we'll rotate the superposition $|\psi\rangle$ in Figure 12-6 counterclockwise, bringing it closer to $|\tau\rangle$ and simultaneously farther away from $|\sigma\rangle$. This process conventionally goes by the alliteratively awkward name of *amplitude amplification*.

Let's run $|\psi\rangle$ from Equation 12.11 through the oracle G_1. The oracle's definition in Equation 12.8 tells us that the first term, $\gamma_\sigma |\sigma\rangle$, will be unchanged. The second term, $\gamma_\tau |\tau\rangle$, will be multiplied by -1. I've written this in symbols in Equation 12.14.

$$G_1 |\psi\rangle = \gamma_\sigma |\sigma\rangle - \gamma_\tau |\tau\rangle \qquad (12.14)$$

We can draw the result as in Figure 12-8.

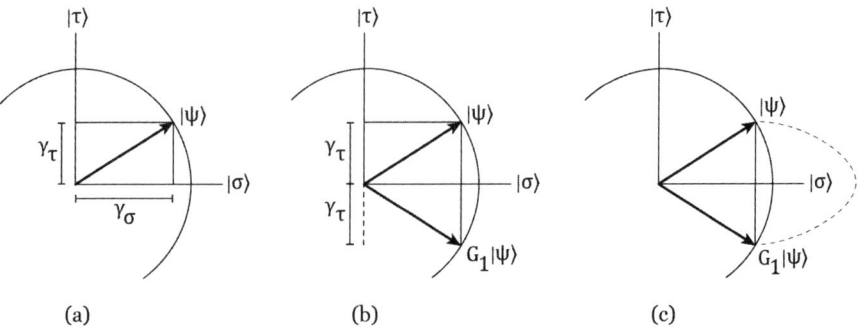

Figure 12-8: The geometry of G_1

Figure 12-8 shows that we can picture the action of the oracle G_1 as a reflection of $|\psi\rangle$ around the $|\sigma\rangle$ axis. Part (a) shows the state $|\psi\rangle$ as a sum of $\gamma_\sigma |\sigma\rangle$ and $\gamma_\tau |\tau\rangle$. In part (b), G_1 negates γ_τ. In part (c), we see that $G_1 |\psi\rangle$ is the reflection of $|\psi\rangle$ around $|\sigma\rangle$. Remember that these diagrams aren't explicit drawings of the states, but rather are geometric versions of their relationships.

We can write the action of G_1 explicitly using Equation 12.3, where $|\psi\rangle$ plays the role of $|\nu\rangle$ and $|\sigma\rangle$ is $|\omega\rangle$. The result of this substitution is shown in Equation 12.15.

$$\begin{aligned} G_1 |\psi\rangle &= \mathrm{Ref}(|\psi\rangle, |\sigma\rangle) \\ &= 2 \langle \sigma | \psi \rangle |\sigma\rangle - |\psi\rangle \end{aligned} \tag{12.15}$$

After all this effort, we're more likely to measure the state $|\tau\rangle$, right? Well, no. Figure 12-9 shows the new situation. The probability of measuring $|\tau\rangle$ hasn't changed *at all*. The amplitude is now negative, so it's $-\sin\theta$, but squaring that to find its probability gives us $\sin^2\theta = 1/N$, just as before.

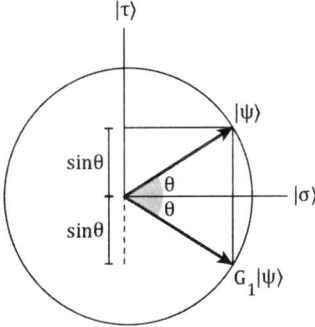

Figure 12-9: The amplitude of $G_1 |\psi\rangle$ is $-\sin\theta$, so its probability $\sin^2\theta$ is the same as the probability of $|\psi\rangle$.

So what was the point of applying G_1?

Earlier, I said that the overall operation of qugate G in Figure 12-1 was made up of two steps: We mark the state that we're looking for, and then we make it more probable. We've just seen how G_1 performs that marking step, turning $|\psi\rangle$ into $G_1 |\psi\rangle$.

Remember that $|\psi\rangle$ is the state of our x register, so the more we can rotate that state toward $|\tau\rangle$, the more likely we are to get back $|\tau\rangle$ when we measure $|\psi\rangle$.

Now we'll adjust the probabilities to make $|\tau\rangle$ more likely to be measured. Let's see how.

G_2: Diffusion

Now things get clever! We want to increase the probability of measuring our marked state. This is done by the next stage of G, called the *diffusion* step. As mentioned earlier, it's just another reflection. But this time, we'll reflect the new state we just made, $G_1 |\psi\rangle$, *around the original* $|\psi\rangle$.

Figure 12-10 shows the geometry.

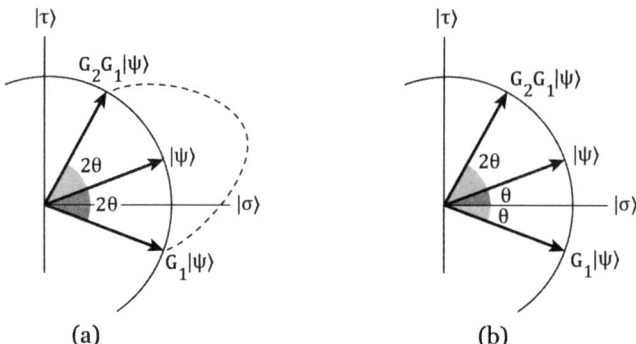

Figure 12-10: (a) G_2 reflects $G_1 |\psi\rangle$ around $|\psi\rangle$. (b) Labeling the angles.

Referring to part (b) of Figure 12-10, the new state $G_2 G_1 |\psi\rangle$ makes an angle of 3θ with the $|\sigma\rangle$ axis, so it has an amplitude of $\sin 3\theta$ on the $|\tau\rangle$ axis.

We've done it! As long as we have four or more qubits, then $\theta < \pi/6$ which means that $\sin 3\theta$ is going to be larger than $\sin\theta$, and we'll be more likely to measure $|\tau\rangle$!

We need θ to be less than $\pi/6$, because otherwise the reflected $|\tau\rangle$ can overshoot the top of the circle, perhaps reducing the probability of measuring it. Happily, as long as we have four or more qubits, the original value of θ meets this criterion (if we have fewer than four qubits, there are only a maximum of eight states to check, so we can skip Grover's algorithm and test them one by one without too much expense).

The step where we apply G_2 is called diffusion because it spreads, or *diffuses*, amplitudes away from the superposition $|\sigma\rangle$ and into the state $|\tau\rangle$. Specifically, the value of γ_τ increases, and the shared amplitude of every other state, γ_σ, decreases.

Applying the oracle and then the diffuser is sometimes called a *Grover iteration* or *Grover operation*. Note that when we increase the probability of measuring t, we're also *decreasing* the probability of measuring anything else, as normalization requires the probabilities of all states to add up to 1.

Happily, we already know how to write down G_2, because we saw it earlier in the chapter. We only need Equation 12.5, using $G_1 |\psi\rangle$ for $|\nu\rangle$ and $|\psi\rangle$ for $|\omega\rangle$. The resulting formula is given in Equation 12.16.

$$
\begin{aligned}
G_2 G_1 |\psi\rangle &= \text{Ref}(|\phi\rangle, |\psi\rangle) \\
&= G_1 |\psi\rangle - 2 \langle\psi|\phi\rangle |\psi\rangle
\end{aligned}
\tag{12.16}
$$

Both G_1 and G_2 are reflections, and thus unitary operations, so they are both qugates that we can implement in real hardware.

When we measure $G_2 G_1 |\psi\rangle$, the probability of measuring t is $\sin^2 3\theta$. That means we have a probability of $\cos^2 3\theta$ of measuring any other state, all of which are collected in $|\sigma\rangle$. All of those states have an equal probability of being measured because they all started out with the same amplitude $1/\sqrt{N}$, and we haven't done anything to any of those states individually.

This combination of reflections is the qugate G, as summarized in Equation 12.17.

$$G = G_2 G_1 \tag{12.17}$$

I mentioned earlier that any sequence of two reflections can be written as a single rotation. Figure 12-10 shows the combined operations $G_2 G_1$ as a single counterclockwise rotation of the original $|\psi\rangle$ by 2θ.

In the preceding figures, I drew θ pretty large so that it would be easy to see. But if there are many qubits, then N will be a big number, so $\sin\theta = 1/N$ will be small. The effect of $G_2 G_1 |\psi\rangle$ will be to rotate $|\psi\rangle$ toward the $|\tau\rangle$ axis, but not by much, as shown in Figure 12-11. Even here, I've exaggerated the geometry a lot for clarity.

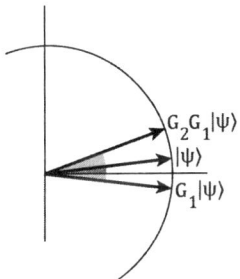

Figure 12-11: Each Grover iteration rotates $|\psi\rangle$ counterclockwise toward the $|\tau\rangle$ axis by an angle of 2θ.

To significantly increase the amplitude $|\psi\rangle$ in this situation, we'll need to apply the Grover operation multiple times. Because we can't build a loop in a quantum circuit, to apply G repeatedly we need to explicitly place it in the circuit multiple times, as I did in Figure 12-1. We can replace the gray box in that diagram with the expanded view in Figure 12-12.

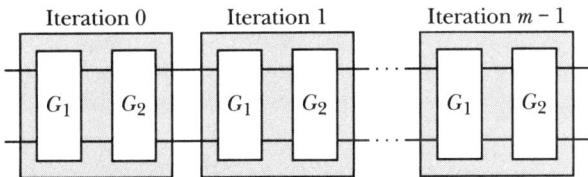

Figure 12-12: The Grover operation repeated m times

This gives us only one loose end to tie up: How many times should we repeat the Grover operation $G_2 G_1$? Let's find out.

Iterating the Grover Oracle

Taken together, we call $G_2 \, G_1$ a single Grover iteration, written G without a subscript. As we just saw, applying G once makes the state we want to measure more probable, but not by a lot. If we repeat the marking and diffusion steps again, measuring the state we're after will become even more probable. We can repeat this process over and over until we've made our goal state very likely to be measured.

How many times should we repeat the Grover operation on $|\psi\rangle$? Ideally, we should repeat it enough times so that the probability γ_τ is 1. Referring back to Figure 12-9, that means we want the angle θ to be $\pi/2$, which is the same as saying that we want $\sin \theta = 1$. The top row of Figure 12-13 shows three different starting angles, corresponding to different numbers of qubits, marked with a heavy line and a black dot. The bottom row of the figure shows the results of applying the Grover iteration many times. The dots from left to right in each plot in the bottom row correspond to the counterclockwise rotation of the starting state $|\psi\rangle$ in the plot above. Note that each column uses a different number of steps.

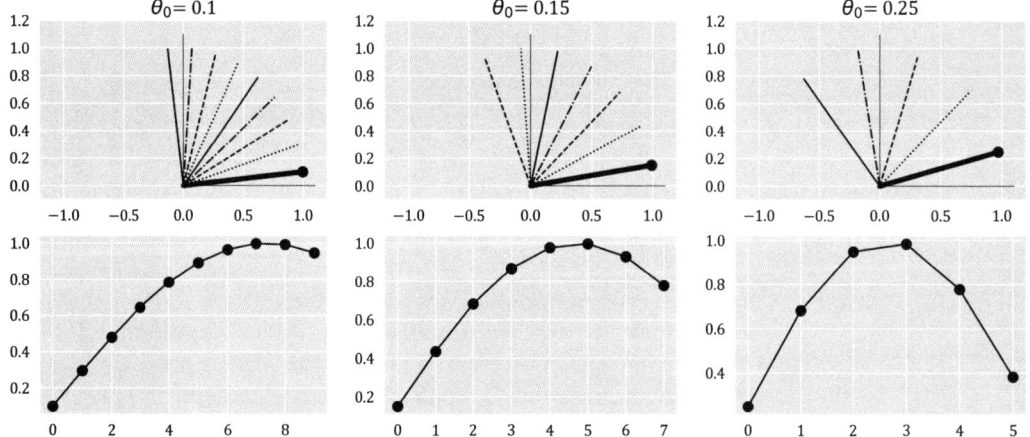

Figure 12-13: Top row: Applying multiple Grover iterations to different starting states $|\psi\rangle$, marked with a heavy line and a dot. Bottom row: Plotting $\sin \theta$ for the angle of each line in the top row.

An interesting result is that Grover's algorithm can *overshoot* the goal. For a while, the output gets closer to $|\tau\rangle$, but then if we keep applying the operation, the output goes past $|\tau\rangle$ and keeps rotating counterclockwise, getting farther away from it. Our goal will be to pick just the right number of iterations so that we get as close as possible to $|\tau\rangle$.

A little algebra tells us how many times to apply the iteration.

Figure 12-10 shows that each time we apply G, we rotate the input state by 2θ, where θ is the original angle of our input state $|\psi\rangle$ with the horizontal axis $|\sigma\rangle$. Therefore, the sequence of angles that we get, starting with the original input, is shown in Equation 12.18.

$$\theta, \, 3\theta, \, 5\theta, \, 7\theta, \, \ldots \qquad (12.18)$$

We can write this using the variable m to refer to the number of iterations. The angle θ_m after m steps is given in Equation 12.19.

$$\theta_m = \theta + 2m\theta = \theta(1 + 2m) \tag{12.19}$$

We'd like to find the value of m that gets this result to $\pi/2$. We can do this by setting $\theta_m = \pi/2$ and solving for m, as in Equation 12.20.

$$\theta(1 + 2m) = \pi/2$$
$$m = \left(\frac{\pi/2}{\theta} - 1\right)/2 \tag{12.20}$$
$$= \frac{\pi}{4\theta} - \frac{1}{2}$$

Recall that we found that $\sin\theta = 1/\sqrt{N}$, so $\theta = \sin^{-1}(1/\sqrt{N})$. Let's plug that into Equation 12.20 for θ, giving us Equation 12.21.

$$m = \frac{\pi}{4\sin^{-1}(1/\sqrt{N})} - \frac{1}{2} \tag{12.21}$$

We could stop here, but that inverse sine is clumsy. When N is big (and it usually will be, as $N = 2^n$ for n qubits), then $\sin^{-1}(1/\sqrt{N})$ is well approximated by just $1/\sqrt{N}$ itself, as shown in Figure 12-14.

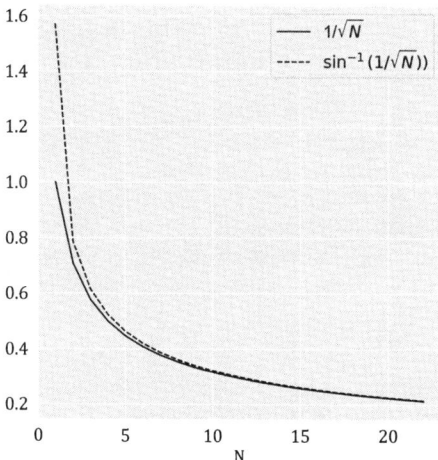

Figure 12-14: As N grows, $1/\sqrt{N}$ (the solid line) becomes a good approximation for $\sin^{-1}(1/\sqrt{N})$ (the dashed line).

So let's replace $\sin^{-1}(1/\sqrt{N})$ in Equation 12.21 with $1/\sqrt{N}$, giving us the more pleasant expression in Equation 12.22.

$$m = \frac{\pi}{4/\sqrt{N}} - \frac{1}{2}$$
$$= \frac{\pi}{4}\sqrt{N} - \frac{1}{2} \tag{12.22}$$

The values of m for different Ns are plotted in Figure 12-15(a). As the values in Figure 12-15(b) show, those results, when rounded to the nearest integer, tell us how many times we need to apply the Grover iteration to get the best chance of measuring the marked bitstring t (which might involve overshooting, as we can see in Figure 12-13).

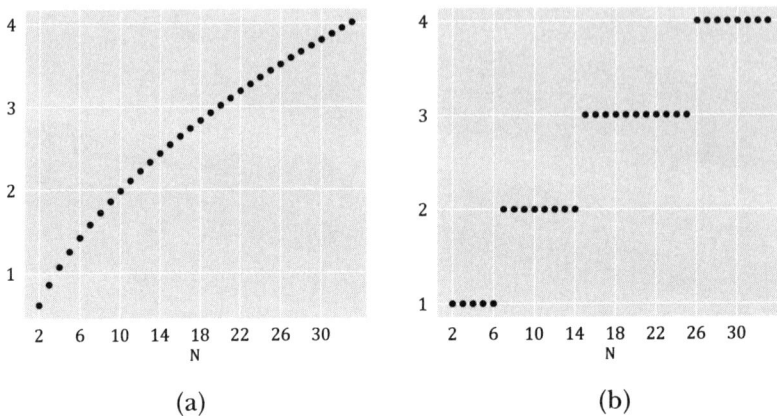

(a) (b)

Figure 12-15: The value of m for different values of N. (a) The results of Equation 12.22. (b) The values on the left rounded to the nearest integer.

All this analysis tells us that to maximize the probability of measuring the state we're looking for, we should repeat the two stages of the Grover iteration m times, where m can be computed from Equation 12.22.

An Example of Grover's Algorithm

Enough theory! Let's implement Grover's algorithm. The G_1 step inverts only the single target state, so it needs to somehow identify when its input qubits match the pattern for that target state. A common way to do this is to use a CX qugate that treats each input line as a control.

Although we've only seen controlled qugates with a single control, we can build up a controlled qugate with any number of control inputs [187]. In practice, most hardware and simulators provide built-in support for only one or two control inputs, so I'll use two qubits in this example. Since there are two controls, we'll name this qugate CCX. When a qugate is controlled by multiple qubits, it's applied only if *all* the controls are $|1\rangle$. Thus, if there are two controls, then *both* must be $|1\rangle$ for the qugate to be applied. If there are three controls, then *all three* must be $|1\rangle$, and so on.

Because we have only two qubits, there are only four states to test. There's no need for Grover's algorithm, since we could test each one of these explicitly. But let's plunge ahead anyway to see how the pieces come together.

Let's say the target bitstring t is 10, so we want the circuit to end up in a superposition where $|10\rangle$ is far more likely than any other state.

To get going, we'll create an equal superposition of the two input qubits by applying H to each. To enable phase kickback in the oracle, we'll also put the auxiliary qubit into the state $|-\rangle$. This initialization step is shown in Figure 12-16.

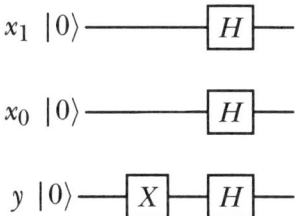

Figure 12-16: The initialization step for our implementation of Grover's algorithm

Now we'll implement the oracle. So far in this book, I've always used controlled qugates where a control state of $|1\rangle$ means the operation is applied and $|0\rangle$ means it's not. We can call this a *1-active* control line. But sometimes it's more convenient to reverse those controls, so $|0\rangle$ means the operation is applied and $|1\rangle$ means it's not, which we can call a *0-active* control line.

We can use a typical, or 1-active, control line as a 0-active line by surrounding it with X qugates. A *CCX* controlled by one control of each type is shown in Figure 12-17(a).

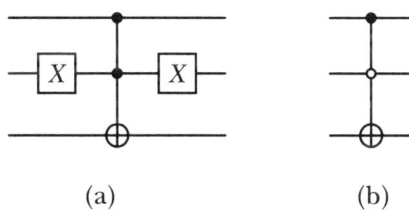

(a)   (b)

Figure 12-17: Two equivalent ways to write an oracle for the target state 10

A notational shortcut for this operation is shown in Figure 12-17(b), where the empty circle represents a 0-active control. The X qugate at the bottom is applied only if *both* control conditions are satisfied. Both versions of this oracle for the target bitstring 10 tell us that the X qugate is applied only when the upper line is $|1\rangle$ *and* the middle line is $|0\rangle$.

The goal of the diffuser is to reflect all states around the state that went into the oracle. To do this, we'll use H qugates to put the superposition state onto the $|0\rangle$ axis, then reflect all of the states that aren't $|0\rangle$ around that axis. Finally, we'll use H qugates to put everything back where it was.

So, we'll start with H qugates to undo the initial round of H qugates. If we apply a *CZ* qugate immediately, that will negate only $|11\rangle$. Therefore, we'll wrap the *CZ* in X qugates so it will apply to all inputs except $|00\rangle$, the original superposition. Then we'll apply H qugates again to restore the superposition before this step. This diffusion step is shown in Figure 12-18.

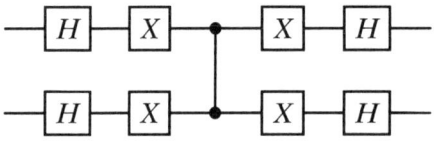

Figure 12-18: A diffusion circuit

The whole circuit is assembled in Figure 12-19.

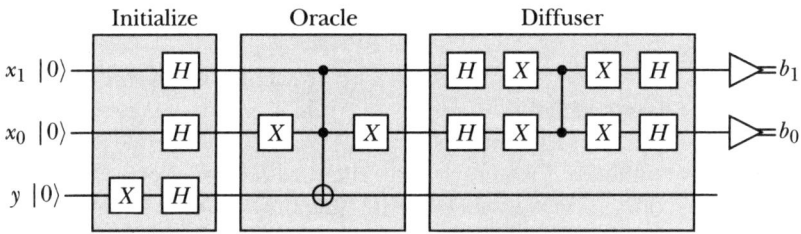

Figure 12-19: Figures 12-16, 12-17, and 12-18 assembled into one circuit

I ran this for 1,024 shots on a five-qubit quantum computer and got the results in Figure 12-20.

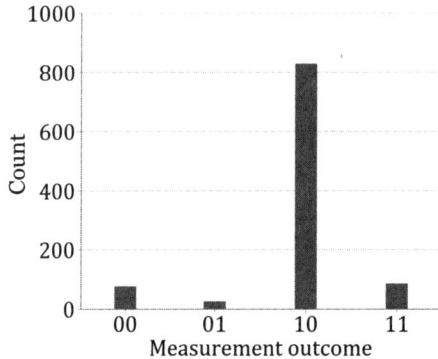

Figure 12-20: The result of running Grover's algorithm in Figure 12-19 for 1,024 shots

Most of the time we measured 10, our target bitstring! Success!

Let's push things a lot further and look for a four-bit marked string. The architecture is about the same as in the previous example, except that we'll need a $CCCCX$ with four controls in the oracle and a $CCCZ$ with three controls in the diffuser. Figure 12-21 shows the circuit for the marked state 1000.

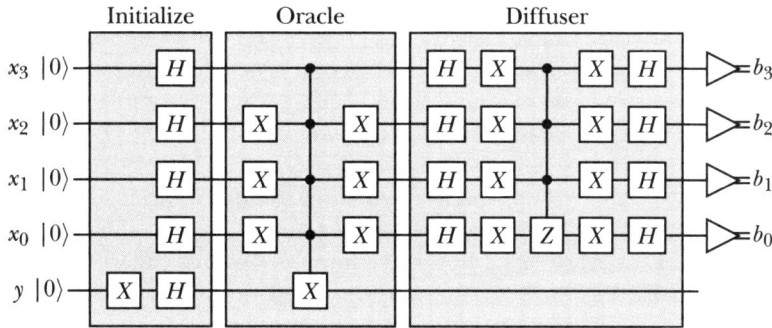

Figure 12-21: Finding the marked state 1000

As mentioned previously, as of 2025, qugates that depend on more than two controls are not native to all quantum simulators and computers. You *can* create these qugates out of simpler qugates, but it can take a lot of those qugates [145] [186]. Some systems automatically expand multiply controlled qugates into equivalent circuits that the simulator or hardware can handle. A direct implementation of such qugates is offered by a demonstration project that runs in the free Wolfram Player [287]. The project is not a general quantum computing library but is hardcoded specifically for Grover's algorithm. Nevertheless, it offers many interactive choices for the number of qubits, the number of iterations, and the identity of the marked state [161].

Using this system, I ran the circuit of Figure 12-21 with just one iteration of the oracle and diffuser. This idealized simulator with no noise returns the target state 1000 about 69 percent of the time. A second iteration gets us to 95 percent, and a third improves this to 98 percent. If we apply the Grover iteration four times, we go right past our optimum and accuracy drops to 76 percent. As $2^4 = 16$, Figure 12-15 accurately tells us that three iterations is the sweet spot.

Observations and Discussion

Let's look at a few aspects of Grover's algorithm and our presentation that can help us better appreciate the technique.

Recall that we often discuss oracle-based quantum algorithms in terms of *oracle complexity*, or the number of times we need to call the oracle (implicitly assuming that everything else makes a negligible impact on the computing resources we need). In Grover's algorithm, we call the oracle once per iteration, so the number of iterations tells us the algorithm's oracle complexity.

We found in Equation 12.22 that the number of iterations we need from Grover's algorithm depends on $\sqrt{N}$. Figure 12-15 shows this behavior graphically. This is more than the single query needed by some of the algorithms we looked at in the previous chapters, but it's way better than the exhaustive search of all $2^n - 1$ states required by conventional algorithms. We say that Grover's algorithm provides a *quadratic speedup* compared to conventional algorithms. This isn't just good; it's the best we can hope to achieve. For this specific searching problem, it's been proven that no quantum algorithm can do better than a quadratic speedup over the conventional approach [16] [99].

We can extend Grover's algorithm to situations where we have more than one marked item [71], and we can tune the algorithm to improve its efficiency [23]. A generalized version of Grover's algorithm, called *amplitude estimation*, is probably used more often now [27].

Finally, because the answer we seek rarely ends up with a probability of 1, Grover's algorithm can sometimes return the wrong result. We can identify this just by testing the answer, which is usually an efficient operation (or at least, more efficient than finding the solution we're testing). If this check shows us that our candidate answer is wrong, we usually just run the algorithm again. Grover's algorithm is great for unstructured data where solutions are hard to find, but easy to check. If there is any kind of structure to the data, or properties that we can exploit, that information can often help us perform search queries more efficiently than using Grover's algorithm.

I've taken a geometric approach to Grover's algorithm. If you go more deeply into the algebra, you can use this approach to explicitly build the reflection operators for any number of dimensions [249].

Summary

In this chapter, we looked at Grover's algorithm. It's a first for us in this book because it doesn't solve a toy, or contrived, problem. Grover's algorithm is a practical step used in many quantum algorithms.

The big idea is that we're given an unstructured database that contains one or more "special" items. In this chapter, we limited ourselves to databases with only one such item. We don't know which item is special, but we're given an oracle that does know. We want to use the oracle to help us ultimately measure that item without having to test every entry in the database, one by one.

As usual, we start with a uniform superposition of n qubits, each state representing the index of one item in a database of 2^n entries. To find the marked entry, we repeat a two-step process.

The first step is to "mark" the special entry. The second step uses this mark (erasing it in the process) to increase the amplitude of that state, which necessarily decreases the amplitudes of all the other states. Repeating these two steps increases the probability of measuring the entry we want at the end of the algorithm.

Both steps can be interpreted geometrically as reflections in the plane. We can build unitary matrix descriptions of these operations, so they can be used as quantum gates. We place several pairs of these qugates in succession to apply the pair of reflections multiple times.

Surprisingly, if we repeat the steps too many times, we actually start to decrease the marked item's amplitude. We found a formula that tells us just how many times we should apply the Grover iteration to maximize the amplitude of the state we want so it will be the most likely one that we'll measure.

13

SHOR'S ALGORITHM

Listen
Do you want to know a secret?
Do you promise not to tell?
Whoa oh oh
Closer
Let me whisper in your ear
Say the words you long to hear
I'm in love with you, ooh
—John Lennon and Paul McCartney, "Do You Want to Know a Secret?," 1964 [123]

Secrets are important. When we have something private to say, we want to trust both the medium we use to share the secret and the person receiving it.

When we share secrets over the internet, such as our credit card numbers, personal medical information, business plans, romances, or anything else we want to keep confidential, we trust the *encryption* algorithms built into our browsers, email systems, and other applications to make certain that even if someone intercepts our messages, they won't be able to read them. The message is replaced by something that looks like nonsense to anyone but the intended recipient, who is the only one with the secret knowledge needed to decrypt it.

Most of the encryption systems for today's web are built on a single technique: a mathematical process that nobody knows how to efficiently reverse.

If anyone does find a way to reverse that process, virtually all of today's secrets will be exposed. This is the Achilles' heel of today's internet security.

You may know where I'm going with this: The algorithm we'll see in this chapter can, if run on a large and reliable enough quantum computer, reverse that operation and break the encryption, laying bare every secret we've shared to any casual reader. That day isn't here yet, because a quantum computer big enough and accurate enough hasn't yet been built, but it surely will be someday.

That's pretty important stuff. Before we look at the algorithm, let's first look at the operation that underlies today's encryption.

Primes and Encryption

In this book, we've been using mathematics as a representation of the real world. Our kets are mathematical objects that follow mathematical rules like matrix multiplication and tensor products, and using them lets us predict what will happen when we perform those operations on real objects. We often call this *applied mathematics*.

However, sometimes mathematics isn't used as a model of the real world but more abstractly, to describe structures and relationships involving purely conceptual objects. We sometimes call such work *theoretical mathematics*, meaning essentially math for its own sake.

Surprisingly often, those two worlds overlap. One famous such case is the form of encryption that is used to protect today's secrets, ranging from credit card numbers to private conversations and even military and industrial information.

Many of these secrets are protected by a form of encryption built around properties of prime numbers. A *prime number* is any number greater than 1 that has no integer divisors other than 1 and itself [245]. For example, the number 6 is not a prime because it has the divisors 2 and 3 (that is, when we divide 6 by either 2 or 3, we get back an integer). We say that both 2 and 3 *divide* 6, meaning that there is no remainder from $6/2$ or $6/3$. Similarly, the number 9 is not a prime because it has a divisor of 3 (that is, 3 divides 9), and 16 is not a prime because it has the divisors 2, 4, and 8. In contrast, 7 is a prime, and so is 13.

The list of primes starts out with $2, 3, 5, 7, 11, 13, 17$ . . . and continues without end [201]. Prime numbers have many properties that are interesting to mathematicians, and they have been studied extensively [235].

A natural question to ask about any integer is whether it's prime. A number that isn't prime is called *non-prime*, or *composite*. Any integer can be in only one category or the other. That is, either it has divisors other than 1 and itself, or it doesn't. If we determine somehow that a given integer is composite, the next question is often to ask which prime numbers divide it.

Suppose we start with 75. This is not a prime, because it's divisible by 3 and 25. But 25 is itself divisible by 5, so $75 = 3 \times 25 = 3 \times 5 \times 5 = 3 \times 5^2$ (using $\times$ for multiplication). Note that in this expression, 3 and 5 are both primes, so we've reduced 75 to its *prime factors* 3 and 5.

A central pillar of the theory of prime numbers is that every non-prime can be written as a product of prime numbers, and this product is *unique*. This property is so important to the theory of mathematics that it's called the *fundamental theorem of arithmetic* [237].

In summary, there is only one possible set of prime numbers that can be multiplied together to produce any composite number.

We're drifting pretty far from quantum computing here! But there's something of great practical importance in all of this. Suppose that we have some huge integer, made up of hundreds of digits (or even more). Also suppose that we know, somehow, that it's composite. Now we'd like to find its prime factors. Nobody knows how to do this efficiently on a classical computer.

By *efficiently*, I mean in any range of time that we'd normally like to wait, like less than a dozen years, or a dozen millennia. There might be an algorithm out there that works more quickly, but a lot of people have worked hard to find one, and nobody has yet. Our best *prime factoring* algorithms are slow enough in practice that people generally consider it impractical to *factor*, or find the prime factors of, integers with huge numbers of digits. In everyday terms, we treat prime factoring of arbitrarily large numbers as something that we just can't do.

This is an asymmetrical situation. If we have two prime numbers, say p and q, then it's entirely practical to find their product $m = pq$. But given an m of hundreds of digits, it's just not feasible to find either p or q. We call the multiplication of two large primes a *one-way function*, meaning that it is, in practical terms, not invertible.

This observation is the basis of RSA cryptography (named for its three inventors, Ron Rivest, Adi Shamir, and Leonard Adleman). Suppose you want to enable people to send you secret messages. In one version of RSA, you pick two big prime numbers, p and q, and multiply them together to find n. Then you publicly publish n, along with another number you derive. This is called a *public key*. People then encrypt their messages using these numbers. You're the only person who can read those messages, because decrypting them requires knowledge of the two primes, p and q. Even though you published their product n, it's impractical for anyone else to find those starting prime numbers, so the message is secure. This description barely scratches the surface of the RSA system, which is a fascinating topic [11].

The key takeaway is that we trust this form of encryption to keep our secrets safe because finding the prime factors of enormous integers is impractical, even on today's biggest and fastest conventional computers.

The impracticality of factoring large prime numbers is used to protect almost everything on the internet today, from state and industrial secrets to banking and healthcare data, individual emails, and even private direct messages. It's the reason we trust that these important secrets are safe on the internet.

The prime and composite numbers used by today's encryption algorithms are often hundreds of digits long. The 2015 US government encryption standards for this technique require the prime factors to have about 310 decimal digits each, resulting in a published composite number of about 617 digits. The expectation is that any algorithm on a classical computer would take decades to find the prime factors of this number and thus crack the encryption [199] [270].

This expectation has been valid for decades, and it probably will remain valid for a while longer. But when quantum computers become sufficiently large and accurate, all of that security will be lost, and all the secrets this encryption method is protecting will be revealed. Let's see why.

Shor's Algorithm

In 1994, Peter Shor published a paper that showed how to use a quantum algorithm as part of a hybrid system to factor a composite number [195]. This algorithm, now known as *Shor's algorithm*, is probably the most famous quantum algorithm because of its implications for breaking the security that many individuals and institutions rely on every day.

Shor's algorithm pierces the key mechanism behind RSA cryptography because it provides an efficient way to factor numbers that are the product of two primes, even when those primes are huge.

The quantum part of Shor's algorithm doesn't directly find these prime numbers. Instead, it solves a related problem whose solution can then be used to determine the prime factors. The algorithm doesn't always succeed, because you start with a number that you guess. But it succeeds often enough that you will almost certainly find the prime factors you're seeking in less than 10 attempts [198, §11.9].

The catch is that the algorithm requires a quantum computer that has many thousands of qubits, along with error-correcting methods to make sure that the computation proceeds reliably. As of 2025, no publicly revealed quantum computers meet these requirements. But quantum computers are getting bigger and more reliable every day, so while using Shor's algorithm to break internet-level cryptography isn't practical today, it seems likely to be practical someday.

I mentioned that Shor's algorithm solves a problem related to prime factoring. This problem is called *period finding*, and it will be our focus in this chapter.

The connection between period finding and prime factoring is complicated. It uses a bunch of ideas from a branch of mathematics called *number theory* [196]. Unfortunately, all of this has little to do with quantum computing, and getting into it would be a huge detour for us. So I'll skip all of that and focus on the problem of period finding, which we'll solve with a hybrid algorithm. If you want to dig into how period finding helps us factor prime numbers, there are lots of great discussions to choose from [11] [47] [71] [137] [198] [294]. The good news is that if you're willing to put in some work, you can build enough working knowledge of all the pieces to see how Shor's algorithm works from beginning to end.

Shor's algorithm is a hybrid technique that uses a conventional computer at the start and end and a quantum computer in the middle. As I said, that quantum step is designed to solve the period-finding problem. If we can efficiently get useful results from the quantum part of the algorithm, we'll have managed the hardest and (by classical standards) most time-consuming part of factoring a composite number. So let's dig in and solve period finding.

Period Finding

Let's start with the big picture of the period-finding problem. Suppose that we have a sequence that is made up of r unique integers. We'll make another sequence that I'll call S by repeating this sequence over and over. Let's say that we repeat all r elements in the original list a total of m times, so S has rm numbers. Equation 13.1 shows an example for a sequence S created from a list of $r = 4$ elements repeated $m = 3$ times, for a list of 12 elements.

$$\text{sequence} = 3, 7, 2, 4$$
$$S = \underbrace{3, 7, 2, 4}, \ \underbrace{3, 7, 2, 4}, \ \underbrace{3, 7, 2, 4} \qquad (13.1)$$

We say that r is the *repeat length*, or *period*, of the sequence that makes up the list, and m is the number of repeats of that list in S.

The period-finding problem starts with a list that is promised to be made up of such a repeating sequence. Our goal is to find r, the period of that sequence.

Classically, we can find the period of the sequence by checking the items of the list one by one, looking for a repeated value. For example, given a sequence S, we start by looking at the integer s_0. Then we look at s_1 and see if it's the same as s_0. If so, the period is 2. Otherwise, we check s_2, then s_3, and so on, until for some integer k we find that $s_k = s_0$. Because we're promised that no entry in the repeating sequence is duplicated, we're done, and we can report that the period is k.

This approach can be practical if nothing gets too big. For example, if we're given the list explicitly, we'd like something that can fit into our computer's memory. It can overflow somewhat, but if the list has many billions of entries, the mere process of retrieving each element may become unacceptably slow. Alternatively, we might not be given the elements of the list at all, but instead a function that takes in a list index, does some computation, and returns the corresponding entry. If that function is time-consuming to evaluate, then the period can't be too long or we'll spend all of our time executing the function instead of comparing items and looking for repeats.

It would be great to bypass these limitations and be able to solve this problem for huge lists, made up of absolutely enormous repeating sequences, however they're represented.

Enter, of course, quantum parallelism!

Before we continue, we need to decide if we're going to tackle this problem in its full generality. The fully general approach starts with $N = 2^n$ qubits and allows the period r to be anything that's $N - 1$ or less.

A special case that makes the math simpler to read and write assumes that r divides N. That is, there are an integer number of complete repeats of the sequence in our N states, or N/r is an integer. For example, in Equation 13.1, we have $N = 12$ and $r = 4$, so $N/r = 3$, an integer.

The quantum circuit for both cases is the same, while the classical part of the algorithm is simpler for the case where r divides N.

As our focus is on the quantum part of the algorithm and not the details handled by the classical computer, I'm going to discuss the special case where r divides N. We'll still see all the key ideas, but our math will be kept to a minimum. The references throughout this chapter show you how to extend the mechanics of the classical part of the algorithm to cover any value of r.

So for us, $N = mr$ for some integer m. We can also write this for m as in Equation 13.2.

$$m = \frac{N}{r} \tag{13.2}$$

Remember that, for us, r always divides N, so m is an integer. To further simplify the math that's coming, I'll create a function f that takes as input any integer x and returns an integer value $x \bmod r$ for some period r. For this chapter, $f(x)$ is defined as in Equation 13.3.

$$f(x) = x \bmod r \tag{13.3}$$

Note that we're implicitly assuming that r is known (in programming terms, this definition assumes that r is a global variable). By leaving r implicit, I can write $f(x)$ rather than the bulkier $f(x, r)$.

Equation 13.3 tells us that inputs $x = 0$ to $x = r - 1$ produce the outputs of $f(x)$ from 0 to $r - 1$, then inputs $x = r$ to $x = 2r - 1$ also produce outputs from 0 to $r - 1$, and so on. Figure 13-1 shows the idea for the period $r = 4$.

m	0	1	2	3	4	5	6	7	8	9	10	11	12	13	14
$m \bmod 4$	0	1	2	3	0	1	2	3	0	1	2	3	0	1	2

Figure 13-1: The value of m mod 4 for increasing values of m

Sometimes I'll write a number x in the form $x = b + jr$ for integers b and j. In this form, j tells us how many full repetitions of the repeating sequence with period r have preceded x, and then b tells us how far we are into the final repetition (numbering, as always, starting with 0). We can solve for j and b by finding $j = \lfloor k/r \rfloor$ and $b = x \bmod r$ (the notation $\lfloor a \rfloor$ means that we take the integer part of a, rounded toward 0).

These definitions don't do a lot for us now, but they will make the math a little easier going forward. A graphical version of this numbering scheme is shown in Figure 13-2, illustrating the values in Figure 13-1.

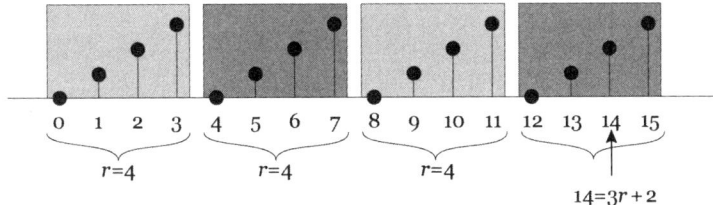

Figure 13-2: The sequence of 4 elements repeats completely 3 times. The entry at index 14 is at index 2 in the repeating sequence.

Figure 13-3 shows the circuit we'll be using to help us find the period r for a given sequence. For now, we'll just look at the big picture. We'll get to the box marked *QFT* shortly.

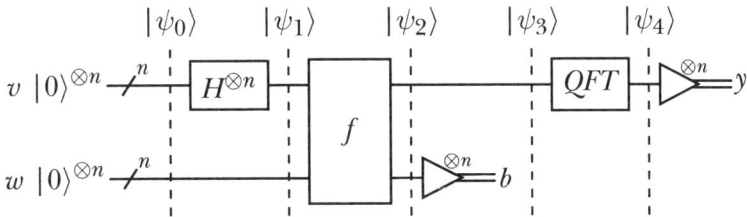

Figure 13-3: Shor's circuit for period finding

A key idea of this algorithm is that the box marked f implements our function $f(x) = x \bmod r$ from Equation 13.3. This sounds like a serious problem, because r is the very thing we're trying to find!

Not to worry. The first steps of Shor's hybrid algorithm, which run on a classical computer, take an integer that we guess and turn that into a candidate for a value of r. Given that value of r, we build the circuit and run it as described in the following section. If we ultimately don't find the prime factors of the number we're trying to factor, we make a new guess, get a new candidate for r, make a new circuit for the qugate marked f, and try again. As I mentioned, the candidates for r will be so good that we will likely never have to guess more than 10 times. You can think of the quantum part of Shor's algorithm as an efficient way to validate, or confirm, that a candidate value for r is correct.

Let's walk through an overview of the circuit. We'll start our process as we often do, by making a register of n qubits, all initialized to $|0\rangle$, and then running them all through Hadamard qugates, creating the uniform superposition $|+\rangle^{\otimes n}$ of all $N = 2^n$ input states. Let's call that the v register. We'll also make a second register, which I'll call the w register, also of n qubits, all initially $|0\rangle$. We'll run these two registers through a big qugate marked f that implements our function $f(x) = x \bmod r$. The top register emerges unchanged, while the lower register emerges with a superposition of all the values of $f(v)$.

This brings us to the location $|\psi_2\rangle$ in Figure 13-3. Though we've been thinking of our qubits as two registers, we can also view them as a single register that combines all inputs x in the top n qubits and all outputs $f(x)$ in the lower n qubits. Thus, by the principle of partial measurement, if we imagine measuring the top register and get an output bitstring b, we can be certain that the lower register, should we choose to measure it, would give us the bitstring $f(y) = b \mod r$.

We can go the other way, too, and note that if we measure the bottom register and get back a bitstring b, then the top register must be in a superposition of all the states that produce an output of $b \mod r$. That is, the upper register is in a superposition of the state $|b\rangle$, and $|b + r\rangle$, and $|b + 2r\rangle$, and so on, up to $|b + (m-1)r\rangle$. So, if we measure the lower register and observe b, then the only states that remain in the upper register are those consistent with observing b in the lower register. This means the upper register holds the superposition of states that can be written in the form $|b + jr\rangle$ for some integer j, and no others.

If we now measured the upper register, we wouldn't learn much, as there are two unknowns: j and b. If we could somehow get rid of one of them, we might have a chance of measuring the other.

To accomplish this, we'll use a quantum version of a staple algorithm in signal processing called the *Fourier transform* [73] [66] [152]. Its quantum version is the *quantum Fourier transform*, or *QFT* [198, §11.10] [249]. Discussing the *QFT* would take us on a long detour, so I'll just use the formula for the *QFT* when we get to it. You can find thorough discussions of the algorithm in the references.

The *QFT* turns a state $|b + jr\rangle$ into the form $e^{\gamma(b+jr)} |k\rangle$ for some complex number γ and state $|k\rangle$. We can write this coefficient as the product $e^{\gamma b} e^{\gamma jr}$. This gives us useful separation of one term involving b and one term involving j and r.

We'll discuss later how collecting a few of these terms enables us to eliminate the j, leaving us with our goal, the original period r.

This overview has introduced a lot of new ideas. Let's see how to turn these concepts into an algorithm.

Analyzing the Circuit

We'll start as we often do, with two sets of quantum registers. In this case, each has n qubits, all initialized to $|0\rangle$. Then we'll apply H qugates to the upper register to place it into a uniform superposition of all $N = 2^n$ states. The next step is to run both registers through a qugate that places $x \mod r$ in the lower register for each input x in the upper register. These steps are shown in Figure 13-4.

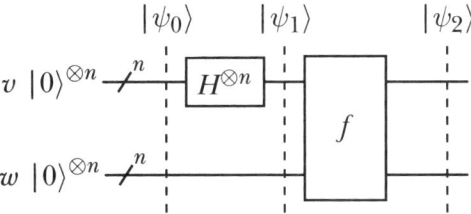

Figure 13-4: The start of a circuit for period finding

Let's write down the three checkpoints in Figure 13-4, starting with $|\psi_0\rangle$ and $|\psi_1\rangle$ in Equation 13.4.

$$|\psi_0\rangle = |0\rangle^{\otimes n} |0\rangle^{\otimes n}$$
$$|\psi_1\rangle = \vee^n \sum_{k\in[N]} |k\rangle |0\rangle \tag{13.4}$$

After applying the function f, we get $|\psi_2\rangle$ in Equation 13.5.

$$|\psi_2\rangle = \vee^n \sum_{k\in[N]} |k\rangle |k \bmod r\rangle \tag{13.5}$$

The w register is thus a superposition of the r states from $|0\rangle$ to $|r-1\rangle$.

There are only two steps to go. First, we measure the w register, as shown in Figure 13-5.

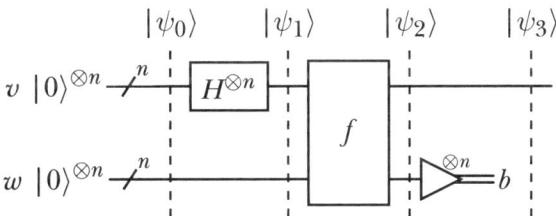

Figure 13-5: Measuring the w register

We'll get a bitstring b from 0 to $r-1$, as those are the only values produced by the function f. Let's interpret the bitstring b as a decimal number, which I'll also write as b.

The principle of partial measurement, which we discussed in Chapter 6, tells us that when we have a superposition of states and we measure some of the qubits, the remaining qubits collapse to values that are consistent with what we measured. Therefore, knowing that the w register has collapsed to the state $|b\rangle$, we also know the entire system will collapse to a superposition of the states from the original superposition that end with this $|b\rangle$. These are the states $|(jr + b)b\rangle$. Thus, the v register is in a superposition of states described by $|jr + b\rangle$ for different values of j.

At this point, the w register has done its job by partially collapsing the v register. We don't need the w register anymore.

Let's think about what's just happened. State $|\psi_2\rangle$ can be viewed as a collection of states, each described by $jr + b$ in the v register and b in the w register. And we just measured b. The value b appears in the w register just once in each set of r states, and there are m such sets. Thus, after measuring b in the w register, we're left with a superposition of m states given by $b + mr$. Because there are m states with equal amplitudes, each must have amplitude $1/\sqrt{m}$ (this makes their probabilities add up to 1, as they must, since $m(1/\sqrt{m})^2 = m/m = 1$). We can write this superposition as in Equation 13.6.

$$|\psi_3\rangle = \frac{1}{\sqrt{m}} \sum_{k \in [m]} |b + kr\rangle \qquad (13.6)$$

Can we find r just from $|\psi_3\rangle$? After all, we now know b, and if we measure $|\psi_3\rangle$, we'll get back some integer $b + mr$. If we can factor out the m, we'll be left with our goal, r.

This might sound a little like brute force. But programmers are fond of saying "Never underestimate the power of brute force," so let's give it a shot.

Suppose that we run the circuit of Figure 13-5 a few thousand times using $N = 100$ and $r = 10$. Then we'll probably (though not definitely) measure some particular value of b at least a few times. Let's call that value b', and throw out all the runs that gave us any other value of b. We could then gather up the corresponding upper register outputs. Let's call them $b' + k_0 r$, $b' + k_1 r$, and so on. Next, we could subtract b' from each, giving us the sequence $k_0 r, k_1 r, \ldots, k_n r$, and from that find r. Success!

For example, suppose $b' = 3$ and $r = 5$. We'll make two measurements, corresponding to $k = 5$ (giving us $3 + (5 \times 5) = 28$) and $k = 11$ (giving us $3 + (5 \times 11) = 58$). Remember that we don't know the values of k, only the measurements 25 and 58. Subtracting $b' = 3$ from each, we get 25 and 55. Their greatest common divisor is 5, and we've recovered our value of r.

This scheme is, unfortunately, impractical when applied to cryptography, where the numbers used are often larger than hundreds of bits [270]. The number of runs of the algorithm we'd need to perform to gather enough measurements would be overwhelming. Brute force, as many would-be conquerors, despots, and authoritarians have discovered, doesn't scale up well.

Let's take a different approach.

Applying the QFT

We're going to leave the conceptual picture now in favor of working the math, because this is where the magic happens. There will be a bunch of equations, so I'll take it one step at a time.

Earlier I said that we would wrangle the term $b + kr$ into an exponent. If we can compute $e^{b+kr} = e^b e^{kr}$, then we might be able to set e^b aside. To get us to that point, I'll apply the *QFT* to $|\psi_3\rangle$ and see what we get from it. Appending the *QFT* to the top register of Figure 13-5 gives us Figure 13-6.

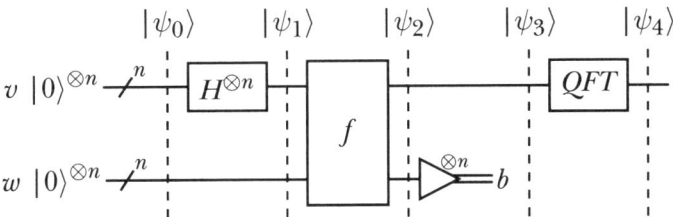

Figure 13-6: Applying the *QFT* to the *v* register

Because we're not going to delve into the workings of the *QFT*, let's just treat it as a normal function. Then the result of applying the *QFT* to $|\psi_3\rangle$ is $|\psi_4\rangle$, given in Equation 13.7 (from here on out, I'll drop the *w* register since we don't need it anymore).

$$|\psi_4\rangle = QFT(|\psi_3\rangle)$$
$$= QFT\left(\frac{1}{\sqrt{m}} \sum_{j \in [m]} |b + jr\rangle\right) \quad \text{From Eq. 13.6} \quad (13.7)$$

The *QFT* is a quantum gate and therefore is linear, so we can move it inside the summation, giving us Equation 13.8.

$$|\psi_4\rangle = \frac{1}{\sqrt{m}} \sum_{j \in [m]} QFT(|b + jr\rangle) \quad \text{Because the } QFT \text{ is linear} \quad (13.8)$$

Now I'll substitute the formula for the *QFT*. This formula is the final step in a logical process, but not one we can summarize in just a few lines. You can find discussions of the *QFT*, and its derivation, in many references [37] [97] [277].

Substituting the definition of the *QFT* into Equation 13.8 gives us Equation 13.9, where the big square brackets contain the result of applying the *QFT* to $|b + jr\rangle$.

$$|\psi_4\rangle = \frac{1}{\sqrt{m}} \sum_{j \in [m]} \left[\frac{1}{\sqrt{N}} \sum_{k \in [N]} \omega^{(b+jr)k} |k\rangle\right] \quad \text{Write out the } QFT \text{ of } |b + jr\rangle \quad (13.9)$$

This expression uses a variable ω, defined in Equation 13.10. As always in this book, $N = 2^n$ for *n* qubits.

$$\omega \stackrel{\Delta}{=} e^{i2\pi/N} \quad (13.10)$$

With this definition, $\omega^{(b+jr)k}$ is a short way of writing the longer (and much messier) term $\exp(i2\pi(b + jr)k/N)$ (remember that $\exp(x)$ is another way to write e^x that saves us the effort of squinting at all those symbols in tiny type as an exponent).

Now that we have the *QFT* applied to $|\psi_4\rangle$, we can juggle things around. I'll reverse the order of the summations and expand the exponential. That will give us a term ω^{bk} that we can pull out of the innermost sum. The steps are in Equation 13.11.

$$|\psi_4\rangle = \frac{1}{\sqrt{m}} \sum_{j\in[m]} \left[\frac{1}{\sqrt{N}} \sum_{k\in[N]} \omega^{(b+jr)k} |k\rangle \right] \qquad \text{Eq. 13.9}$$

$$= \frac{1}{\sqrt{mN}} \sum_{j\in[m]} \sum_{k\in[N]} \omega^{(b+jr)k} |k\rangle \qquad \text{Pull } 1/\sqrt{N} \text{ to the front}$$

$$= \frac{1}{\sqrt{mN}} \sum_{k\in[N]} \sum_{j\in[m]} \omega^{(b+jr)k} |k\rangle \qquad \text{Reverse order of sums} \qquad (13.11)$$

$$= \frac{1}{\sqrt{mN}} \sum_{k\in[N]} \sum_{j\in[m]} \omega^{bk}\, \omega^{jrk} |k\rangle \qquad \text{Expand exponential}$$

$$= \frac{1}{\sqrt{mN}} \sum_{k\in[N]} \omega^{bk} \sum_{j\in[m]} \omega^{jrk} |k\rangle \qquad \text{Move } \omega^{bk} \text{ out of final summation}$$

Recall that the thorn in our brute-force approach was that it depended on obtaining the same b multiple times, and that became increasingly unlikely for large values of N. By using the *QFT*, we've gotten the term involving b outside of the inner loop. This means we don't have to hope that we'll luckily measure many states with the same b. Next, we'll get rid of the b entirely, giving us a clear path to our goal, finding the value of r.

Finding the Probability of $|k\rangle$

We'll now find the probability of measuring one particular, but arbitrary, state, and we'll see how that leads us to r.

I'll start by isolating the amplitude of just one specific state $|k\rangle$ from Equation 13.11, which I'll square to get the probability of measuring k, written $\Pr(k)$. This is shown in Equation 13.12. I've removed the loop over k, as we're now interested in only one specific value of k.

$$\Pr(k) = \left| \frac{1}{\sqrt{mN}} \omega^{bk} \sum_{j\in[m]} \omega^{jrk} \right|^2 \qquad (13.12)$$

Let's clean this up. First, notice that ω^{bk} is now a constant. Thus, we can pull it out front, giving us Equation 13.13.

$$\Pr(k) = |\omega^{bk}|^2 \left| \frac{1}{\sqrt{mN}} \sum_{j\in[m]} \omega^{jrk} \right|^2 \qquad (13.13)$$

Recall that ω^{bk} is a shorthand for $\exp(i2\pi(bk)/N)$. Although the phase is complicated, this is a complex number with a magnitude of 1, so its square is also 1. As multiplying the probability by 1 doesn't change that probability, we can just drop this term.

Whoa, we did it; we just got rid of b!

The resulting simpler expression is shown in Equation 13.14.

$$\Pr(k) = \left| \frac{1}{\sqrt{mN}} \sum_{j \in [m]} \omega^{jrk} \right|^2 \tag{13.14}$$

Since b doesn't show up at all anymore, there's no need to run the algorithm multiple times to get a bunch of outputs with the same b. Thanks to the *QFT*, we've replaced brute force with finesse.

We can keep simplifying, and things get even better. First, let's pull the fraction at the start out of the expression. Noting that $(1/\sqrt{mN})^2 = 1/(mN)$, we get Equation 13.15.

$$\Pr(k) = \frac{1}{mN} \left| \sum_{j \in [m]} \omega^{jrk} \right|^2 \tag{13.15}$$

We'll soon find it useful to get the $1/m$ part of the opening fraction into the squared expression. Let's get started by rewriting the fraction $1/mN$ as in Equation 13.16.

$$\frac{1}{mN} = \frac{1}{m^2 r} \qquad \text{Use } N = mr \text{ from Eq. 13.2}$$
$$= \frac{1}{r} \left(\frac{1}{m} \right)^2 \tag{13.16}$$

The fraction $1/m$ is getting squared while $1/r$ isn't, so we can keep the $1/r$ out front and move $1/m$ inside the absolute value. The result of this move is Equation 13.17.

$$\Pr(k) = \frac{1}{r} \left| \frac{1}{m} \sum_{j \in [m]} \omega^{jrk} \right|^2 \tag{13.17}$$

Let's expand the ω term using Equation 13.10, and then substitute N with rm as shown in Equation 13.18.

$$\Pr(k) = \frac{1}{r} \left| \frac{1}{m} \sum_{j \in [m]} \exp\left(i2\pi jrk/N \right) \right|^2 \qquad \text{Expand } \omega$$

$$= \frac{1}{r} \left| \frac{1}{m} \sum_{j \in [m]} \exp\left(i2\pi jrk/(rm) \right) \right|^2 \qquad \text{Substitute } N = rm \tag{13.18}$$

$$= \frac{1}{r} \left| \frac{1}{m} \sum_{j \in [m]} \exp\left(i2\pi jk/m \right) \right|^2 \qquad \begin{array}{l} \text{Cancel the } r \text{ values} \\ \text{in the exponent} \end{array}$$

This is nice. We've now isolated r at the start of this expression. If we can figure out the value of the term between the absolute value bars, we'll know r. We're just about done!

Having been down this road before, I can report back that if we examine the probability of measuring some specific values of k, we're going to learn something useful. In particular, suppose that $k = cm$ for some integer c. That is, k is a multiple of m, the number of repeats of our pattern in the original input.

Let's plug $k = cm$ into Equation 13.18 and see what happens. The steps are shown in Equation 13.19.

$$\Pr(k = cm) = \frac{1}{r}\left|\frac{1}{m}\sum_{j\in[m]}\exp(i2\pi j(cm)/m)\right|^2 \quad \text{Use } k = cm$$

$$= \frac{1}{r}\left|\frac{1}{m}\sum_{j\in[m]}\exp(i2\pi jc)\right|^2 \quad \text{Cancel the } m \text{ terms}$$

$$= \frac{1}{r}\left|\frac{1}{m}\sum_{j\in[m]}1\right|^2 \quad \text{Since } j, c \in \mathbb{Z}, \exp(i2\pi jc) = 1 \tag{13.19}$$

$$= \frac{1}{r}\left|\frac{1}{m}m\right|^2 \quad \text{Add up } m \text{ values of 1 to } m$$

$$= \frac{1}{r} \quad \text{The probability of measuring } k = cm$$

The next-to-last step is why I put the $1/m$ term back inside the squared expression. When we measure $|\psi_4\rangle$, the probability that we'll get back a value cm is $1/r$. This tells us that in our running example, when $|\psi_4\rangle$ has four states, it can be drawn as in Figure 13-7.

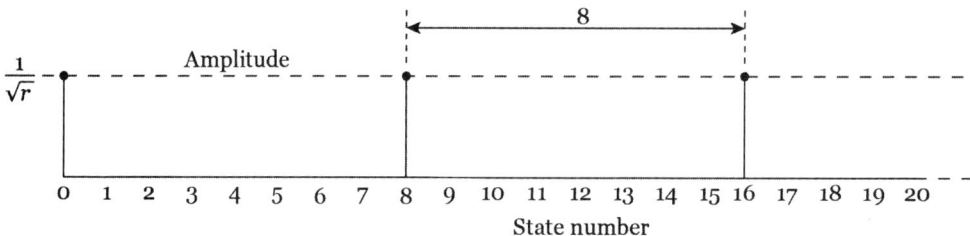

Figure 13-7: The superposition $|\psi_4\rangle$ resulting from applying the QFT to $|\psi_3\rangle$

How many of those values are there? They start with $c = 0$, then $c = 1$, and so on, up to some largest nonzero value of c that I'll call $c_{\max}$. Whatever this is, we know that $c_{\max}m = N$, so $c_{\max} = N/m = r$. So when we measure $|\psi_4\rangle$, we can get back any one of r different states cm, for c ranging from 0 to $r - 1$.

But hold on. If there are r possible such states, and each has a probability of $1/r$ of being measured, then the probability of measuring one of these states is $r(1/r) = 1$.

This is huge news! It means that when we measure $|\psi_4\rangle$, we are *guaranteed* to get back a state cm. We will *never* get back any other state, because there's no probability left over for any other state to take on.

We're definitely going to measure a state $cm = cN/r$. We know N, so collecting enough values of cm will let us recover r. The beauty of the QFT in this algorithm is that it gets us to this conclusion *regardless of what we measure in the lower register*.

The QFT transformed $|\psi_3\rangle$ into a form that let us eliminate any dependence on b, and ultimately led us to measure a very specific superposition that we could use to find r.

And that's how we use a quantum computer to find the period r of some input sequence. The whole algorithm is shown in one place in Figure 13-8. Our final measurement y in this circuit will always be some number cN/r, where we know N and we also know c is an integer.

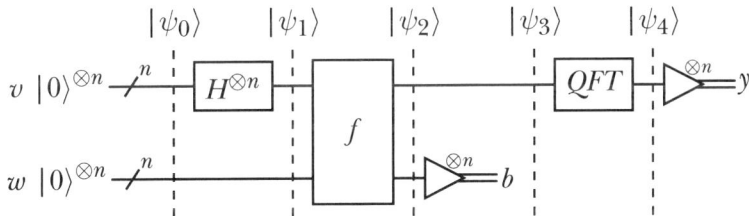

Figure 13-8: The full quantum circuit for period finding

As I said at the start of this discussion, I've been assuming that the initial sequence repeats an integer number of times, or that N/r is an integer. We used that assumption several times to simplify the math. Assuming that r divides N is a major assumption, however, and it's unlikely to be true for most sequences we'd work with in practice.

Ifwe relax this assumption, we can still carry out the circuit of Figure 13-8, and we still get $|\psi_4\rangle$ as in Equation 13.11 and the probability of measuring k in Equation 13.12. But after that, things get more complicated. Specifically, it's the classical part of the algorithm that comes after this quantum step that gets more complicated, because we can't isolate r as neatly as we did in Equation 13.19.

Generally speaking, even when r doesn't divide N, the closer a state is to an integer value of N/r, the more likely that state is to be measured. By running the circuit multiple times and combining multiple measurements, we can still recover r during the classical step by using a few more mathematical techniques [196] [47] [71] [198].

Connecting to Prime Factoring

I said earlier that connecting period finding to prime factoring would be too big a detour for this discussion. But we can still consider the big picture, so let's look at Shor's algorithm in overview and demonstrate it in action. I'll just state the steps here, and you can find justification for everything from the references.

We'll use two ideas from number theory in this process. The first is modulo arithmetic, which we've already seen.

The second is the *gcd*, which stands for the *greatest common divisor*. Given two positive integers a and b, their greatest common divisor, written $\gcd(a, b)$, is the largest integer that divides both a and b. For example, $\gcd(12, 16) = 4$, because while both 4 and 2 divide both 12 and 16, the larger of these divisors is 4. Using some larger numbers, $\gcd(35, 40) = 5$, because 35 has the divisors 5 and 7 and 40 has the divisors 2, 4, 5, 8, 10, and 20, so the largest divisor they have in common is 5.

If the gcd of two numbers is 1, like in $\gcd(7, 15)$, then we say that the two numbers are *relatively prime* (we also say that they are *coprime* to one another, or that they are *coprimes*). This doesn't mean that either number is a prime number (for example, $\gcd(9, 16) = 1$ even though neither 9 nor 16 is prime), but only that they have no prime factors in common.

The gcd can be quickly computed with a process called the *Euclidean algorithm* [234], which is available as a function in almost every classical math library. The process is deterministic, so there's no guessing, and it's simple enough that it doesn't require any external reference data, such as a table of prime numbers [265].

Shor's algorithm is a hybrid sandwich: It starts and ends with classical computing and performs the quantum algorithm we saw earlier in the middle. I'll outline the basic process with a step-by-step approach [198]. The quantum part is in step 5:

1. **Setup:** We're given some integer M that we're promised is the product of two prime numbers, p and q. We want to find these two primes (we really only need one, say p, as it's practical to compute $q = M / p$ on a classical computer).

2. **Pick circuit size:** Find the smallest n such that $2^n > M^2$. That n will be the number of qubits in each register of our network, representing $N = 2^n$ states.

3. **Guess a:** Guess a random integer a in the range $[2, \sqrt{M}]$.

4. **See if we're lucky:** Find $\gcd(a, M)$. If this is not 1, then we were lucky: a and M have a common factor greater than 1, and we've achieved our goal. We stop here and return $\gcd(a, M)$. Otherwise, we continue to step 5.

5. **Go quantum:** Run the quantum algorithm of Figure 13-8, where the function $f(x)$ is $x \bmod a$. Measure the registers.

6. **Compute a candidate:** Because we generally won't have an integer number of repeats (that is, $m = N / r$ won't be an integer), we use some math we haven't discussed to process the measurement y, along with any previous measurements, into a guess for r.

7. **Check this** r**:** Before we go further, we check that this guess for r satisfies two conditions. Specifically, if r is odd or if the expression $a^{r/2} + 1 \bmod M = 0$, we go back to step 3 and try a new guess for a. Otherwise, we continue.

8. **Try to factor** M**:** Find $\gcd(a^{r/2} - 1, M)$ and $\gcd(a^{r/2} + 1, M)$. If either of these is not 1, we've found a prime factor, so we return it. Otherwise, we go back to step 3 and try a new choice for a.

We're doing some guessing in this algorithm, so it falls under the category of a *probabilistic algorithm*. Using some analysis of each guess, we can compute that the probability of failing to find a prime number after 10 times through this loop is about 10^{-10}, or 1 in 10 billion [198, §11.9].

An Example of Shor's Algorithm

Let's run through Shor's algorithm, using the popular example of factoring $M = 15$ [146, §5.4].

I'm going to pick $a = 7$ for the example, so let's look at the pattern $7^x \bmod 15$ for increasing values of x, as shown in Table 13-1.

Table 13-1: The function $7^x \bmod 15$ for the first few values of x

x	7^x	$7^x \bmod 15$
0	1	1
1	7	7
2	49	4
3	343	13
4	2,401	1
5	16,807	7
6	117,649	4
7	823,543	13
8	5,764,801	1
9	40,353,607	7
10	282,475,249	4
11	1,977,326,743	13

This tells us that the period r has the value 4, as $7^x \bmod 15$ repeats every four terms.

I'm going to pick $n = 11$ qubits, giving us $2^{11} = 2,048$ states. Conveniently, $N / r = 2,048 / 4 = 512$, which is an integer, so we have exactly $m = 512$ repetitions of the sequence $(1, 7, 4, 13)$ in our 2,048 starting states.

Now let's go through the same steps outlined in the previous section using these values:

1. **Setup:** We're given $M = 15$.

2. **Pick circuit size:** Because $M^2 = 225$, any value of $n \geq 8$ will do, as $2^8 = 256 > 225$. Let's arbitrarily pick $n = 11$.

3. **Pick a:** Arbitrarily pick $a = 7$.

4. **See if we're lucky:** Find the gcd of a and M. Because $\gcd(7, 15) = 1$, we weren't lucky, so we continue.

5. **Go quantum:** Build the quantum gate that implements the function $x \bmod 7$. Run the quantum algorithm of Figure 13-8, where the function $f(x)$ is $7^x \bmod 15$. When we measure the second register, Table 13-1 tells us that we could get 1, 7, 4, or 13. I'd like this example to complete successfully, and I know that getting 4 will get us there, so let's suppose we measure 4 (if we didn't, one of the later steps would fail, and we'd return to step 3). Then we know that $|\psi_3\rangle$ is a superposition of all 512 states $|x\rangle$ where $7^x \bmod 15 = 4$, as shown in Equation 13.20 (the first few values come from Table 13-1).

$$
\begin{aligned}
|\psi_3\rangle &= \frac{1}{\sqrt{512}} \sum_{k \in [512]} |4k + 2\rangle \\
&= \frac{1}{\sqrt{512}} \Big(|2\rangle + |6\rangle + |10\rangle + \cdots + |2,046\rangle \Big)
\end{aligned}
\tag{13.20}
$$

This superposition has a nice graphical interpretation, shown in Figure 13-9.

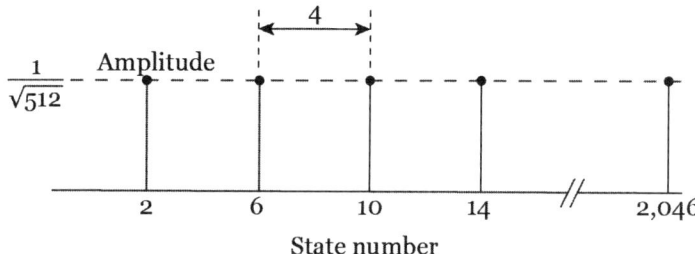

Figure 13-9: The superposition $|\psi_3\rangle$. Each of these 512 states has a probability of $1/512$.

Now we apply the *QFT*. The upper register turns into an equal superposition of the four values cm, or $512c$, for $c \in [r]$, or $0, 1, 2, 3$.

Thus, $|\psi_4\rangle$ is given in Equation 13.21.

$$|\psi_4\rangle = \frac{1}{\sqrt{4}} \sum_{k \in [4]} |512k\rangle$$

$$= \frac{1}{\sqrt{4}} \Big(|0\rangle + |512\rangle + |1{,}024\rangle + |1{,}536\rangle \Big)$$

(13.21)

Once again, we can draw this superposition graphically, as shown in Figure 13-10.

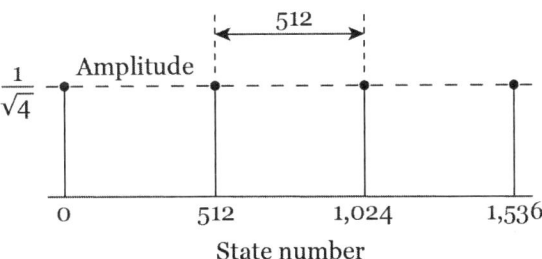

Figure 13-10: The superposition $|\psi_4\rangle$. Each of the 4 states has a probability of 1/4.

6. **Compute a candidate:** Suppose when we measure $|\psi_4\rangle$ we get 1,536. Then, using a technique called *continued fractions*, from a starting point of $1{,}536 / 2{,}048 = 3/4$, we get a guess of $r = 4$.

7. **Check this r:** First, $r = 4$ is even, so we pass the first test. Second, $7^{4/2} + 1 \bmod 15 = 50 \bmod 15 = 5$, and that's not 0, so we pass that test as well and continue to step 8.

8. **Try to factor M:** We now find $\gcd(7^{4/2} - 1, 15) = \gcd(48, 15) = 3$ and $\gcd(7^{4/2} + 1, 15) = \gcd(50, 15) = 5$. These are both not 1, so we've found the prime factors of 15, or $15 = 3 \times 5$. Encryption cracked!

Discussion

I mentioned earlier that we can almost surely factor a prime number using Shor's algorithm in just 10 runs. So that's the end of internet security, right?

Not quite. The first problem is simply one of size. To crack security systems used in 2025, we'd need a reliable, low-noise quantum computer of somewhere between 5,000 and 24,000 qubits [67]. As of 2025, the largest publicly announced quantum computer is 1,121 qubits [69]. Quantum computers are growing in size and reliability, but getting up to many thousands of useful qubits may take a while.

Another issue, though more easily accommodated, is that we have to create a new quantum circuit for each value of a that we want to test, because each choice of a requires a new circuit to evaluate $f(x) = x \bmod a$. That might take some work [297].

To counter the possibility of a large quantum computer opening up all the secrets now protected by the RSA system, a new field of *quantum cryptography* is developing ways to protect information. The goal is to find practical methods that cannot be efficiently cracked by a quantum computer [11].

New encryption methods don't change the fact that Shor's algorithm could be used retroactively, to decrypt everything that is, or was, encrypted with a system like RSA. A lot of this stuff won't matter, like everyday credit card transactions from decades ago. But some of those old secrets might still be important, such as military or industrial plans, or people's private health histories, or their diaries and photos, videos, and other personal, confidential records. Even if you re-encrypt these with some new kind of quantum cryptography, if the old versions still exist in some archives, or on old hard drives, or anywhere else, they will still be vulnerable.

This isn't a new situation. The history of cryptography is filled with techniques that were eventually broken and replaced with better methods, leaving the old secrets open to discovery by anyone with the means of accessing them. Even so, there are some famous examples of secret messages, ciphers, and codes that have not been cracked. Some are ancient and presumably were actually used in practice (like the Linear A script found on tablets in Crete), while others may be one-off mysteries (or hoaxes) that might never be undone [54].

It's probably best to always treat any encryption scheme as temporary and find additional ways to protect information you really don't want anyone to be able to read for a long time. Almost nothing is forever, including encryption.

Summary

Shor's algorithm is probably the most famous quantum algorithm. It can solve a problem that is so time-consuming for classical computers that it is used as the core of encryption techniques around the world.

The idea behind many of those techniques begins with two huge prime numbers that are multiplied together to get a new, even bigger number. Given that bigger number, we'd like to reverse the process and find its two prime factors.

Shor's algorithm solves this problem by breaking it down into pieces. First, the work is shared between classical and quantum computers, so the complete system forms a hybrid algorithm. Second, through some clever number theory that we didn't discuss, the factoring problem is replaced with a different problem that asks us to find the period of a repeating sequence in a sequence of numbers. We start the process by making a guess, and if our guess doesn't lead us to the prime numbers we seek, we make a different guess and try again.

In this chapter, we assumed that the period, or length of the repeating sequence divided the total length of the input. This meant that we have an integer number of full instances of the repeating sequence in the input. With a more complicated classical algorithm, this restriction can be removed.

Shor's algorithm is not yet a practical tool for breaking the widespread RSA encryption system and its variants. Cracking RSA would require a reliable quantum computer with more qubits than anyone's been able to build. But we will probably have such computers one day, and then the secrets protected today by encryption algorithms such as RSA, which depend on the difficulty of factoring prime numbers, will be exposed.

To address this, researchers are working on new encryption algorithms, native to quantum computers, that aren't vulnerable to Shor's algorithm.

14

NEXT STEPS

If the quantum circuit model really does capture everything that happens in physics, then we'll be able to use quantum computers in the future to explore fundamental physics at a very deep level. But if not, that's even more exciting. It means that Nature will ultimately allow even more powerful information processors than the quantum computers we currently foresee.
—John Preskill, *Quantum Computing 40 Years Later*, 2021 [160]

Study hard what interests you the most in the most undisciplined, irreverent and original manner possible.
—Richard Feynman, *The Quotable Feynman*, 1965 [62]

We've covered a lot of material in this book, but there is much more to be discovered in the field of quantum computing. To help you launch yourself in new directions and embrace new ideas, in this final chapter I'll present some advanced ideas, ways for you to run your own quantum programs on real and simulated quantum computers, and a bunch of cool applications to inspire your own programs.

Further Ideas

Let's start by looking at a few topics that you might find useful to explore if you decide to pursue quantum computing in more depth.

Superdense Coding

The technique of *superdense coding* (also called *dense coding*) is kind of the flip side of the quantum teleportation protocol we discussed in Chapter 7.

Rather than using entanglement and two classical bits to transfer a quantum state from one quantum object to another, superdense coding is a way to communicate two classical bits by exchanging one quantum bit [278].

In this scenario, we imagine a world in which Alice and Bob once again want to exchange some information. As in quantum teleportation, I'll assume that they each share one qubit from a previously entangled pair.

The goal is to enable Alice to send two classical bits to Bob. Perhaps she's worried that if she sent those bits using some classical method, like radio or in print, an eavesdropper could intercept them. Even if they were encrypted, there's some chance the eavesdropper could crack that encryption and learn the information in the bits.

To prevent this possibility, Alice can securely communicate her two classical bits to Bob by physically sending him a single quantum bit. Then, Bob can use that qubit to recover the classical bits.

The circuit is shown in Figure 14-1.

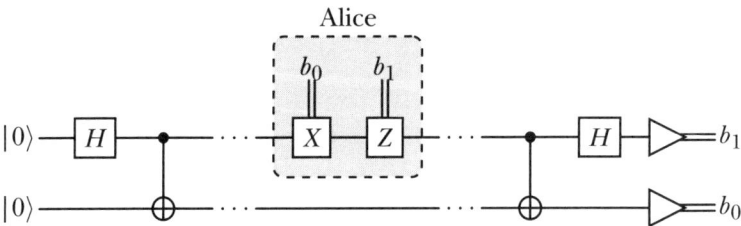

Figure 14-1: A quantum circuit for superdense coding

We start as in teleportation, creating the Bell pair $|\beta_s^+\rangle$, as we saw in Figure 7-3. Now Alice and Bob go their separate ways, each with one of the entangled pair of qubits. Some time elapses, as shown by the dots on the qubit lines.

Later, Alice has two bits, named b_1 and b_0, that she wants to send to Bob. Recall that she can't (or doesn't want to) send her two bits to Bob. Instead, as shown in the box marked "Alice," she uses her classical bits as controls on two qugates applied to her qubit, a controlled-X and then a controlled-Z.

Now Alice hands her qubit to a courier, who physically delivers it to Bob. This also takes some time, represented by the dots.

When Bob receives this qubit, he applies a CX and then an H qugate to the system of qubits that are both now in his possession. Bob then measures the two particles and recovers Alice's two classical bits.

Superdense coding offers a secure way for Alice to send her information to Bob. If an eavesdropper (conventionally given the delightful name Eve) intercepts the quantum particle that Alice sends to Bob, she is lacking Bob's half of the entangled pair and thus can't extract Alice's information from the intercepted qubit.

Like quantum teleportation, superdense coding has been proven to work in real hardware.

POVM and the Density Matrix

Our discussion of measurement focused on projection operators, because they let us make measurements without requiring us to cover additional math. There is a more general approach to measurement that instead uses *positive operator-valued measures (POVMs)* [194] [276].

You might need to learn a few new mathematical tools to master this approach to measurement, like the trace operator, which adds up the elements along the main diagonal of a matrix. But then you'll be able to measure systems that are less well defined than the states we've talked about in this book.

Let's see what this approach offers us, starting with an analogy. Suppose that you work with a bookstore that sends out curated boxes of books to different organizations once a month. Each box has its own set of books tailored to that type of organization, so there might be one box of books selected for children's hospitals, a different box of selections for high school libraries, another box for senior community centers, and so on. They've sent you a list of the contents of each of this month's boxes, and you've selected and ordered one box for your group.

Each month, the bookstore creates a summary sheet listing how many boxes of each kind were ordered and emails that to everyone along with their invoice. Unfortunately, the bookstore employee who filled the boxes this month forgot to label them and sent each recipient a random box. As a result, when you receive your order, you won't know which selection of books you've received until you open the box and look.

Now let's replace the books with quantum states and the boxes with superpositions. Suppose you're working with a provider who creates systems of qubits, where each system is in one of a number of known superpositions of different states. Each month, they provide a summary that describes the different superpositions their customers ordered and indicates how many systems of each superposition were sent out. But, like the bookstore worker, this month the person doing the packaging forgot to add the labels and sent out the boxes containing the qubits haphazardly, so you don't know which superposition describes the qubits you received.

In other words, you have a system of qubits, and you know it's in one of the superpositions in the list. You also know how many systems in each superposition were sent out. However, you don't know which superposition you have. It would be nice to be able to represent this scenario in mathematical terms, so you can work with the probabilities of the different superpositions you might have received and the different states you might measure.

You can mathematically model this situation with a *density matrix* [19] [257]. A nice advantage of the density matrix approach is that it doesn't assume we're working with isolated quantum systems. Let's see what that means.

In this book, I've assumed that our qubits (and the systems we made from them) are *isolated* from the environment. In fact, that condition was part of Postulate 1 in Chapter 2. Isolation means that qubits don't interact with one another (unless we specifically cause them to), they don't interact with the electronics or mechanical devices that make up the quantum computer, and they're not affected by heat or vibration or gravity waves or anything else. They are isolated from the rest of the universe. We did briefly discuss how these outside phenomena introduce noise in our measurements, but we didn't try to model that noise.

The density matrix formulation lets us represent our quantum bits and systems more realistically, as part of the universe we all inhabit. We don't have to make the assumption that they are isolated, in their own little pocket universes. With this mathematical model, we can explicitly acknowledge the universe's effects on our qubits and make the universe a part of our complete quantum system. We can then use mathematical techniques to remove the parts of the universe we don't want to deal with at any given moment, thus isolating just the parts that interest us.

Quantum Encryption

We saw in Chapter 13 that Shor's algorithm can break public-key encryption schemes such as RSA encryption (or at least, it will be able to do so when quantum computers get big and reliable enough).

Recall that RSA is based on the observation that nobody knows of an efficient way to use a classical computer to factor a huge composite number into its prime factors. But just because nobody's found a way to do this yet doesn't mean nobody will ever find a way.

In response to these issues, a field of research known as *post-quantum cryptography* (or sometimes just *quantum cryptography*) has emerged [11]. The goal of the field is to find encryption schemes that are safe from being broken even by quantum computers. In other words, quantum cryptographers aim to develop algorithms that are known (or at least, strongly believed) to be theoretically impractical (or even impossible) for both classical or quantum computers to crack, not just for some cases, but in every use.

An important problem is that the field of cryptography has shown us time and again that no matter how clever an encryption system is, there is almost always someone equally or more clever who can break it. So unless a scheme can be rigorously *proven* to be uncrackable, there's always the possibility that someone will come along and find a way to read the secrets it protects.

Nevertheless, as the RSA and other public-key schemes have shown us, having cryptography that works well today is definitely better for protecting secrets than no cryptography at all.

Quantum encryption is becoming an increasingly real technology. In late 2024, the US National Institute of Standards and Technology (NIST) announced a set of new encryption standards that are intended to resist decryption even by quantum computers [149].

Quantum Error Correction

As we've seen, quantum computers make mistakes, whether due to imperfections in the computers themselves, unintended interactions among the qubits, or the influence of some event or phenomenon happening somewhere in the universe where the quantum computers are located. No man, and no quantum computer, is an island.

Taken together, all of these effects that disrupt the ideal functioning of a quantum computer are called *noise*.

We will probably never be free of noise and its effects. The universe is a busy place, and things happen everywhere, all the time.

Noise isn't unique to quantum computers. Classical computers make lots of errors and are also subject to events happening in the universe. For example, cosmic rays are particles resulting from the explosions of stars, perhaps many millions of years ago. If a cosmic ray smashes into a classical computer chip, it can flip one or more bits [159]. As programmers, we rarely have to deal with this problem because engineers and scientists have spent decades improving hardware resistance to cosmic rays and other events and building software tools to automatically detect and correct errors when they do happen.

Quantum computers are newer, and we don't have these robust error management tools yet. This is why people sometimes say that we're currently in the noisy intermediate-scale quantum (NISQ) era [273] (the "intermediate-scale" part is a deliberately vague term that refers to the number of reliable qubits we can currently build in a single quantum computer, which is more than a handful but less than thousands).

To help us deal with errors, there's a robust field of research and development collectively known as *quantum error correction* (*QEC*) [177]. People are working hard to bring quantum error detection and correction up to the same level of reliability that we're used to with classical computers, so that users of quantum computers don't have to think about these kinds of errors.

Thanks to these efforts, quantum computers are becoming more reliable. Google's Willow quantum computing chip, announced in late 2024, demonstrated exponential quantum error correction, with the number of errors actually decreasing as the qubits are assembled into increasingly larger clusters [147].

Other Diagrams

In this book, I've used the well-known circuit diagram approach for drawing quantum circuits. This technique is popular because it shows every step in detail.

There's another graphical representation for quantum circuits that takes a more abstract view. They're called *ZX-diagrams*, associated with a conceptual approach called the *ZX-calculus* [222] [142]. The technique offers a nice visual way to present the conceptual structure of some quantum calculations while hiding many of their details.

Quantum Advantage

As of 2025, quantum computers and their software are still in their relatively early days of development compared to classical computers. Quantum computers today are not able to solve a lot of practical problems faster or more cheaply than classical computers.

The term *quantum advantage* refers to a change in this situation marked by the demonstration of a quantum computer solving a problem that no classical computer can solve with reasonable amounts of time and memory (this is sometimes called *quantum supremacy*).

There have been claims of particular systems reaching quantum supremacy [9]. However, they are often met with caution, because it's hard to know if there are more efficient classical solutions waiting to be discovered.

In fact, quantum advantage has become something of a cat-and-mouse game between quantum and classical researchers. When the former claim to have demonstrated a quantum computation that beats all classical approaches, the latter often develop new classical solutions that run not just faster than previous classical algorithms, but faster than the quantum demonstration [204]. Then the quantum researchers improve their hardware and software to beat the classical versions (perhaps on other problems), and the classical hardware and software are improved to beat the quantum solutions, and so it goes.

Like many people, I believe that eventually quantum computers will someday achieve useful quantum advantage. That is, they will produce results for a variety of practical problems more quickly than any classical computer could hope to match, perhaps even theoretically. Nobody knows when that day will come.

But it might not be too long, because quantum computing technology is developing rapidly. One popular measure of the power of a quantum computer is how many qubits it contains. The largest publicly announced quantum computer as of early 2025 is the IBM Condor, which offers 1,121 qubits [69]. Regardless of the particular hardware, superpositions are essential to quantum computing. But superpositions are delicate and notoriously difficult to maintain, with lifetimes that are often measured in tiny fractions of a second. Recently, a group of researchers used a collection of ytterbium isotopes to build and maintain an isolated superposition state for a whopping 23 minutes [155] [291]. The representations of qubits themselves are also improving, with new and advancing technologies. Surprisingly, an entirely mechanical qubit has even been developed [290]. New methods for representing qubits are still being discovered and developed [295].

Further Reading

There are lots of places to go to learn more about everything I've talked about in this book. In addition to the references included in the text, here are a few pointers to get you started.

Quantum Mechanics Books

Quantum computing is built on the theory of quantum mechanics. There are many fine introductory books on this subject, including a few that I particularly like for non-physicists who want an introduction to the field [213] [81] [188]. There are so many books on this topic that it's worth taking the time to look at a few to find the best fit for you.

Quantum Computing Books

The king of quantum computing textbooks is *Quantum Computation and Quantum Information*, by Michael Neilsen and Isaac Chuang [146]. It has all the benefits and drawbacks of being a comprehensive, detailed, university-level textbook. A more approachable book is *Quantum Computer Science: An Introduction*, by N. David Mermin [137]. While authoritative and clear, it's also idiosyncratic in its notation and style of figures.

There are many other books on quantum computing that take a variety of different approaches, from casual to rigorous. An online search (or better yet, a visit to your local technical bookstore) is the best way to browse the books that are available and find one that suits your experience and interests.

Quantum Computing Lecture Notes

Some professors have put complete course notes for their university-level courses online, and they can be a treasure trove of readable, accessible information. I referred to three sets of lecture notes while writing this book [71] [198] [293].

Quantum Computing Online

There are many websites, videos, blogs, and other online media related to quantum computing. Their styles, content, and accuracy vary considerably. And while new resources pop up frequently, I find that they also often disappear without warning. If you want to pursue these online resources, spend some time with your favorite search engine and sample the options that are available to you now.

The most important thing to attend to, in my experience, is that the material is explained clearly and completely, at a technical level that matches your current expertise. If something is too hard for you now, bookmark it and come back to it later (if it's still there!). There's no need to get frustrated trying to comprehend material that is opaque to you when there are so many other options out there.

One site that I've found that has been stable and offers good information is IBM's Qiskit project [96].

Quantum Computing Resources

There are some terrific resources for quantum computing available online. These include simulators, real hardware, and tools for drawing circuits.

Simulators

You can write and run quantum algorithms today on a wide variety of simulators, some of which can even run on your home computer. Big lists of simulators for many languages and platforms can be found online [166] [44].

As of 2025, well-known environments for building and simulating quantum circuits include Qiskit [96] and the IBM Quantum Platform [95] (the latter of which offers a high-quality simulator capable of up to 32 qubits).

Time on bigger simulators can be purchased from a variety of online vendors. A web search will return your current options and what they cost.

Other Software

There are several free libraries that let you create and manipulate quantum objects, operators, and programs in any way you choose. Be cautious, because while these systems are designed for flexibility and accuracy, you're the one in control, and most of them will let you casually break the laws of physics if you're not careful!

As of 2025, just a few of the popular quantum libraries for the Python language that appear to be under active development include Qiskit [164], Cirq [78], QCpy [65], and QuTiP [169] [103] [102].

A terrific hub for quantum computing software and information is the Quantum Open Source Foundation (QOSF) [168]. Among its projects, the QOSF maintains an extensive and diverse list of open source projects in quantum computing for many different programming languages and operating systems [167]. Browse for any project that might speak to you or could help you in your work. If you find some software you love on that list, it's probably written and maintained by volunteers, so consider chipping in and helping out!

Real Hardware

Though this isn't a book on the hardware of quantum computers, it can be interesting to see how different organizations are building these machines [249] [160].

Challenges remain on many fronts. For example, classical computers are dependent on random-access memory, but it's not clear when (or even if) we'll have quantum versions of this ubiquitous storage mechanism [101].

There are companies that own real, working quantum computers and will sell you time on their systems. IBM's and Amazon's services have been around for a while, and as of 2025 they both offer their customers 10 minutes of free quantum computation per month, usually on machines with single-digit numbers of qubits [95] [6]. Other offerings seem to come and go, and prices and plans change frequently. If you need more computing time or power than the free tiers offer, I urge you to do an online search and find a provider who offers services and prices that meet your needs.

There's nothing like running your code on a real, live quantum computer to be sure that you know what you're doing!

Drawing Circuits

A great way to share your quantum circuits is to draw them.

There are several packages that are designed to make good-looking circuits in LaTeX, such as quantikz [110], qcircuit [185], and yquant [46].

All of these are built on LaTeX, so they're as brittle and unforgiving as LaTeX itself and inherit that system's problematic error messaging. In this book (which I wrote in LaTeX), I used quantikz version 2 to make all of the circuit diagrams, because I like its simple syntax and have found that I can usually (though not always) coerce it into drawing what I want. I build up my circuits one small step at a time, recompiling the LaTeX after each change, so I can isolate and fix errors as they happen.

Getting Help

Everyone gets lost or stuck sometimes. Maybe some idea just doesn't click, or two things don't go together properly, or an algorithm doesn't produce the answers you expect. You may find yourself wanting help. If you don't personally know an expert in quantum computing, being lost can feel lonely, stressful, and difficult.

At these times, it can seem appealing to talk to AI systems such as digital assistants and chatbots. They're friendly and responsive, are available at any time of day or night, and can provide comprehensible answers.

But I urge caution. Compared to many other topics, there is not a lot of data for these systems to train on, or even consult in real time. Making things worse, as of 2025, most AI systems are poor at the kind of math that we use in quantum computing. As a result of these limitations, they make a lot of mistakes, ranging from major conceptual blunders regarding basic principles to subtle missteps in carrying out the mechanics.

In other words, when we ask most of today's AI systems about quantum computing, they frequently get things a little wrong, or even wildly wrong. Or they speak nonsense. And they do it all with total confidence!

I've tried it myself, and I've found that asking these systems for help with quantum computing is often a frustrating and fruitless experience. Even when you know they're wrong, and tell them, they cheerfully agree (and might even apologize), and then typically give you a different answer that's wrong in a new way.

So what can you do if you get stuck? Throughout this book there are references relevant to the topics I've discussed. If you have questions those references don't answer, or the answers are unclear to you and a search for better answers doesn't help, there are online boards where you can ask questions and often get back expert responses.

In particular, I've found that Stack Exchange [207] is a great place to start. Among its many specialized boards, you can find experts on mathematics [208], physics [209], and quantum computing [210]. It's free to sign up, search answers to previous questions, and post your own questions. My experience is that if you respectfully and clearly ask one specific question per post, and explain what you've already done to answer it for yourself, there's a good chance that you'll get back a helpful response. A little friendliness and humility go a long way.

The Philosophy of Quantum Mechanics

> *The universe is not only queerer than we suppose, but queerer than we can suppose.*
>
> —J.B.S. Haldane, *Possible Worlds and Other Essays*, 1927 [87]

The philosophy of quantum mechanics, and by extension quantum computing, is fascinating. What is a superposition, really? How does entanglement actually work? Who can make a measurement? How does measurement cause a superposition to collapse? Exactly how does quantum parallelism work? Arguments over answers to these and many other questions have been raging for over a century, with no sign of stopping or even slowing down.

The different ways of trying to understand what quantum mechanics "means" are called *interpretations* of quantum mechanics.

Not everyone thinks that looking for an interpretation of quantum mechanics is a worthwhile pursuit. There is a contingent of physicists who believe that our mathematical theory has been shown to make correct predictions about the results of experiments, and that's the end of the story. The extreme version of this view is that any attempt to find a deeper meaning, or understanding, of the math is pointless. The math works, and that's not just all that we know, it's all that we *can* know, ever. This is usually called the *Copenhagen interpretation* [59]. Some adherents of this approach urge us to therefore ignore what quantum mechanics might mean, or why the quantum world behaves as it does, and instead just "shut up and calculate" [136].

We can follow this line of thought further and conclude that mathematics doesn't just provide an accurate description of nature, or even a model of nature, but is the essential stuff of nature itself. This can lead us to the conclusion that "our universe is not just described by mathematics—it *is* mathematics" (emphasis added) [218].

An attractive quality of this philosophy is that it easily wins the Occam's razor argument: The simplest explanation for why mathematics has been so successful at describing the universe is because mathematics is the universe.

I can see the appeal of this argument. There is much beauty in mathematics, from the elegant precision of a graceful equation to the sophisticated relationships among abstract concepts. Sometimes I feel that's enough, or even far more than enough, to make a universe as deep and meaningful as I could ever want.

But I doubt that the universe actually is mathematics. Even getting past our arbitrary symbols and syntax, I feel that there's more to existence than patterns wheeling and forming and breaking. There's a gap between the soaring abstractions of invisible patterns and the graspable magic of the everyday, real world that I feel that I inhabit. I admit that feeling could be an illusion, and my perception of myself could be merely an accidental side effect created by the flow and turbulence of a mathematical universe. But my sense of conscious awareness is so strong, and so appealing, that I'll part with it only if convincing evidence forces me to. So far, I haven't seen such evidence.

Thus, I like to think of the math as a description of the universe, a way to see its footprints or hear the echoes of its calls, perhaps to other universes. The math works because we have, seemingly with incredible luck, found a way to use it to track some aspects of the reality that unfolds around us. This idea fits nicely with Plato's allegory of the cave, which suggests that we perceive only the shadows of reality [205]. In this story, the shadows tell us much of reality, but there is so much that we are unable to perceive, or perhaps even comprehend, that we will be forever unaware of the universe's true nature.

I find that thinking about interpretations of quantum mechanics is great fun. It's letting your fantasy and imagination run wild, building on a foundation of experimental evidence.

Personally, my favorite interpretations involve time travel. There's something so cool about a causally bidirectional universe, the future and the past cocreating everything dynamically. Sometimes we refer to a model where current events can alter what's happened in the past as *retrocausal*. One way to use retrocausality is to imagine that every state in a superposition has its own corresponding reality, and an observation tomorrow changes yesterday to bring about the universe we experience today.

One good way to get started with interpretations is to read ideas from the physicists who discovered (or invented) quantum mechanics. Some were interested in interpretations and discussed them lucidly. This is a great way to see how to play the game, and then you can branch out into the theories that came later [55] [215] [230] [189].

The *many-worlds* interpretation of quantum mechanics has recently become part of popular culture. This interpretation says that when someone makes a quantum observation, they split into multiple copies of themselves, one copy for each possible outcome of the measurement. Each of these copies of the observer lives in their own newly created universe, which is identical to all the others except for that single measurement. These new universes are now independent, and each evolves on its own from that instant. This splitting happens every time a quantum observation is made, creating an infinite branching tree of independent observers and universes. So each time you look at something, and thereby observe a photon, you branch into multiple copies of yourself, each in their own reality.

What a breathtaking idea! The original paper presenting this idea is an exciting read that you're now well equipped to understand and enjoy [58].

An extension of this idea, called the *multiverse*, combines the many-worlds hypothesis with ideas from computation, evolution, and the philosophy of science. The result is a wild conception of many different, independent universes existing simultaneously [48]. As with so many of the wild ideas in quantum mechanics, some people think that the multiverse is real, and others don't [29].

From these starting points, you can find nearly limitless variations and new interpretations, and you can even make up your own. Go nuts, but don't get carried away into pseudoscience! The explanations that are the most fun are the ones that are solidly rooted in science, anyway; after all, the whole incentive is to forge a closer relationship with the amazing universe we inhabit.

Be cautious with videos, blog posts, and other media that are not based on mathematics. As I discussed in the Introduction, metaphors are inherently incomplete and inaccurate, so reasoning from them can lead to drawing seemingly reasonable conclusions that are more fantasy than true reflections of the world we inhabit. The only description of quantum phenomena that has proven itself to be reliably accurate and predictive is mathematical. Interpretations that are not based on the math are built on quicksand and risk telling us more about the person doing the interpretation than the universe we share.

Inventing new interpretations of quantum phenomena is a rewarding activity that feels endless, partly because it naturally leads us to all the other great unsolved questions: What is reality? What is consciousness? What does it mean to lead a good life? Who am I? Why are we here?

Applications

Let's look at a few uses of quantum computing for science and fun.

Maze Solving

Hedge mazes (also called corn mazes) are great fun. They're usually set in a big field, with tall hedges (or corn plants) planted to form a person-sized maze. You walk into the entrance and are immediately facing a corridor lined with opaque hedges on both sides. Soon the corridor splits, and you need to pick a new direction. The goal is to find the exit and emerge from the maze. Some of these mazes even offer bridges, tunnels, and other variations to make the solving process more interesting and fun [163].

There are lots of classical strategies for solving a maze. One of the simplest, and most famous, is the *right-hand rule* [107]. Just put your right hand on the wall to your right as you enter the maze, and never lose contact. You'll eventually find the exit (it's not a fun way for people to solve a maze, but it's effective if you're a computer).

Can a quantum computer solve a maze? One way to approach this problem is to ignore the geometry, or layout, of the maze, and instead think of it as a bunch of straight corridors joined by intersections. We can draw this abstractly as a *graph* of lines and vertices, mark two vertices as the "entrance" and "exit," and see if a quantum computer can get from the entrance to the exit. The answer is yes, it can, but surprisingly, it's kind of useless if you're out there walking around in the maze. The problem is that while a quantum algorithm can find the exit, it cannot tell you what path it took to get there [36]! It seems that the algorithms we have now must "forget" the path in the course of finding the exit.

Quantum computing is full of these kinds of surprising results.

Ray Tracing

An appealing application of quantum computing involves the computer graphics algorithm called *ray tracing* [10]. The idea is to create a synthetic image, perhaps for a game, TV show, or film, using the principles of classical optics. In this approach, we follow light rays through a mathematical version of a 3D environment, using the color of the light they carry to build up a description of what we would see if we were actually viewing that scene.

Because light rays in this approach don't interact with one another, the computations that work out the colors of the light rays can be carried out independently for each ray. We sometimes say that ray tracing is *embarrassingly parallel*.

Since modern images and animated sequences can require many millions (or even billions) of these independent rays to be traced, and quantum computers can evaluate information about vast numbers of rays simultaneously, using quantum computers to accelerate ray tracing sounds like a great match!

You can read a variety of approaches to this problem in the graphics literature [157] [74] [75] [76] [104] [5] [128] [183] [296].

Suppose that the input to one of these quantum algorithms is a superposition of all the rays we want to trace for a scene, and our circuit can find the intersections of rays and objects. The allure is that if we could intersect every ray with every object in parallel, then we could process this list of intersections to find the first object hit by each ray. But we're stymied by the same problem that everyone runs into when trying to use quantum parallelism to accelerate conventional algorithms: We can measure only a single outcome at a time. That is, even if the quantum circuit can intersect 10 or 100 million rays with a million objects *simultaneously*, the information in the resulting superposition exists only inside the computer. Any measurement at the end will return a single ray–object intersection pair. Quantum ray tracing remains appealing because of the hope that somehow we'll figure out how to extract something more useful from that huge bundle of information, which is tantalizingly close but just out of reach.

Games

We can learn a lot from games! Games can offer highly structured environments that reward clever insight and problem solving, and they frequently provide clear metrics for evaluating performance and quantifying improvements in skill.

You might enjoy getting into a quantum algorithm for the board game *Battleship* [94]. If you'd rather play a quantum version of *Bulls and Cows* (sold commercially as *Mastermind*, and more recently reinvented in word form as *Wordle*), you can play that as well [124]. One of the first video games ever, *Pong*, is still fun, and you can download source code for its quantum version, *QPong* [228].

A couple of nice roundups of some basic and more sophisticated games that can be solved by, or played on, quantum computers offer a nice way to get into the subject [56] [127]. There's also a fun personal essay describing one person's work on quantum games, including another version of *Battleship* and the wonderfully named *Quantum Awesomeness* [288].

Other Applications

The majority of published quantum algorithms have been focused on mathematical and scientific topics. A great list called the *Quantum Algorithm Zoo* contains descriptions and pointers to many of these algorithms [106]. The diverse topics currently listed include cryptography, mathematics, physics, and AI. By clicking on the links provided in the descriptions, you can read the original papers and learn all the details.

There has also been work on finding quantum algorithms that are versatile enough to solve a range of real-world problems [154] [13].

It's fascinating and instructive to read through the descriptions of quantum algorithms and see the variety of clever and imaginative ways that people have applied quantum computing to diverse tasks.

Wrapping Up

This brings us to the end of our journey.

This book has been a multiyear labor of love. I poured endless hours and energy into creating this book because I find this topic absolutely fascinating and beautiful, and I wanted to help you see that beauty as I do.

We're only at the beginning of what I'm sure will turn out to be an amazing adventure. I look forward to learning about the incredible things you'll do with quantum computing!

APPENDIX: NOTATION

Nothing is more easy than the invention of notation, and nothing of worse example and consequence than the confusion of mathematical expressions by unknown symbols.
—Augustus De Morgan, *A Treatise on the Calculus of Functions*, 1836 [45]

Oh no. No. No. I meant no scheme. I merely posed a little academic accounting theory. It was just a thought.
—Mel Brooks, *The Producers*, 1967 [28]

Table A-1: Constants

e	Euler's number, $2.7182\ldots$
i	Imaginary number, defined as $\sqrt{-1}$
π	The ratio of a circle's circumference to its radius, $3.1415\ldots$
$\vee$	$1/\sqrt{2} = \sqrt{1/2} = \sqrt{2}/2$

Table A-2: Sets

$\mathbb{B}$	The set of bits (0 and 1)
$\mathbb{C}$	The set of complex numbers
$\mathbb{N}$	The set of natural (or counting) numbers
$\mathbb{R}$	The set of real numbers
$\mathbb{Z}$	The set of integer numbers
$\mathbb{R} \times \mathbb{R}$	The set of all pairs of real numbers
$\mathbb{R}^n$	The set of all tuples of n real numbers
$\{a, b, c \ldots\}$	A set of elements
$a \in \mathbb{X}$	a belongs to set $\mathbb{X}$
$\mathbb{B}^n$	A bitstring of n elements

Table A-3: Sequences and ranges

$[a, b]$	The integers from a up to and including b
(a, b)	The integers from $a + 1$ up to and including $b - 1$
$[d]$	The sequence of integers from 0 up to and including $d - 1$

Table A-4: Common symbols

$\oplus$	Exclusive OR
$\otimes$	Tensor product
$A^{\otimes n}$	An operator A tensored with itself n times
$\lvert\psi\rangle^{\otimes n}$	A state $\lvert\psi\rangle$ tensored with itself n times
0^n	A bitstring of n bits, all 0s
1^n	A bitstring of n bits, all 1s
$a \cdot b$	For bitstrings a and b, $a_0 b_0 \oplus a_1 b_1 \oplus \ldots \oplus a_{n-1} b_{n-1}$
$\delta_{j,k}$	Kronecker delta for integers j, k; returns 1 if $j = k$, else 0

Table A-5: Common variables

d	The dimensionality (or number of elements) of a vector
n	Often the number of qubits in a circuit
N	2^n for a given n
k	A generic integer
θ	When discussing geometry, a generic angle
$\lvert\psi\rangle, \lvert\phi\rangle$	Generic kets
$\langle\psi\rvert, \langle\phi\rvert$	Generic bras
ω	$e^{i2\pi/N}$
λ	$e^{i2\pi}$

Table A-6: Common variable types by typography

b	Roman lowercase italic b: a bit or bitstring
r	Roman lowercase italic: a real or integer
A	Roman uppercase italic: a list, set, operator, or matrix
α	Greek lowercase: a complex number
$\mathbf{v}$	Roman lowercase bold: a vector

Table A-7: Numbering and counting

1	First number when counting
0	First number when numbering
Top down	Order for numbering elements in a column vector
Left to right	Order for numbering elements in a row vector
Top down	Order for tensoring operators in a circuit diagram
Top down	Order for tensoring states in a circuit diagram

Table A-8: XOR properties

$x \oplus x = 0$	X1
$x \oplus 0 = x$	X2
$(x \oplus y) \oplus z = x \oplus (y \oplus z)$	X3

Table A-9: Complex numbers

$\overline{\alpha}$	The complex conjugate of α
α^*	An alternative form of $\overline{\alpha}$
$a + bi$	The Cartesian form of a complex number
$re^{i\theta}$	The polar form of a complex number

Table A-10: Vectors, quantum states, and operations

$\mathbf{a} \cdot \mathbf{b}$	The dot product of $\mathbf{a}$ and $\mathbf{b}$
$\langle \mathbf{a}, \mathbf{b} \rangle$	The inner product of $\mathbf{a}$ and $\mathbf{b}$
$|\psi\rangle$	A ket (a complex column matrix)
$\langle\psi|$	A bra (a complex row matrix)
$\langle\phi|\psi\rangle$	A braket (or inner product) of $|\phi\rangle$ and $|\psi\rangle$
$|\phi\rangle\langle\psi|$	A ketbra (or outer product) of $|\phi\rangle$ and $|\psi\rangle$
$\langle onto, from \rangle$	Projection using the inner product

Table A-11: Properties of vectors

$|\mathbf{v}|$	Magnitude of a vector
$\dim(\mathbf{v})$	Dimensionality of a vector

Table A-12: Vector and linearity properties

$\sigma\mathbf{v} = \sigma \begin{bmatrix} \nu_0 \\ \nu_1 \\ \nu_2 \end{bmatrix} = \begin{bmatrix} \sigma\nu_0 \\ \sigma\nu_1 \\ \sigma\nu_2 \end{bmatrix}$	V1
$\mathbf{v} + \mathbf{w} = \begin{bmatrix} \nu_0 \\ \nu_1 \\ \nu_2 \end{bmatrix} + \begin{bmatrix} \omega_0 \\ \omega_1 \\ \omega_2 \end{bmatrix} = \begin{bmatrix} \nu_0 + \omega_0 \\ \nu_1 + \omega_1 \\ \nu_2 + \omega_2 \end{bmatrix}$	V2
$f(\sigma\mathbf{v}) = \sigma f(\mathbf{v}), \quad \sigma \in \mathbb{C}$	L1
$f(\mathbf{v} + \mathbf{w}) = f(\mathbf{v}) + f(\mathbf{w})$	L2

Table A-13: Matrix operations

$\overline{A}$	The complex conjugate of matrix A
A^{T}	The transpose of matrix A
$A^{\dagger}$	The adjoint of matrix A
A^{-1}	The inverse of matrix A
AB	Ordinary matrix multiplication of matrices A and B

Table A-14: Tensor product properties

$\overline{A \otimes B} = \overline{A} \otimes \overline{B}$	Conjugate distributivity
$(A \otimes B)^{-1} = A^{-1} \otimes B^{-1}$	Tensor inverse property
$(A \otimes B)^T = A^T \otimes B^T$	Tensor transpose property
$(A \otimes B)^\dagger = A^\dagger \otimes B^\dagger$	Tensor adjoint property
$A \otimes (B + C) = (A \otimes B) + (A \otimes C)$	Tensor product is distributive
$A \otimes (B \otimes C) = (A \otimes B) \otimes C$	Scalar tensor property
$c(A \otimes B) = (cA) \otimes B$ $= A \otimes (cB)$	For any $c \in \mathbb{R}$
$(A \otimes B)(C \otimes D) = AC \otimes BD$	Mixed-product property

Table A-15: Bitstring operations

$a \cdot b = \left(\sum_{k \in [d]} a_n \oplus b_n \right) \bmod 2$	Bitstring dot product
$a \oplus b = a_{n-1} \oplus b_{n-1}, \ldots, a_0 \oplus b_0$	Bitstring XOR

Table A-16: Common states

$$|0\rangle = \begin{bmatrix} 1 \\ 0 \end{bmatrix}, \quad |1\rangle = \begin{bmatrix} 0 \\ 1 \end{bmatrix}, \quad |+\rangle = \vee \begin{bmatrix} 1 \\ 1 \end{bmatrix}, \quad |-\rangle = \vee \begin{bmatrix} 1 \\ -1 \end{bmatrix}$$

Table A-17: Integer bras and kets

$|k\rangle$	For integer k, a ket of all 0 elements except a 1 at index k
$\langle k|$	A bra of all 0 elements except a 1 at index k, or $|k\rangle^\dagger$

Table A-18: Superpositions

$\alpha |\phi\rangle + \beta |\psi\rangle$	Superposition of 2 states, $|\alpha|^2 + |\beta|^2 = 1$
$\sum_{k \in [n]} \alpha_k |\phi_k\rangle$	Superposition of n states, $\sum_{k \in [n]} |\alpha_k|^2 = 1$
$\vee^n \sum_{k \in [n]} |k\rangle$ $= H^{\otimes n} |0\rangle^{\otimes n}$ $= |+\rangle^{\otimes n}$	Uniform superposition of 2^n states

Table A-19: Operations on quantum states

$\big|\,\lvert\psi\rangle\,\big|$	The magnitude of $\lvert\psi\rangle$
$\big|\,\langle\phi\rvert\,\big|$	The magnitude of $\langle\phi\rvert$
$\langle\phi\rvert\,M\,\lvert\psi\rangle$	A bramket of $\lvert\phi\rangle$, M, and $\lvert\psi\rangle$
$\langle r\rvert\,A\,\lvert c\rangle$	Element $a_{r,c}$ for integers r and c
$\langle\phi\rvert\,A\,\lvert\psi\rangle = (\langle\phi\rvert\,A)\,\lvert\psi\rangle$ $= \langle\phi\rvert\,(A\,\lvert\psi\rangle)$	Matrix products are associative
$\lvert\psi\phi\rangle = \lvert\phi\rangle\,\lvert\psi\rangle = \lvert\phi\rangle \otimes \lvert\psi\rangle$	Ways to write a product state

Table A-20: Hadamard operations on common states

$H\,\lvert 0\rangle = \lvert +\rangle$

$H\,\lvert 1\rangle = \lvert -\rangle$

$H\,\lvert +\rangle = \lvert 0\rangle$

$H\,\lvert -\rangle = \lvert 1\rangle$

Table A-21: Algebraic expressions for $H\,\lvert j\rangle$

$H\,\lvert j\rangle = \vee \sum_{k\in\mathbb{B}} (-1)^{j\cdot k}\,\lvert k\rangle$	For $j \in \mathbb{B}$
$H\,\lvert j\rangle = \vee(\lvert 0\rangle + e^{i2\pi(j/2)}\,\lvert 1\rangle)$	For $j \in \mathbb{B}$
$H\,\lvert j\rangle = \vee(\lvert 0\rangle + \lambda^{j/2}\,\lvert 1\rangle))$	For $j \in \mathbb{B}$

Table A-22: Bell states

$\lvert\beta_s^{+}\rangle = \lvert\Psi^{+}\rangle = \lvert\beta(0,0)\rangle = \vee\left(\lvert 00\rangle + \lvert 11\rangle\right)$

$\lvert\beta_m^{+}\rangle = \lvert\Phi^{+}\rangle = \lvert\beta(0,1)\rangle = \vee\left(\lvert 01\rangle + \lvert 10\rangle\right)$

$\lvert\beta_s^{-}\rangle = \lvert\Psi^{-}\rangle = \lvert\beta(1,0)\rangle = \vee\left(\lvert 00\rangle - \lvert 11\rangle\right)$

$\lvert\beta_m^{-}\rangle = \lvert\Phi^{-}\rangle = \lvert\beta(1,1)\rangle = \vee\left(\lvert 01\rangle - \lvert 10\rangle\right)$

Table A-23: Measurement operators

Π_k	Projection matrix $\lvert k\rangle\langle k\rvert$
Π_0	Projection matrix $\lvert 0\rangle\langle 0\rvert$
Π_1	Projection matrix $\lvert 1\rangle\langle 1\rvert$
$\Pr_{\lvert\psi\rangle}(k)$	Probability that measuring $\lvert\psi\rangle$ will return k

Table A-24: Eigenthings

ε_k	Eigenvector k
λ_k	Eigenvalue k

Table A-25: Common qugate matrices in the computational basis

I	The identity of any size; for example, $I_2 = \begin{bmatrix} 1 & 0 \\ 0 & 1 \end{bmatrix}$	
$CX, CNOT$	Controlled-X or controlled-NOT,	$\begin{bmatrix} 1 & 0 & 0 & 0 \\ 0 & 1 & 0 & 0 \\ 0 & 0 & 0 & 1 \\ 0 & 0 & 1 & 0 \end{bmatrix}$
CX'	CX with control below target,	$\begin{bmatrix} 1 & 0 & 0 & 0 \\ 0 & 0 & 0 & 1 \\ 0 & 0 & 1 & 0 \\ 0 & 1 & 0 & 0 \end{bmatrix}$
H	Hadamard, $\vee \begin{bmatrix} 1 & 1 \\ 1 & -1 \end{bmatrix}$	
P	Phase, $\begin{bmatrix} 0 & 1 \\ 1 & e^{i\theta} \end{bmatrix}$	
S	$\pi/2$ phase, $\begin{bmatrix} 0 & 1 \\ 1 & e^{i\pi/2} \end{bmatrix}$	
T	$\pi/4$ phase, $\begin{bmatrix} 0 & 1 \\ 1 & e^{i\pi/4} \end{bmatrix}$	
$SWAP$	Swap, $\begin{bmatrix} 1 & 0 & 0 & 0 \\ 0 & 0 & 1 & 0 \\ 0 & 1 & 0 & 0 \\ 0 & 0 & 0 & 1 \end{bmatrix}$	
U	Generic unitary, $\begin{bmatrix} \cos\left(\frac{\theta}{2}\right) & -e^{i\lambda}\sin\left(\frac{\theta}{2}\right) \\ e^{i\phi}\sin\left(\frac{\theta}{2}\right) & e^{i(\phi+\lambda)}\cos\left(\frac{\theta}{2}\right) \end{bmatrix}$	

Table A-26: Pauli matrices

X, NOT	Pauli X, also σ_x, $\begin{bmatrix} 0 & 1 \\ 1 & 0 \end{bmatrix}$
Y	Pauli Y, also σ_y, $\begin{bmatrix} 0 & -i \\ i & 0 \end{bmatrix}$
Z	Pauli Z, also σ_z, $\begin{bmatrix} 1 & 0 \\ 0 & -1 \end{bmatrix}$

Table A-27: Common geometrical matrices

Rot(θ)	Rotation by θ counterclockwise, $\begin{bmatrix} \cos\theta & -\sin\theta \\ \sin\theta & \cos\theta \end{bmatrix}$
Ref(θ)	Reflection by line θ counterclockwise from X axis, $\begin{bmatrix} \cos 2\theta & \sin 2\theta \\ \sin 2\theta & -\cos 2\theta \end{bmatrix}$

Table A-28: Circuit symbols: general

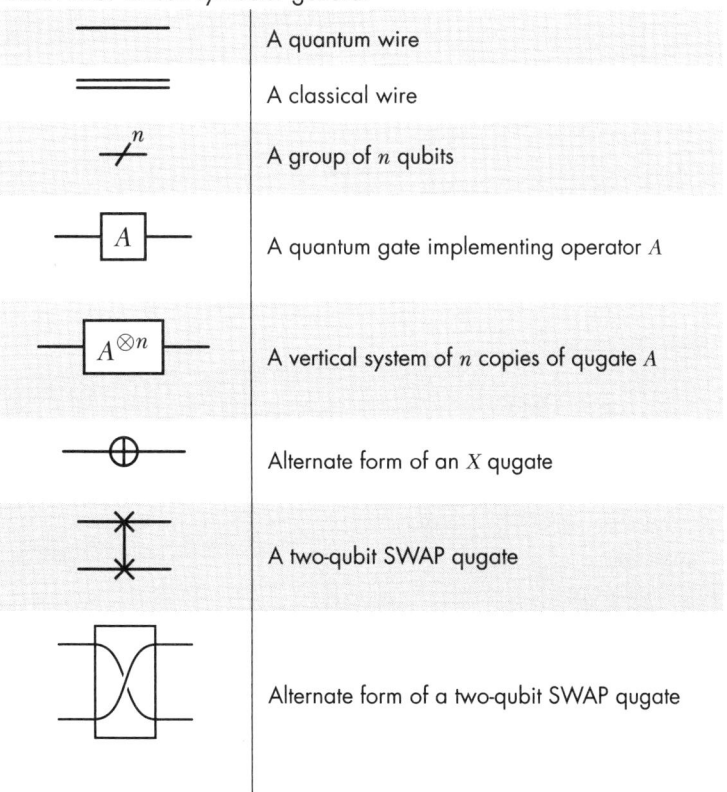

	A quantum wire
	A classical wire
n	A group of n qubits
A	A quantum gate implementing operator A
$A^{\otimes n}$	A vertical system of n copies of qugate A
$\oplus$	Alternate form of an X qugate
	A two-qubit SWAP qugate
	Alternate form of a two-qubit SWAP qugate

Table A-29: Circuit symbols: measurement and entanglement

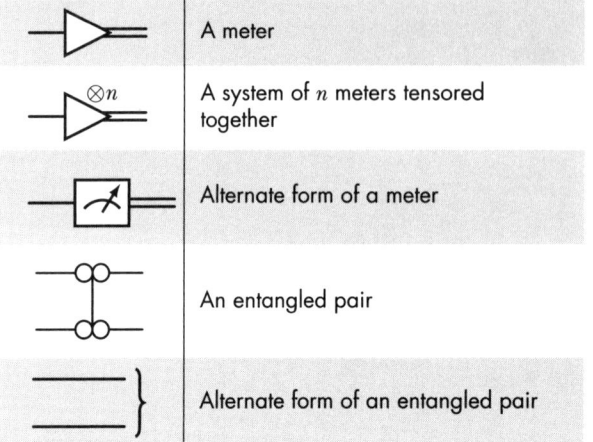	A meter
	A system of n meters tensored together
	Alternate form of a meter
	An entangled pair
	Alternate form of an entangled pair

Table A-30: Circuit symbols: controlled qugates

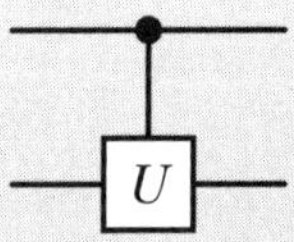	A controlled-U qugate enabled only if the control is $|1\rangle$
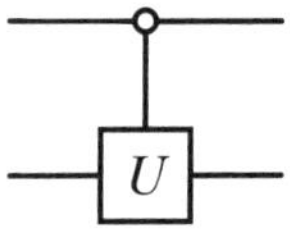	A controlled-U qugate enabled only if the control is $|0\rangle$
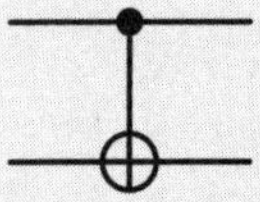	Common shortcut for a CX qugate
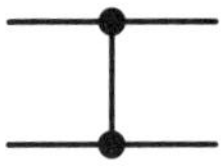	Common shortcut for a CZ qugate
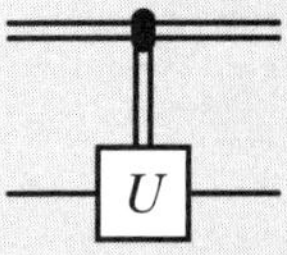	A U qugate controlled by a classical bit
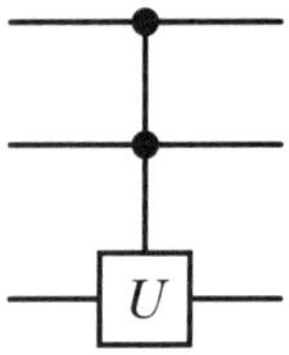	A controlled-U qugate enabled only if both controls are $|1\rangle$

BIBLIOGRAPHY

I dabbled in bird-watching once. Well, by dabbled, I mean I bought a bird guide and put it up on the shelf and looked up at it sometimes.

—Scott Jacobson, *Bob's Burgers*, "Crows Encounters of the Bird Kind," 2023 [100]

Whenever possible, I've tried to use free and accessible online references. I cited books and other traditional references when they are the superior choice or offer a unique perspective.

[1] Ackerman, Jennifer, and Eleanor Grant. "Born to Move." *https://www.imdb.com/title/tt1776626*, 2010.

[2] Adams, Douglas. *Dirk Gently's Holistic Detective Agency*. Pocket Books, 1987.

[3] Adams, Douglas. *The Hitchhiker's Guide to the Galaxy*. Pan Books, 1978.

[4] Agrawal, Pankaj, and Arun K. Pati. "Probabilistic Quantum Teleportation." *https://arxiv.org/pdf/quant-ph/0210004*, 2002.

[5] Alves, Carolina, Luís Paulo Santos, and Thomas Bashford-Rogers. "A Quantum Algorithm for Ray Casting Using an Orthographic Camera." In *2019 International Conference on Graphics and Interaction*, 2019. *https://repositorium.sdum.uminho.pt/bitstream/1822/71756/3/Quantum_Algorithm.pdf*.

[6] Amazon Web Services. "Amazon Braket." *https://aws.amazon.com/braket/*, 2025.

[7] American Kennel Club. "Breed Weight Chart." *https://www.akc.org/expert-advice/nutrition/breed-weight-chart/*, 2023.

[8] Answers.com. "How Much Does a Ladybug Weigh?" *https://www.answers.com/Q/How_much_does_a_ladybug_weigh*, 2023.

[9] Arute, Frank, Kunal Arya, Ryan Babbush, et al. "Quantum Supremacy Using a Programmable Superconducting Processor." *Nature* 574 (2019): 505–10. *https://www.nature.com/articles/s41586-019-1666-5*.

[10] Arvo, James, Robert L. Cook, Andrew Glassner, Eric Haines, Pat Hanrahan, Paul S. Heckbert, and David Kirk. *An Introduction to Ray Tracing*. Morgan Kaufmann, 1989.

[11] Aumasson, Jean-Philippe. *Serious Cryptography: A Practical Introduction to Modern Encryption*. No Starch Press, 2017.

[12] Azad, Kalid. "Understanding the Birthday Paradox." *Better Explained* blog, 1997. *https://betterexplained.com/articles/understanding-the-birthday -paradox/*.

[13] Babbush, Ryan, Dominic W. Berry, Robin Kothari, Rolando D. Somma, and Nathan Wiebe. "Exponential Quantum Speedup in Simulating Coupled Classical Oscillators." *Physical Review X* 13 (2023): 041041.

[14] Bacon, Dave. "CSE 599d—Quantum Computing: Simon's Algorithm." University of Washington, 2006. *https://courses.cs.washington.edu/ courses/cse599d/06wi/lecturenotes8.pdf*.

[15] Bartlett, Albert A. "The Exponential Function—Part 1." *The Physics Teacher* 14, no. 7 (1976): 393–401. *https://doi.org/10.1119/1.2339436*.

[16] Bennett, Charles H., Ethan Bernstein, Gilles Brassard, and Umesh Vazirani. "Strengths and Weaknesses of Quantum Computing." *https:// arxiv.org/pdf/quant-ph/9701001*, 1997.

[17] Berger, Cynthia. "True Colors: How Birds See the World." *https:// www.nwf.org/Magazines/National-Wildlife/2012/AugSept/Animals/Bird -Vision*, 2012.

[18] Bernstein, Ethan, and Umesh Vazirani. "Quantum Complexity Theory." *SIAM Journal on Computing* 26, no. 5 (1997): 1411–73. *http://wpage.unina.it/pieroandrea.bonatti/didattica/complexity/slides/ Bernstein-Vazirani.pdf*.

[19] Bertlmann, Reinhold, and Nicolai Friis. "Density Matrices." In *Modern Quantum Theory: From Quantum Mechanics to Entanglement and Quantum Information*, 321–49. Oxford University Press, 2023.

[20] Bethesda Game Studios. "Starfield." *https://www.imdb.com/title/ tt8545540*, 2023.

[21] Beutlich, Robert E. "Radionics and the Little Black Box." *https://altered -states.net/barry/update183/radionics2.htm*.

[22] Bouwmeester, Dik, Jian-Wei Pan, Klaus Mattle, Manfred Eibl, Harald Weinfurter, and Anton Zeilinger. "Experimental Quantum Teleportation." *https://arxiv.org/pdf/1901.11004*, 1997.

[23] Boyer, Michel, Gilles Brassard, Peter Høyer, and Alain Tapp. "Tight Bounds on Quantum Searching." *Fortschritte der Physik* 46, no. 4–5 (1998): 187–99. *https://www.researchgate.net/publication/227992557 _Tight_Bounds_on_Quantum_Searching*.

[24] Bradley, Tai-Danae. "The Tensor Product, Demystified." *https://www .math3ma.com/blog/the-tensor-product-demystified*, 2018.

[25] Brahic, Catherine. "Birds Can 'See' the Earth's Magnetic Field." *New Scientist*, 2008. *https://www.newscientist.com/article/dn13811-birds-can-see -the-earths-magnetic-field/*.

[26] Brassard, Gilles, and Peter Høyer. "An Exact Quantum Polynomial-Time Algorithm for Simon's Problem." *https://arxiv.org/pdf/quant-ph/9704027.pdf*, 1997.

[27] Brassard, Gilles, Peter Høyer, Michele Mosca, and Alain Tapp. "Quantum Amplitude Amplification and Estimation." In *AMS Contemporary Mathematics Series, vol. 305*. American Mathematical Society, 2002. *https://arxiv.org/pdf/quant-ph/0005055*.

[28] Brooks, Mel. *The Producers*, 1967. *https://www.imdb.com/title/tt0063462*.

[29] Brooks, Michael. "How a Quantum Innovation May Quash the Idea of the Multiverse." *New Scientist*, 2025. *https://www.newscientist.com/article/mg26435252-200-how-a-quantum-innovation-may-quash-the-idea-of-the-multiverse/*.

[30] Cai, Guangya, and Daowen Qiu. "Optimal Separation in Exact Query Complexities for Simon's Problem." *Journal of Computer and System Sciences* 97 (2018): 83–93. *https://doi.org/10.1016/j.jcss.2018.05.001*.

[31] Cartwright, Jon. "The Quantum World: A Concise Guide to the Particles That Make Reality." *https://www.newscientist.com/article/2367423-the-quantum-world-a-concise-guide-to-the-particles-that-make-reality/*, 2023.

[32] Carver, George Washington. "Quotefancy." Approx. 1934. *https://quotefancy.com/george-washington-carver-quotes*.

[33] Castelvecchi, Davide. "Magnetic Sense Shows Many Animals the Way to Go." *Scientific American*, 2012. *https://www.scientificamerican.com/article/the-compass-within/*.

[34] Cherney, David, Tom Denton, and Andrew Waldron. "7.1: Linear Transformations and Matrices." In *Linear Algebra*, LibreTexts. *https://math.libretexts.org/Bookshelves/Linear_Algebra/Map%3A_Linear_Algebra_(Waldron_Cherney_and_Denton)/07%3A_Matrices/7.01%3A_Linear_Transformations_and_Matrices*.

[35] Chesser, Preston. "Oracle of Delphi." *https://ehistory.osu.edu/articles/oracle-delphi*.

[36] Childs, Andrew M., Matthew Coudron, and Amin Shiraz Gilani. "Quantum Algorithms and the Power of Forgetting." *https://arxiv.org/pdf/2211.12447*, 2022.

[37] Clark, Bryan. "Quantum Fourier Transform." University of Illinois Urbana-Champaign, 2022. *https://courses.physics.illinois.edu/phys498cmp/sp2022/QC/QFT.html*.

[38] Cleve, Richard. "Classical Lower Bound for Simon's Problem." University of Waterloo, 2011. *https://cs.uwaterloo.ca/~cleve/courses/F11CS667/SimonClassicalLB.pdf*.

[39] Cleve, Richard, Artur Ekert, Chiara Macchiavello, and Michele Mosca. "Quantum Algorithms Revisited." *Proceedings of the Royal Society of London A* 454 (1998): 339–54. *https://doi.org/10.1098/rspa.1998.0164*.

[40] Cohen, Danny. "On Holy Wars and a Plea for Peace." *IEEE Computer* 14, no. 10 (1981): 49–54. *https://gwern.net/doc/cs/algorithm/1981-cohen.pdf*.

[41] Conrad, Keith. "Tensor Products." *https://kconrad.math.uconn.edu/blurbs/linmultialg/tensorprod.pdf*.

[42] Cycling '74. "What Is Max?" *https://cycling74.com/products/max*.

[43] Daniels, Greg, Brent Forrester, and Ricky Gervais. "A.A.R.M." (*The Office*, Season 9, Episode 17). *https://www.imdb.com/title/tt2669740*, 2013.

[44] Dargan, James. "Top 63 Quantum Computer Simulators for 2024." *https://thequantuminsider.com/2022/06/14/top-63-quantum-computer-simulators-for-2022/*.

[45] De Morgan, Augustus. *A Treatise on the Calculus of Functions*, 1836.

[46] Desef, Benjamin. "Yquant." *https://github.com/projekter/yquant*.

[47] de Wolf, Ronald. "Quantum Computing: Lecture Notes." QuSoft, CWI and University of Amsterdam, 2023. *https://arxiv.org/pdf/1907.09415*.

[48] Deutsch, David. *The Fabric of Reality*. Viking, 1997.

[49] Deutsch, David. "Quantum Theory, the Church–Turing Principle and the Universal Quantum Computer." *Proceedings of the Royal Society A.* 85, no. 400 (1985): 97–117. *https://doi.org/10.1098/rspa.1985.0070*.

[50] Dirac, Paul. "A New Notation for Quantum Mechanics." In *Mathematical Proceedings of the Cambridge Philosophical Society*, vol. 35, 416–18, 1939. *https://doi.org/10.1017/S0305004100021162*.

[51] Dobrijevic, Daisy. "How Hot Is the Sun?" *https://www.space.com/17137-how-hot-is-the-sun.html*, 2023.

[52] Duan, Lu-Ming, and Guang-Can Guo. "Probabilistic Cloning and Identification of Linearly Independent Quantum States." *https://arxiv.org/pdf/quant-ph/9804064*, 1997.

[53] Duan, Lu-Ming, and Guang-Can Guo. "Two Non-Orthogonal States Can Be Cloned by a Unitary-Reduction Process." *https://arxiv.org/pdf/quant-ph/9704020*, 1987.

[54] Dunin, Elonka. "Famous Unsolved Codes and Ciphers." *https://elonka.com/UnsolvedCodes.html*, 2023.

[55] Einstein, Albert, Boris Podolsky, and Nathan Rosen. "Can Quantum Mechanical Description of Physical Reality Be Considered Complete?" *Physical Review* 47 (1935): 777–80. *https://doi.org/10.1103/PhysRev.47.777*.

[56] Etim, Inemesit. "10 Quantum Games That Can Help You Learn the Field of Quantum Computing." *https://quantumzeitgeist.com/10-quantum-games-that-can-help-you-learn-the-field-of-quantum-computing/*, 2022.

[57] Euler, Leonhard. *Elements of Algebra*. Longman, Rees, Orme, and Co., 1765. 2015 update of 1828 translation by John Hewlett of the

original *Vollständige Anleitung zur Algebra. https://ia800308.us.archive .org/2/items/ElementsOfAlgebraLeonhardEuler2015/ElementsOfAlgebra _LeonhardEuler_Edition2015.pdf.*

[58] Everett III, Hugh. ""Relative State" Formulation of Quantum Mechanics." *Reviews of Modern Physics* 29, no. 3 (1957): 454–62. *https://typeset .io/papers/relative-state-formulation-of-quantum-mechanics-3crdqguh8o.*

[59] Faye, Jan. "Copenhagen Interpretation of Quantum Mechanics." In *Edward N. Zalta (ed.), The Stanford Encyclopedia of Philosophy (Winter 2019 Edition). https://plato.stanford.edu/archives/win2019/entries/qm -copenhagen/.*

[60] Feldman, Joel. "Trig Identities—Cosine Law and Addition Formulae." *https://personal.math.ubc.ca/~feldman/m100/trigId.pdf,* 2000.

[61] Fermilab. "Physics Questions People Ask Fermilab." *https://www.fnal .gov/pub/science/inquiring/questions/atoms.html,* 2014.

[62] Feynman, Richard P. "From a Letter to Ms. J. M. Szabados of Victoria, Australia." *https://en.wikiquote.org/wiki/Richard_Feynman,* 1965.

[63] Fisher, Trevor, Derek Funk, and Rachel Sams. "The Birthday Problem and Generalizations." *https://d31kydh6n6r5j5.cloudfront.net/uploads/ sites/66/2019/04/birthday_comps.pdf,* 2013.

[64] Florio, G., and D. Picca. "Quantum Implementation of Elementary Arithmetic Operations." *https://arxiv.org/pdf/quant-ph/0403048,* 2004.

[65] Freeze, Brennan T., Aundre Barras, Paris Osuch, Soren Sevier Richenberg, and Suzanne Rivoire. "QC.py: Quantum Computing Simulation and Visualization Suite." In *Proceedings of the 54th ACM Technical Symposium on Computer Science Education,* vol. 2, 2023. *https://doi.org/10.1145/3545947.3576334.*

[66] Gabel, Robert A., and Richard A. Roberts. *Signals and Linear Systems* (2nd ed.). John Wiley & Sons, 1980.

[67] Gagliardoni, Tommaso. "Quantum Attack Resource Estimate: Using Shor's Algorithm to Break RSA vs DH/DSA vs ECC." *https://research .kudelskisecurity.com/2021/08/24/quantum-attack-resource-estimate-using -shors-algorithm-to-break-rsa-vs-dh-dsa-vs-ecc/,* 2021.

[68] Galilei, Galileo. *Il Saggiatore.* 1623. Quoted at *https://mathshistory.st -andrews.ac.uk/Biographies/Galileo/quotations/.*

[69] Gambetta, Jay. "The Hardware and Software for the Era of Quantum Utility Is Here." IBM Quantum Research Blog, December 2023. *https:// www.ibm.com/quantum/blog/quantum-roadmap-2033.*

[70] Gauss, Carl Friedrich. *Theoria Residiorum Biquadraticorum, Commentario Secunda.* Dieterich, 1832. Quoted in Robert Edouard Moritz. *Memorabilia Mathematica* (Macmillan, 1914). *https://www.gutenberg.org/ cache/epub/44730/pg44730-images.html.*

[71] Gharibian, Sevag. "Introduction to Quantum Computation." Paderborn University, 2021. *https://groups.uni-paderborn.de/fg-qi/courses/UPB_INTRO_QUANTUM/S2021/notes/IQC_Masterfile.pdf.*

[72] Gide, André. *The Counterfeiters: A Novel.* Knopf, 1927.

[73] Glassner, Andrew. "An Introduction to the Fourier Transform." In *SIGGRAPH Courses '24*, 2024. *https://dl.acm.org/doi/10.1145/3664475.3664537.*

[74] Glassner, Andrew. "Quantum Computing, Part 1." *https://www.glassner.com/wp-content/uploads/2014/04/CG-CGA-PDF-01-07-Quantum-Computing-1-July01.pdf*, 2001.

[75] Glassner, Andrew. "Quantum Computing, Part 2." *https://www.glassner.com/wp-content/uploads/2014/04/CG-CGA-PDF-01-09-Quantum-Computing-2-Sept01.pdf*, 2001.

[76] Glassner, Andrew. "Quantum Computing, Part 3." *https://www.glassner.com/wp-content/uploads/2014/04/CG-CGA-PDF-01-11-Quantum-Computing-3-Nov01.pdf*, 2001.

[77] Gleick, James. *Chaos: Making a New Science.* Viking, 1987.

[78] Google Quantum AI. "Cirq." *https://quantumai.google/cirq.*

[79] Gowers, Timothy. "How to Lose Your Fear of Tensor Products." *https://www.dpmms.cam.ac.uk/~wtg10/tensors3.html.*

[80] Graham, Ian. "What Is the Geological Time Scale?" *https://australian.museum/learn/australia-over-time/evolving-landscape/the-geological-time-scale/*, 2020.

[81] Griffiths, David J., and Darrell F. Schroeter. *Introduction to Quantum Mechanics* (3rd ed.). Cambridge University Press, 2018.

[82] Grover, Lov K. "A Fast Quantum Mechanical Algorithm for Database Search." *https://arxiv.org/pdf/quant-ph/9605043*, 1996.

[83] Grover, Lov K. "Quantum Computers Can Search Arbitrarily Large Databases by a Single Query." *Physical Review Letters* 79 (1997): 4709–12. *https://arxiv.org/pdf/quant-ph/9706005.pdf.*

[84] Guinness World Records. "Heaviest Watermelon." *https://www.guinnessworldrecords.com/world-records/heaviest-watermelon.*

[85] Haber, Howard E. "Coordinates, Matrix Elements and Changes of Basis." University of California, Santa Cruz, 2006. *http://scipp.ucsc.edu/~haber/archives/physics116A06/basis.pdf.*

[86] Haener, Thomas, Mathias Soeken, Martin Roetteler, and Krysta M. Svore. "Quantum Circuits for Floating-point Arithmetic." *https://arxiv.org/pdf/1807.02023*, 2018.

[87] Haldane, J.B.S. *Possible Worlds and Other Essays.* Chatto & Windus, 1927.

[88] Hamilton, Kirk. "'Soul Vaccination' by Tower of Power." *Strong Songs* podcast, 2023. *https://strongsongspodcast.com.*

[89] Harper, David, and L.M. Stockman. "The History of Fortran." *https://www.obliquity.com/computer/fortran/history.html*, 2013.

[90] Hatfield, Gary. "René Descartes." *https://plato.stanford.edu/entries/descartes/*, 2023.

[91] Hoexum, Eline Sophie. "Revisiting the Proof of the Complexity of the Sudoku Puzzle." Master's thesis, Rijksuniversiteit Groningen, June 2020. *https://fse.studenttheses.ub.rug.nl/22745/1/bMATH_2020_HoexumES.pdf.pdf*.

[92] Hughes, John. Personal communication, 2023.

[93] Hunt, Leigh. As quoted in *The Farmer's Wife*, 36, 1933. *https://en.wikiquote.org/wiki/Leigh_Hunt*.

[94] Hunziker, Markus, David A. Meyer, Jihun Park, James Pommersheim, and Mitch Rothstein. "The Geometry of Quantum Learning." *Quantum Information Processing* 9, no. 3 (2003): 321–41. *https://arxiv.org/abs/quant-ph/0309059*.

[95] IBM. "IBM Quantum Platform." *https://quantum.ibm.com*.

[96] IBM. "Qiskit Documentation." *https://www.ibm.com/quantum/qiskit*.

[97] IBM. "Quantum Fourier Transform." *https://github.com/Qiskit/textbook/blob/main/notebooks/ch-algorithms/quantum-fourier-transform.ipynb*, 2022.

[98] IBM. "Single Qubit Gates." *https://qiskit.org/textbook/ch-states/single-qubit-gates.html*, 2023.

[99] IBM Quantum Learning. "Grover's Algorithm." *https://learning.quantum.ibm.com/tutorial/grovers-algorithm*.

[100] Jacobson, Scott. "Crows Encounters of the Bird Kind" (*Bob's Burgers*, Season 13, Episode 17). *https://www.imdb.com/title/tt27003633*, 2023.

[101] Jaques, Samuel, and Arthur G. Rattew. "QRAM: A Survey and Critique." *https://arxiv.org/pdf/2305.10310.pdf*, 2023.

[102] Johansson, J.R., P.D. Nation, and Franco Nori. "QuTiP 2: A Python Framework for the Dynamics of Open Quantum Systems." *Computer Physics Communications* 184, no. 4 (2013): 1234–40. *https://doi.org/10.1016/j.cpc.2012.11.019*.

[103] Johansson, J.R., P.D. Nation, and Franco Nori. "QuTiP: An Open-Source Python Framework for the Dynamics of Open Quantum Systems." *Computer Physics Communications* 183, no. 8 (2012): 1760–72. *https://doi.org/10.1016/j.cpc.2012.02.021*.

[104] Johnston, Eric. "An Exploratory Study in Quantum Acceleration of Ray Tracing." *https://www.machinelevel.com/qc/doc/Quantum%20Ray%20Tracing.pdf*, 2015.

[105] Jordan, Louis, and Billy Austin. "Is You Is or Is You Ain't My Baby?" *https://open.spotify.com/track/0IzMBujKJ24egd9HIK3OhG?si=a2a3cd66a3d54515*, 1943.

[106] Jordan, Stephen. "Quantum Algorithm Zoo." *https://quantum algorithmzoo.org*, 2025.

[107] Joyce, Alex. "Why Right-Hand Rule for Mazes Works." *https://www .nytimes.com/1989/09/06/opinion/l-why-right-hand-rule-for-mazes-works -075389.html*, 1989.

[108] Juster, Norton. *The Phantom Tollbooth*. Epstein & Carroll, 1961.

[109] Kay, Alastair (DaftWullie). "How Do I Show That a Two-Qubit State Is an Entangled State?" *https://quantumcomputing.stackexchange .com/questions/2263/how-do-i-show-that-a-two-qubit-state-is-an-entangled -state*, 2018.

[110] Kay, Alastair. "Tutorial on the Quantikz Package." *https://arxiv.org/ pdf/1809.03842*, 2023.

[111] Kay, Alastair (DaftWullie). "Understanding Oracles for Simon's Algorithm." *https://quantumcomputing.stackexchange.com/questions/32360/ understanding-oracles-for-simons-algorithm*, 2023.

[112] Kernighan, Brian W., and Dennis M. Ritchie. *The C Programming Language*. Pearson, 1978.

[113] Khan Academy. "Intro to the Trigonometric Ratios." *https://www .khanacademy.org/math/geometry-home/right-triangles-topic/intro-to-the -trig-ratios-geo/v/basic-trigonometry*, 2011.

[114] Kun, Jeremy. "How to Conquer Tensorphobia." *https://jeremykun.com/ 2014/01/17/how-to-conquer-tensorphobia/*, 2014.

[115] Kun, Jeremy. "Tensorphobia and the Outer Product." *https://jeremy kun.com/2016/03/28/tensorphobia-outer-product/*, 2014.

[116] Kuttler, Ken. "Multiplication of Matrices." *https://math.libretexts.org/ Bookshelves/Linear_Algebra/A_First_Course_in_Linear_Algebra_(Kuttler)/, 02%3A_Matrices/2.02%3A_Multiplication_of_Matrices*, 2023.

[117] Kuttler, Kenneth. *Linear Algebra, Theory and Applications*. 2012. *https:// resources.saylor.org/wwwresources/archived/site/wp-content/uploads/2012/ 02/Linear-Algebra-Kuttler-1-30-11-OTC.pdf*.

[118] Kwan, Dan, and Daniel Scheinert. *Everything Everywhere All at Once*. *https://www.imdb.com/title/tt6710474*, 2022.

[119] Lanese, Nicoletta. "Short-Necked Giraffe Relative Discovered in China. It Used Its Helmet Head to Bash Rivals." *https://www.live science.com/extinct-giraffe-relative-fossils*, 2022.

[120] Lankham, Isaiah, Bruno Nachtergaele, and Anne Schilling. "9.1: Inner Products." *https://math.libretexts.org/Bookshelves/Linear_Algebra/ Book%3A_Linear_Algebra_(Schilling_Nachtergaele_and_Lankham)/09% 3A_Inner_product_spaces/9.01%3A_Inner_Products*.

[121] LEGO. "MINDSTORMS Reference." *https://makecode.mindstorms.com/ reference*, 2018.

[122] Leiber, Jerry, and Mike Stoller. "Searchin'." *https://genius.com/The -coasters-searchin-lyrics*, 1957.

[123] Lennon, John, and Paul McCartney. "Do You Want to Know a Secret?" *https://music.youtube.com/watch?v=uRQ7ecvU56k*, 1964.

[124] Li, Lvzhou, Jingquan Luo, and Yongzhen Xu. "Playing Mastermind on Quantum Computers." *https://arxiv.org/pdf/2207.09356*, 2023.

[125] LibreTexts. "1.5: Composition of Functions." In *Precalculus 2e* (Open-Stax). *https://math.libretexts.org/Bookshelves/Precalculus/Precalculus_2e _(OpenStax)/01%3A_Functions/1.05%3A_Composition_of_Functions*.

[126] Lincoln, Don. "Quantum Foam, Virtual Particles, and Other Curiosities." *https://www.pbs.org/wgbh/nova/article/quantum-foam-virtual -particles-and-other-curiosities*, 2012.

[127] Livingston, Morgan, and Richard Liu. "Games Round Up: Quantum Computing." *https://www.wilsoncenter.org/blog-post/games-round -quantum-computing*, 2022.

[128] Lu, Xi, and Hongwei Lin. "A Framework for Quantum Ray Tracing." *https://arxiv.org/pdf/2203.15451*, 2022.

[129] Ludwig, Howard. "Why Is the Set of All Integers Denoted by $\mathbb{Z}$?" *https://www.quora.com/Why-is-the-set-of-all-integers-denoted-by-mathbb-Z*, 2019.

[130] Luff, Christine. "How Long Does It Take to Run a Marathon?" *https://www.verywellfit.com/how-long-does-it-take-to-run-a-marathon -2911423*, 2024.

[131] Martinez, Josu Etxezarreta. "How to Construct Matrix of Regular and 'Flipped' 2-qubit CNOT?" *https://quantumcomputing.stackexchange.com/ questions/5179/how-to-construct-matrix-of-regular-and-flipped-2-qubit-cnot*, 2019.

[132] Mathoma. "Geometric Algebra in 2D—Two Reflections Is a Rotation." *https://www.youtube.com/watch?v=Hy2gbdbrJZ8*, 2016.

[133] Matsos, Helen. "How Animals Sense Earth's Magnetic Field." *https:// phys.org/news/2020-05/animals-earth-magnetic-field.html*, 2020.

[134] McAllister, Willy. "Basic Electrical Quantities: Current, Voltage, Power." *https://www.khanacademy.org/science/physics/circuits-topic/ circuits-resistance/a/ee-voltage-and-current*.

[135] Merino, Orlando. "A Short History of Complex Numbers." *https:// www.math.uri.edu/ ~merino/spring06/mth562/ShortHistoryComplex Numbers2006.pdf*, 2006.

[136] Mermin, N. David. "Could Feynman Have Said This?" *Physics Today* 57, no. 5 (2004): 10–11. *https://doi.org/10.1063/1.1768652*.

[137] Mermin, N. David. *Quantum Computer Science: An Introduction.* Cambridge University Press, 2007.

[138] Merriam-Webster. "Black Box." *https://www.merriam-webster.com/dictionary/black%20box*.

[139] Merriam-Webster. "Oracle." *https://www.merriam-webster.com/dictionary/oracle*.

[140] Microsoft Azure Quantum Documentation. "Dirac Notation in Quantum Computing." *https://learn.microsoft.com/en-us/azure/quantum/concepts-dirac-notation*, 2025.

[141] Microsoft Azure Quantum Documentation. "Understanding Quantum Oracles." *https://learn.microsoft.com/en-us/azure/quantum/concepts-oracles*, 2024.

[142] Miller-Bakewell, Hector, and John van de Wetering. "The ZX-calculus." *https://zxcalculus.com*.

[143] Nahin, Paul J. *Dr. Euler's Fabulous Formula: Cures Many Mathematical Ills*. Princeton University Press, 2017.

[144] Najera, Jesus. "The History of Euler's Number (*e*)." *https://www.cantorsparadise.org/the-history-of-eulers-number-e-8c982994a39b/*, 2020.

[145] Nakanishi, Ken M., Takahiko Satoh, and Synge Todo. "Quantum-Gate Decomposer." *https://arxiv.org/pdf/2109.13223*, 2021.

[146] Neilsen, Michael A., and Isaac L. Chuang. *Quantum Computation and Quantum Information* (10th anniv. ed.). Cambridge University Press, 2011.

[147] Neven, Hartmut. "Meet Willow, Our State-of-the-Art Quantum Chip." *https://blog.google/technology/research/google-willow-quantum-chip/*, 2024.

[148] Nicholas, Jackie, and Peggy Adamson. "Introduction to Trigonometric Functions." *https://www.sydney.edu.au/content/dam/students/documents/mathematics-learning-centre/introduction-trigonometric-functions.pdf*, 1998.

[149] NIST. "NIST Releases First 3 Finalized Post-Quantum Encryption Standards." *https://www.nist.gov/news-events/news/2024/08/nist-releases-first-3-finalized-post-quantum-encryption-standards*, 2024.

[150] Nykamp, Duane Q. "Linear Transformations." *https://mathinsight.org/linear_transformation_definition_euclidean*.

[151] O'Connor, J.J., and E.F. Robertson. "Jean Robert Argand." *https://mathshistory.st-andrews.ac.uk/Biographies/Argand/*, 2019.

[152] Oppenheim, Alan V., and Ronald W. Schafer. *Digital Signal Processing*. Prentice-Hall, 1975.

[153] Osgood, Brad. "Lecture Notes for EE 261: The Fourier Transform and Its Applications." *https://see.stanford.edu/materials/lsoftaee261/book-fall-07.pdf*, 2007.

[154] Padavic-Callaghan, Karmela. "Quantum Computers Have Finally Arrived, But Will They Ever Be Useful?" *New Scientist*, 2025. *https://www.newscientist.com/article/2467128-quantum-computers-have-finally-arrived-but-will-they-ever-be-useful/*.

[155] Padavic-Callaghan, Karmela. "Quantum 'Schrödinger's Cat' Survives for a Stunning 23 Minutes." *New Scientist*, 2024. *https://www.newscientist.com/article/2453356-quantum-schrodingers-cat-survives-for-a-stunning-23-minutes/*.

[156] Pais, A. "Einstein and the Quantum Theory." *Reviews of Modern Physics* 51 (1979): 863–914. *https://eclass.aegean.gr/modules/document/file.php/511165/quantum_mechanics/einstein_quantum.pais%281979%29.pdf*.

[157] Pevzner, Vadim, and Karl Hess. "Quantum Ray Tracing: A New Approach to Quantum Transport in Mesoscopic Systems." In K. Hess, J.P. Leburton, and U. Ravaioli (eds.), *Computational Electronics*, 227–30. The Springer International Series in Engineering and Computer Science, vol. 113. Springer, 1991. *https://doi.org/10.1007/978-1-4757-2124-9_45*.

[158] Pirandola, Stefano, Jens Eisert, Christian Weedbrook, Akira Furusawa, and Samuel L. Braunstein. "Advances in Quantum Teleportation." *Nature Photonics* 9 (2015): 641–52. *https://arxiv.org/pdf/1505.07831*.

[159] Pomeroy, Ross. "Did Your Computer Crash? It Might Have Been a Cosmic Ray." *https://bigthink.com/hard-science/cosmic-rays-computer-crash/*, 2022.

[160] Preskill, John. "Quantum Computing 40 Years Later." *https://arxiv.org/pdf/2106.10522.pdf*, 2021.

[161] Prokopenya, Alexander. "Quantum Circuit Implementing Grover's Search Algorithm." *https://demonstrations.wolfram.com/QuantumCircuitImplementingGroversSearchAlgorithm/#popup1*, 2011.

[162] ProofWiki. "Transpose of Matrix Product." *https://proofwiki.org/wiki/Transpose_of_Matrix_Product*, 2024.

[163] Puzzling World. "The Great Maze." *https://www.puzzlingworld.co.nz/attractions/the-great-maze*.

[164] Qiskit authors. "Qiskit." *https://pypi.org/project/qiskit/*.

[165] Quantiki. "Bell State." *https://quantiki.org/wiki/bell-state*, 2015.

[166] Quantiki. "List of QC Simulators." *https://quantiki.org/wiki/list-qc-simulators*, 2024.

[167] Quantum Open Source Foundation. "Open-Source Quantum Software Projects." *https://github.com/qosf/awesome-quantum-software*, 2025.

[168] Quantum Open Source Foundation. "Public Repository." *https://github.com/qosf*, 2025.

[169] QuTiP developers and contributors. "QuTiP: Quantum Toolbox in Python." *https://qutip.org*, 2024.

[170] Rauch, Earl Mac. *The Adventures of Buckaroo Banzai Across the 8th Dimension. https://www.imdb.com/title/tt0086856*, 1984.

[171] Reany, P. "The No-Cloning Theorem Made Easy." *https://advancedmath .org/Physics/Quantum/QM1/No-Cloning-Theorem.pdf*, 2023.

[172] Richman, Lynne. "The History and Evolution of Numbers." *https:// sites.math.rutgers.edu/~zeilberg/math436/projects/RichmanP.pdf*, 2017.

[173] Robbins, Tom. *Even Cowgirls Get the Blues*. Houghton Mifflin, 1976.

[174] Roberts, Molly. "Numbers With Cool Names: Amicable, Social, Friendly." *https://tomrocksmaths.com/2023/05/10/numbers-with-cool -names-amicable-sociable-friendly/*, 2023.

[175] Robinson, Andrew. "Did Einstein Really Say That?" *Nature* 557 (2018). *https://www.nature.com/articles/d41586-018-05004-4*.

[176] Roddenberry, Gene, and Jerome Bixby. "Mirror, Mirror" (*Star Trek*, Season 2, Episode 4). *https://www.imdb.com/title/tt0708438*, 1970.

[177] Roffe, Joschka. "Quantum Error Correction: An Introductory Guide." *Contemporary Physics* 60, no. 3 (2019): 226−45. *https://arxiv.org/pdf/ 1907.11157*.

[178] Rogers, Leo. "The History of Negative Numbers." *https://nrich.maths .org/articles/history-negative-numbers*, 2011.

[179] The Royal Swedish Academy. "Nobel Prize in Physics 2022: Popular Information." *https://www.nobelprize.org/prizes/physics/2022/popular -information/*, 2023.

[180] Rui, Pinshu, Wen Zhangx, Yanlin Liao, and Ziyun Zhang. "Part Probabilistic Cloning of Linearly Dependent States." *https://arxiv.org/pdf/ 1603.07036*, 2016.

[181] Running Level. "1k Run Times." *https://runninglevel.com/running -times/1k-times*.

[182] Salton, Grant, Daniel Simon, and Cedric Lin. "Exploring Simon's Algorithm With Daniel Simon." *https://aws.amazon.com/blogs/quantum -computing/simons-algorithm/*, 2021.

[183] Santos, Luís Paulo, Thomas Bashford-Rogers, João Barbosa, and Paul Navrátil. "Towards Quantum Ray Tracing." *https://arxiv.org/pdf/2204 .12797*, 2022.

[184] Scale of Universe. "How Big Is the Smallest Object Visible to the Naked Eye?" *https://scaleofuniverse.com/en/universe/smallest-object -visible-to-the-naked-eye*, 2023.

[185] Scholten, Travis L., Bryan Eastin, and Steven Flammia. "Qcircuit— Macros to Generate Quantum Circuits." *https://www.ctan.org/pkg/ qcircuit*, 2018.

[186] Schuch, Norbert. "How to Construct a Multi-qubit Controlled-Z From Elementary Gates?" *https://quantumcomputing.stackexchange.com/ questions/4078/how-to-construct-a-multi-qubit-controlled-z-from-elementary -gates*, 2018.

[187] Schuch, Norbert, and Jens Siewert. "Programmable Networks for Quantum Algorithms." *Physical Review Letters* 91, no. 2 (2003): 027902. *https://arxiv.org/pdf/quant-ph/0303063*.

[188] Schwichtenberg, Jakob. *No-Nonsense Quantum Mechanics*. No Nonsense Books, 2020.

[189] Schwitzgebel, Eric. "How to Wrap Your Head around the Most Mind-bending Theories of Reality." *New Scientist*, 2024. *https://www.newscientist.com/article/mg26134833-100-how-to-wrap-your-head-around-the-most-mind-bending-theories-of-reality/*.

[190] Scratch Consortium. "Scratch Reference." *https://scratch.mit.edu*.

[191] Seneca the Younger. "Moral and Political Essays." *https://en.wikiquote.org/wiki/Seneca_the_Younger*, 40.

[192] Shakespeare, William. *The Tragedy of Coriolanus*. 1605. *https://www.folger.edu/explore/shakespeares-works/coriolanus/read/*.

[193] Shannon, C.E. "A Mathematical Theory of Communication." *Bell System Technical Journal* 27, no. 3 (1948): 379–423. *https://people.math.harvard.edu/~ctm/home/text/others/shannon/entropy/entropy.pdf*.

[194] Shor, Peter. "18.435/2.111 POVM Lecture." Massachusetts Institute of Technology, 2008. *https://math.mit.edu/~shor/18.435/POVM-lecture.pdf*.

[195] Shor, Peter. "Algorithms for Quantum Computation: Discrete Logarithms and Factoring." In *Proceedings, 35th Annual Symposium on Foundations of Computer Science*, 124–34. IEEE Computer Society Press, 1994. *https://math.mit.edu/~shor/papers/algsfqc-dlf.pdf*.

[196] Shor, Peter. "Lecture Notes for 8.360/18.435 Quantum Computation." Massachusetts Institute of Technology, 2022. *https://math.mit.edu/~shor/435-LN/*.

[197] SideFX. "Houdini." *https://www.sidefx.com/products/houdini/*, 2025.

[198] Simha, Rahul. "Introduction to Quantum Computing." *https://www2.seas.gwu.edu/~simhaweb/quantum/modules.html*, 2024.

[199] Simmons, Gustavus J. "RSA Encryption." *https://www.britannica.com/topic/RSA-encryption*, 2022.

[200] Simon, Daniel R. "On the Power of Quantum Computation." *SIAM Journal on Computing* 26, no. 5 (1997): 1474–83. *https://doi.org/10.1137/S0097539796298637*.

[201] Sloane, N. J. A. "The On-Line Encyclopedia of Integer Sequences." *https://oeis.org/A000040*, 2025.

[202] Smith, Kurt. "Descartes' Theory of Ideas." *https://plato.stanford.edu/Archives/fall2013/entries/descartes-ideas/*, 2013.

[203] Solow, Daniel. *How to Read and Do Proofs* (6th ed.). Wiley, 2013.

[204] Sparkes, Matthew. "Google's Claim of Quantum Supremacy Has Been Completely Smashed." *New Scientist*, 2024. *https://www.newscientist.com/article/2437886-googles-claim-of-quantum-supremacy-has-been-completely-smashed/*.

[205] Sparknotes authors. "The Allegory of the Cave." *https://www.sparknotes.com/philosophy/republic/themes/*.

[206] Spivery, William. "Operator." *https://www.youtube.com/watch?v=WEAtDmTbpu4*, 1944.

[207] Stack Exchange. "All Sites." *https://stackexchange.com/sites#*.

[208] Stack Exchange. "Mathematics Stack Exchange." *https://math.stackexchange.com*.

[209] Stack Exchange. "Physics Stack Exchange." *https://physics.stackexchange.com*.

[210] Stack Exchange. "Quantum Computing Stack Exchange." *https://quantumcomputing.stackexchange.com*.

[211] Strang, Gilbert. *Introduction to Linear Algebra* (6th ed.). MIT Press, 2023.

[212] Strogatz, Steven. "How Infinite Series Reveal the Unity of Mathematics." *https://www.quantamagazine.org/how-infinite-series-reveal-the-unity-of-mathematics-20220124/*, 2022.

[213] Susskind, Leonard, and Art Friedman. *Quantum Mechanics: The Theoretical Minimum*. Basic Books, 2014.

[214] Swift, Jonathan. *Gulliver's Travels into Several Remote Nations of the World*. Benjamin Motte, 1726. *https://www.gutenberg.org/files/829/829-h/829-h.htm*.

[215] Szabó, László E. "The Einstein—Podolsky—Rosen Argument and the Bell Inequalities." *https://arxiv.org/pdf/0712.1318*, 2007.

[216] Tagore, Rabindranath. *Stray Birds*. Macmillan, 1916. *https://www.gutenberg.org/files/6524/6524-h/6524-h.htm*.

[217] Tavares, Américo. "Understanding Imaginary Exponents." *https://math.stackexchange.com/questions/9770/understanding-imaginary-exponents*, 2010.

[218] Tegmark, Max. "Shut Up and Calculate." *https://arxiv.org/pdf/0709.4024*, 2007.

[219] Tomayko, James E. *Computers in Spaceflight: The NASA Experience*. 1988. *https://history.nasa.gov/computers/Source2.html*.

[220] Treil, Sergei. *Linear Algebra Done Wrong*. 2017. *https://www.math.brown.edu/streil/papers/LADW/LADW_2017-09-04.pdf*.

[221] University of Alberta, Centre for Teaching and Learning. "Why Reflect on Your Teaching?" *https://www.ualberta.ca/centre-for-teaching-and-learning/teaching-support/reflection/why-reflect-on-your-teaching/index.html*.

[222] van de Wetering, John. "ZX-calculus for the Working Quantum Computer Scientist." *https://arxiv.org/pdf/2012.13966*, 2020.

[223] van Gogh, Vincent. Letter to Theo van Gogh. *https://vangoghletters .org/vg/letters/let274/letter.html*, 1882.

[224] Vazirani, Umesh V. "Lecture 3: Hilbert Spaces, Tensor Products, Teleportation." *https://people.eecs.berkeley.edu/~vazirani/f04quantum/notes/ lec3.pdf*, 2004.

[225] Vazirani, Umesh V. "Lecture 4: Quantum Circuit Model, Solovay–Kitaev Theorem, BQP." *https://people.eecs.berkeley.edu/~vazirani/ f04quantum/notes/lec4.pdf*, 2004.

[226] Vernimmen, Tim. "Sharks Can Navigate Via Earth's Magnetic Field, Study Confirms for the First Time." *https://www.nationalgeographic .com/animals/article/sharks-can-navigate-via-earths-magnetic-field-study -confirms-for-the-first-time*, 2021.

[227] Waits, Tom. "Foreign Affair." *https://www.youtube.com/watch?v= -iTLk3gjEec*, 1977.

[228] Weaver, James, Huang Junye, Jarrod Reilly, and Anastasia Jeffery. "QPong," 2019. *https://github.com/QPong/QPong*, *https://github.com/ QPong/QPong-Unity*, and *www.youtube.com/watch?v=a1NZC5rqQD8*.

[229] Weinersmith, Zach. "Saturday Morning Breakfast Cereal." *https://www .smbc-comics.com/comic/contrived*, 2023.

[230] Weisberger, Mindy. ""God Plays Dice With the Universe," Einstein Writes in Letter About His Qualms With Quantum Theory." *https:// www.livescience.com/65697-einstein-letters-quantum-physics.html*, 2019.

[231] Weisstein, Eric W. "Argand Diagram." *https://mathworld.wolfram.com/ ArgandDiagram.html*.

[232] Weisstein, Eric W. "Birthday Problem." *https://mathworld.wolfram.com/ BirthdayProblem.html*.

[233] Weisstein, Eric W. "Bracket." *https://mathworld.wolfram.com/Bracket.html*.

[234] Weisstein, Eric W. "Euclidean Algorithm." *https://mathworld.wolfram .com/EuclideanAlgorithm.html*.

[235] Weisstein, Eric W. "Euclid's Theorems." *https://mathworld.wolfram .com/EuclidsTheorems.html*.

[236] Weisstein, Eric W. "Exponential Function." *https://mathworld.wolfram .com/ExponentialFunction.html*.

[237] Weisstein, Eric W. "Fundamental Theorem of Arithmetic." *https:// mathworld.wolfram.com/FundamentalTheoremofArithmetic.html*.

[238] Weisstein, Eric W. "Gaussian Elimination." *https://mathworld.wolfram .com/GaussianElimination.html*.

[239] Weisstein, Eric W. "Hilbert Space." *https://mathworld.wolfram.com/ HilbertSpace.html*.

[240] Weisstein, Eric W. "Kronecker Delta." *https://mathworld.wolfram.com/KroneckerDelta.html*.

[241] Weisstein, Eric W. "Linear Function." *https://mathworld.wolfram.com/LinearFunction.html*.

[242] Weisstein, Eric W. "List." *https://mathworld.wolfram.com/List.html*.

[243] Weisstein, Eric W. "Natural Number." *https://mathworld.wolfram.com/NaturalNumber.html*.

[244] Weisstein, Eric W. "Operator." *https://mathworld.wolfram.com/Operator.html*.

[245] Weisstein, Eric W. "Prime Number." *https://mathworld.wolfram.com/PrimeNumber.html*.

[246] Weisstein, Eric W. "Set." *https://mathworld.wolfram.com/Set.html*.

[247] Weisstein, Eric W. "Square Root." *https://mathworld.wolfram.com/SquareRoot.html*.

[248] Weisstein, Eric W. "Vector Space." *https://mathworld.wolfram.com/VectorSpace.html*.

[249] Whitfield, James D., Jun Yang, Weishi Wang, Joshuah T. Heath, and Brent Harrison. "Quantum Computing 2022." *https://arxiv.org/pdf/2201.09877.pdf*, 2022.

[250] Wikipedia. "Bell State." *https://en.wikipedia.org/wiki/Bell_state*.

[251] Wikipedia. "Bernstein–Vazirani Algorithm." *https://en.wikipedia.org/wiki/Bernstein%E2%80%93Vazirani_algorithm*.

[252] Wikipedia. "Bloch Sphere." *https://en.wikipedia.org/wiki/Bloch_sphere*.

[253] Wikipedia. "Bohr Radius." *https://en.wikipedia.org/wiki/Bohr_radius*.

[254] Wikipedia. "Born Rule." *https://en.wikipedia.org/wiki/Born_rule*.

[255] Wikipedia. "Bra–ket Notation." *https://en.wikipedia.org/wiki/Bra%E2%80%93ket_notation*.

[256] Wikipedia. "Class (Computer Programming)." *https://en.wikipedia.org/wiki/Class_(computer_programming)*.

[257] Wikipedia. "Density Matrix." *https://en.wikipedia.org/wiki/Density_matrix*.

[258] Wikipedia. "Deutsch–Jozsa Algorithm." *https://en.wikipedia.org/wiki/Deutsch-Jozsa_algorithm*.

[259] Wikipedia. "*e* (Mathematical Constant)." *https://en.wikipedia.org/wiki/E_(mathematical_constant)*.

[260] Wikipedia. "Euler's Formula." *https://en.wikipedia.org/wiki/Euler%27s_formula*.

[261] Wikipedia. "Felix Bloch." *https://en.wikipedia.org/wiki/Felix_Bloch*.

[262] Wikipedia. "Fortran." *https://en.wikipedia.org/wiki/Fortran*.

[263] Wikipedia. "Fredkin Gate." *https://en.wikipedia.org/wiki/Fredkin_gate*.

[264] Wikipedia. "Gaussian Elimination." *https://en.wikipedia.org/wiki/ Gaussian_elimination.*

[265] Wikipedia. "Greatest Common Divisor." *https://en.wikipedia.org/wiki/ Greatest_common_divisor.*

[266] Wikipedia. "'Hello, World!' Program." *https://en.wikipedia.org/wiki/ %22Hello,_World!%22_program.*

[267] Wikipedia. "Invertible Matrix." *https://en.wikipedia.org/wiki/Invertible _matrix.*

[268] Wikipedia. "Jaques Hadamard." *https://en.wikipedia.org/wiki/Jacques _Hadamard.*

[269] Wikipedia. "John von Neumann." *https://en.wikipedia.org/wiki/John _von_Neumann.*

[270] Wikipedia. "Key Size." *https://en.wikipedia.org/wiki/Key_size.*

[271] Wikipedia. "Matrix Multiplication." *https://en.wikipedia.org/wiki/ Matrix_multiplication.*

[272] Wikipedia. "Natural Number." *https://en.wikipedia.org/wiki/Natural _number.*

[273] Wikipedia. "Noisy Intermediate-Scale Quantum Era." *https://en .wikipedia.org/wiki/Noisy_intermediate-scale_quantum_era.*

[274] Wikipedia. "NP (Complexity)." *https://en.wikipedia.org/wiki/NP _(complexity).*

[275] Wikipedia. "Oracle Machine." *https://en.wikipedia.org/wiki/Oracle _machine.*

[276] Wikipedia. "POVM." *https://en.wikipedia.org/wiki/POVM.*

[277] Wikipedia. "Quantum Fourier Transform." *https://en.wikipedia.org/ wiki/Quantum_Fourier_transform.*

[278] Wikipedia. "Superdense Coding." *https://en.wikipedia.org/wiki/ Superdense_coding.*

[279] Wikipedia. "Toffoli Gate." *https://en.wikipedia.org/wiki/Toffoli_gate.*

[280] Wikipedia. "Triple Modular Redundancy." *https://en.wikipedia.org/ wiki/Triple_modular_redundancy.*

[281] Wikipedia. "Unitary Matrix." *https://en.wikipedia.org/wiki/Unitary _matrix.*

[282] Wikipedia. "Vector (Mathematics and Physics)." *https://en.wikipedia .org/wiki/Vector_(mathematics_and_physics).*

[283] Wikipedia. "Vector Space." *http://en.wikipedia.org/wiki/Vector_space.*

[284] Wikipedia. "Wolfgang Pauli." *https://en.wikipedia.org/wiki/Wolfgang _Pauli.*

[285] Wikipedia. "Zero to the Power of Zero." *https://en.wikipedia.org/wiki/ Zero_to_the_power_of_zero.*

[286] Wilson, Robin. *Euler's Pioneering Equation: The Most Beautiful Theorem in Mathematics*. Oxford University Press, 2018.

[287] Wolfram. "Wolfram Player." *https://www.wolfram.com/player/*.

[288] Wootten, James. "Making Games With Quantum Computers." *https://decodoku.medium.com/games-computers-and-quantum-84bfdd2c0fe0*, 2020.

[289] Yang, Jianhao M. "Probabilistic Quantum Teleportation." *https://arxiv.org/pdf/1601.02501*, 2017.

[290] Yang, Yu, Igor Kladaric, Maxwell Drimmer, et al. "A Mechanical Qubit." *Science* 386, no. 6723 (2024): 783–88. *https://www.science.org/doi/10.1126/science.adr2464*.

[291] Yang, Y.A., W.-T. Luo, J.-L. Zhang, S.-Z Wang, Chang-Ling Zou, T. Xia, and Z.-T. Lu. "Minutes-scale Schrödinger-cat State of Spin-5/2 Atoms." *Nature Photonics* 19 (2024): 89–94. *https://arxiv.org/pdf/2410.09331v1*.

[292] Young, Peter. "Simon's Algorithm." *https://web.archive.org/web/20231004205621/https://young.physics.ucsc.edu/150/simon.pdf*, 2019.

[293] Young, Peter. "An Undergraduate Course on Quantum Computing" (4th ed.). University of California, Santa Cruz, 2024. *https://bpb-us-e1.wpmucdn.com/sites.ucsc.edu/dist/7/1905/files/2025/03/phys_150_all.pdf*.

[294] Young, Peter. "Using Period Finding to Factor an Integer." *https://web.archive.org/web/20221028100132/https://young.physics.ucsc.edu/150/period.pdf*, 2019.

[295] Zhang, Naiyuan J., Ron Q. Nguyen, Navketan Batra, et al. "Excitons in the Fractional Quantum Hall Effect." *Nature* 637 (2025): 327–32. *https://doi.org/10.1038/s41586-024-08274-3*.

[296] Zhang, Yingwen, Antony Orth, Duncan England, and Benjamin Sussman. "Ray Tracing With Quantum Correlated Photons to Image a Three-dimensional Scene." *Physical Review A* 105 (2022): L011701. *https://doi.org/10.1103/PhysRevA.105.L011701*.

[297] Zhu, Sebastian, William Yue, and Vincent Fan. "Shor's Algorithm and the Period Finding Problem." *https://math.mit.edu/research/highschool/primes/materials/2019/December/Fan-Yue-Zhu_Shor_s_Algorithm_and_the_Period_Finding_Problem.pdf*, 2019.

INDEX

The fonts used in *Quantum Computing* are New Baskerville, Futura, The Sans Mono Condensed, and Dogma. The book was typeset with LaTeX 2_ε package nostarch by Boris Veytsman with many additions by Alex Freed, Miles Bond, and other members of the No Starch Press team.

The book was created on a Mac Studio. I wrote the text in iTerm2 windows running the vi text editor. I proofed the results, and tweaked the mathematical typesetting from LaTeX 2_ε, using TeXStudio and TeXShop. Circuit diagrams were made with the quantikz library. The outputs of quantum programs on real hardware were generated by running those programs using the IBM Quantum service. Local simulations were mostly performed in Python using the Qiskit library. I drew the figures with Adobe Illustrator and Photoshop, using a Wacom tablet. Other programming was done in Python, using primarily the numpy, scipy, sympy, scikit-image, and matplotlib libraries.

RESOURCES

Visit *https://nostarch.com/quantum-computing* for errata and more information.

More no-nonsense books from **NO STARCH PRESS**

DEEP LEARNING
A Visual Approach
BY ANDREW GLASSNER
768 PP., $99.99
ISBN 978-1-7185-0072-3
full color

COMPUTER ARCHITECTURE
From the Stone Age to the Quantum Age
BY CHARLES FOX
560 PP., $59.99
ISBN 978-1-7185-0286-4

HOW AI WORKS
From Sorcery to Science
BY RONALD T. KNEUSEL
192 PP., $29.99
ISBN 978-1-7185-0372-4

MATH FOR DEEP LEARNING
What You Need to Know to Understand Neural Networks
BY RONALD T. KNEUSEL
344 PP., $49.99
ISBN 978-1-7185-0190-4

MATH FOR PROGRAMMING
Learn the Math, Write Better Code
BY RONALD T. KNEUSEL
504 PP., $49.99
ISBN 978-1-7185-0358-8

THE MANGA GUIDE TO LINEAR ALGEBRA
BY SHIN TAKAHASHI, IROHA INOUE, *and* TREND-PRO CO., LTD.
264 PP., $24.99
ISBN 978-1-59327-413-9

PHONE:
800.420.7240 OR
415.863.9900

EMAIL:
SALES@NOSTARCH.COM
WEB:
WWW.NOSTARCH.COM